A GUIDE TO CLASSICAL AND MODERN MODEL THEORY

TRENDS IN LOGIC
Studia Logica Library

VOLUME 19

SCOPE OF THE SERIES

Trends in Logic is a bookseries covering essentially the same area as the journal *Studia Logica* – that is, contemporary formal logic and its applications and relations to other disciplines. These include artificial intelligence, informatics, cognitive science, philosophy of science, and the philosophy of language. However, this list is not exhaustive, moreover, the range of applications, comparisons and sources of inspiration is open and evolves over time.

Volume Editor
Ryszard Wojácki

The titles published in this series are listed at the end of this volume.

A GUIDE TO CLASSICAL AND MODERN MODEL THEORY

by

ANNALISA MARCJA
University of Florence, Italy

and

CARLO TOFFALORI
University of Camerino, Italy

KLUWER ACADEMIC PUBLISHERS
DORDRECHT / BOSTON / LONDON

A C.I.P. Catalogue record for this book is available from the Library of Congress.

ISBN 1-4020-1330-2 (HB)
ISBN 1-4020-1331-0 (PB)

Published by Kluwer Academic Publishers,
P.O. Box 17, 3300 AA Dordrecht, The Netherlands.

Sold and distributed in North, Central and South America
by Kluwer Academic Publishers,
101 Philip Drive, Norwell, MA 02061, U.S.A.

In all other countries, sold and distributed
by Kluwer Academic Publishers,
P.O. Box 322, 3300 AH Dordrecht, The Netherlands.

Printed on acid-free paper

Printed in the Netherlands.

Preface

This book deals with Model Theory. So the first question that a possible, recalcitrant reader might ask is just: What is Model Theory? Which are its intents and applications? Why should one try to learn it? Another, more particular question might be the following one. Let us assume, if you like, that Model Theory deserves some attention. Why should one use this book as a guide to it?
The answer to the former question may sound problematic, but it is quite simple, at least in our opinion. For, Model Theory has been developing, since its birth, a number of methods and concepts that do have their intrinsic relevance, but also provide fruitful and notable applications in various fields of Mathematics. We could mention here its role in Algebra and Algebraic Geometry, for instance the analysis of differentially closed fields (and the results on the differential closure of a differential field), or p-adic fields (and the asymptotic solution of Artin's Conjecture), as well as the recent Hrushovski's model theoretic approach to classical problems, like Mordell-Lang's Conjecture or Manin-Mumford's Conjecture.
So Model Theory is today a lively, sprightly and fertile research area, which surely deserves the attention of the mathematical world and, consequently, its own references. This recalls the latter question above. Actually there do exist some excellent textbooks explaining Model Theory, such as [56] and [57]. Also Poizat's book [131] should be mentioned; it was written more than ten years ago, but it is still up-to-date, and it has been recently translated in English. In addition more specialistic references treat adequately some particular fields in Model Theory, such as stability theory, simplicity theory, o-minimality, classification theory and so on.
Nevertheless, we believe that this book has its own role and its own originality in this setting. Indeed we wish to address this work not only to the experts of the area, but also, and mainly, to young people having a basic knowledge of model theory and wishing to proceed towards a deeper analysis, as well as to mathematicians which are not directly involved in Model

Theory but work in related and overlapping fields, such as Algebra and Geometry. Accordingly we will emphasize the frequent and fruitful connections between Model Theory and these branches of Mathematics (differentially closed fields, Artin's Conjecture, Mordell-Lang's Conjecture and so on). In each case, we aim at giving a detailed report or, at least, at sketching the main ideas and techniques of the model theoretic approach.
Our book wishes also to follow a historical perspective in introducing Model Theory. Of course, this does not mean to provide a full history of Model Theory (although such a project could be interesting and worthy of some attention), but just to insert any basic concept in the historical framework where it was born, and so to better clarify the reasons why it was introduced.
Hence, after shortly recalling in Chapter 1 basic Model Theory (structures and theories, compactness and definability), we deal in Chapter 2 with quantifier elimination, in particular with the work of Alfred Tarski on algebraically closed fields and real closed fields. We will discuss the role of quantifier elimination in Model Theory, but we will treat briefly also its intriguing role in the $P = NP$ problem within the new models of computation (such as the Blum-Shub-Smale approach, and so on).
Chapter 3 will be concerned with Abraham Robinson's ideas: model completeness, model companions, existentially closed structures. We will consider again algebraically closed fields and real closed fields, but we will illustrate also other crucial classes, like differentially closed fields, separably closed fields, p-adically closed fields and, finally, existentially closed difference fields (a rather recent matter, with some remarkable applications to Algebraic Geometry).
Chapter 4 deals with imaginary elements. They are essentially classes of definable equivalence relations in a structure $\mathcal{A}$, so elements in some quotient structure. We describe Shelah's construction of $\mathcal{A}^{eq}$, englobing these classes as new elements in the whole structure, and we show that these imaginary elements can be sometimes eliminated, because the corresponding quotients can be simulated by some suitable definable subsets of $\mathcal{A}$.
Chapters 5 and 6 are devoted to Morley's Theorem on uncountable categoricity. Actually its proof will be given only in Chapter 7, but here we describe Morley's ideas -algebraic closure, totally transcendental theories, prime models, an so on- and we illustrate their richness and their applications.
We will be led in this way to one of the main topics in Model Theory, namely the Classification Problem. We will explain in Chapter 7 the more relevant ideas in the formidable work of Shelah on this matter (simplicity, stability, superstability, modularity), and we will discuss their significance in some

important algebraic classes, like differential fields, difference fields, and so on. We wish also to deal with the Zilber program of classifying structures up to biinterpretability, in particular with Zilber's Conjecture on strongly minimal sets, and its brilliant solution due to Hrushovski.
Also Chapter 8 largely owes to Hrushovski. In fact, after illustrating in more detail the natural connection between Model Theory and Algebraic Geometry, we will describe the Hrushovski proof of Mordell-Lang conjecture; we will refer very quickly also to the Hrushovski solution of the related Manin-Mumford conjecture. In particular we will realize how deeply Model Theory, actually both pure Model Theory and Model Theory applied to algebra are involved in these proofs.
The final Chapter is devoted to a (comparatively) recent and fertile area in Model Theory: o-minimality. We will expound the basic results on o-minimal theories, and we will discuss some intriguing developments, including Wilkie's solution of a classical problem of Tarski on the exponentiation in the real field.

We assume some familiarity with the basic notions of Algebra, Set Theory and Recursion Theory. [65], [66] or [78], and [121] respectively are good references for these areas. Incidentally, let us point out that we are working within the usual Zermelo - Fraenkel axiomatic system, including the Axiom of Choice. We also assume some acquaintance with basic Model Theory, such as it is usually proposed in any introductory course. However, Chapter 1 is devoted, as already said, to a short and somewhat informal sketch of these matters.

As its title states, this book aims at being only a guide. We do not claim to provide an exhaustive treatment of Model Theory; indeed our omissions are likely to be much more numerous and larger than the topics we deal with. But we have aimed at giving an almost complete report of at least two crucial subjects (ω-stability and o-minimality), and at providing the basic hints towards some conspicuous generalizations (such as superstability, stability, and so on).
In a similar way, we have treated in detail some key algebraic examples (algebraically closed fields, real closed fields, differentially closed fields in characteristic 0), but we have provided at least some basic information on other relevant structures (like p-adic fields, existentially closed fields with an automorphism, differentially and separably closed fields in a prime characteristic). In conclusion, we do hope that the outcome of our work is a sufficiently clear and terse picture of what Model Theory is, and provides a report as homogeneous and general as possible. Incidentally, let us say

that this book is not a literal translation of the former italian version [108]; all the material was revised and rewritten; our treatment of some topics, like quantifier elimination and model completeness, are entirely new; and we have added some relevant matters, such as prime models and Morley's Theorem on uncountable categorical theories.

Contents

Chapter 1

Structures

1.1 Structures

The aim of this chapter is to sketch out basic model theory. We wish to summarize some key facts for people already acquainted with them, but also, at the same time, to introduce them to people unfamiliar to logic, and perhaps disliking too many logical details. Accordingly we will use a rather colloquial tone. The fundamental question to be answered is: what is Model Theory? As we will see in more detail in Section 1.2, Model Theory is -or, more precisely, was at its beginning- the study of the relationship between mathematical formulas and structures satisfying or rejecting them. But, in order to fully appreciate this matter, it is advisable for us preliminarily to recall what a structure is, and which kind of formulas we are dealing with. This section is devoted to the former concept.

Structures are an algebraic notion. Actually, since Galois, Algebra is not only the solving of equations, or literal calculus, but becomes the science of structures (groups, rings, fields, and so on). This new direction gets clearer at the beginning of the last century, with Steinitz's work on fields and, later, the publication of the Van der Waerden book. What is a structure? Basically, it is a non-empty set A, with a collection of distinguished elements, operations, and relations. For instance, the set $\mathbf{Z}$ of integers with the usual operations of addition $+$ and multiplication $\cdot$ is a structure, as well as the same set $\mathbf{Z}$ with the order relation $\leq$. Note that, in these examples, the underlying set is the same (the integers), but, of course, the structure changes: in the former case we have the ring of integers, in the latter the integers as an ordered set. To make this kind of difference among structures clearer, we have to choose a *language*, in other words to specify how many distinguished elements, how many n-ary operations and relations (for every natural $n \neq 0$) we want to involve in building our structure. So, when we discuss the integral domain of integers, our language needs two binary operations (for addition and multiplication), while, in the latter case, a binary relation (for the order) is enough. Notice that the language of the ring case works as well for all the structures admitting two binary operations, and hence possibly for structures which are not rings; for instance, the reals with the functions

$$f(x, y) = sin(x - y), \quad g(x, y) = e^{x \cdot y}$$

for all x and y in $\mathbf{R}$ provide a new structure for our language, but, of course, the algebraic features of this structure are very far from the basic properties of integral domains. Accordingly it is advisable, from a general point of view, to distinguish the constant, operation and relation symbols of a language L and the elements, operations and relations embodying these symbols in a given structure for L. Symbols are something like the characters in a tragedy (like Hamlet), while their interpretations in a structure are the actors playing on the stage (Laurence Olivier, or Kenneth Branagh, or your favourite "Hamlet").
In this framework, we can at last provide a sharp definition of *structure*. We fix a language L. For simplicity, we assume that L is countable, hence either finite or denumerable (but most of what we shall say can be extended without problems to uncountable languages).

Definition 1.1.1 *A structure $\mathcal{A}$ for L is a pair consisting of a non empty set A, called the* **universe** *of $\mathcal{A}$, and a function mapping*

(i) *every constant c of L into an element $c^{\mathcal{A}}$ of A,*

and, for any positive integer n,

(ii) *every n-ary operation symbol f of L into an n-ary operation $f^{\mathcal{A}}$ of A (hence a function from A^n into A),*

(iii) *every n-ary relation symbol R of L into an n-ary relation $R^{\mathcal{A}}$ of A (hence a subset of A^n).*

The structure $\mathcal{A}$ is usually denoted as follows

$$\mathcal{A} = (A, (c^{\mathcal{A}})_{c \in L}, (f^{\mathcal{A}})_{f \in L}, (R^{\mathcal{A}})_{R \in L}).$$

Let us propose some examples, which will be useful later in this book.

Examples 1.1.2 1. A graph is a non empty set A with a binary relation P both irreflexive and symmetric. Hence a graph can be viewed as a structure $\mathcal{A}$ in the language L consisting of a unique binary relation symbol R, with $R^{\mathcal{A}} = P$. Also a non empty set A partially ordered by some relation $\leq$ can be regarded as a structure in the same language L; this time, $R^{\mathcal{A}} = \leq$.

2. A (multiplicative) group $\mathcal{G}$ is a structure of the language $L = \{1, \cdot, ^{-1}\}$, where 1 is a constant, $\cdot$ and $^{-1}$ are operation symbols of arity 2 and 1 respectively. $1^{\mathcal{G}}$ represents the identity element in $\mathcal{G}$, while $\cdot^{\mathcal{G}}$ and $^{-1^{\mathcal{G}}}$ denote the product and the inverse operation in $\mathcal{G}$. Actually one might enrich L with some additional symbols; for instance, one might introduce a new binary operation symbol $[\ ,\ ]$ corresponding to the commutator operation in $\mathcal{G}$. But, for a and b in G, $[a, b]$ is just $a \cdot b \cdot a^{-1} \cdot b^{-1}$, so $[\ ,\ ]$ is not really new, and is implicitly defined by L. Actually we will prefer L later; but it is noteworthy that L can capture and express some further operations (and relations and constants) of $\mathcal{G}$ besides those literally interpreting its symbols.

3. A field $\mathcal{K}$ is a structure of the language $L = \{0, 1, +, -, \cdot\}$ where 0 and 1 are constant, and $+$, $-$ and $\cdot$ are operation symbols (each having an obvious interpretation in $\mathcal{K}$). Alternatively, $\mathcal{K}$ can be viewed as a structure in the language $L' = L \cup \{^{-1}\}$ with a new operation symbol $^{-1}$; obviously, $^{-1}$ has to be interpreted within $\mathcal{K}$ in the inverse operation for nonzero elements of K. However, according to the general definition of structure given before, $^{-1^{\mathcal{K}}}$ should denote a 1-ary operation with domain K. So we run into the problem of defining 0^{-1}; this can be overcome by agreeing, for instance, $0^{-1} = 0$, but this solution may sound slightly artificial. So we will prefer to adopt below the language L when dealing with fields. Indeed, when a and b are two elements in a field $\mathcal{K}$, then $a = b^{-1}$ can be equivalently expressed by saying $a \cdot b = 1$.

4. An ordered field is a structure in the language $L = \{+, -, \cdot, 0, 1, \leq\}$ obtained by adding a new binary relation symbol $\leq$. Its interpretation in a given ordered field is clear: the order relation in the field.

5. Let N denote the set of natural numbers. 0 is an element of N; the successor s (mapping each natural n into $n + 1$) is a 1-ary function from N to N. Giuseppe Peano pointed out that the Induction Principle (together with the auxiliary conditions that s is $1-1$ but 0 is not in

its image) fully characterizes $(\mathbf{N}, 0, s)$. A suitable language to discuss this structure should include a constant symbol and a 1-ary operations symbol.

6. Let $\mathcal{K}$ be a (countable) field. A vectorspace $\mathcal{V}$ over $\mathcal{K}$ can be regarded as a structure in the language $L_K = \{0, +, -, r\ (r \in K)\}$, where 0 is a constant, + and − are operation symbols with arity 2 and 1 respectively, and, for every $r \in K$, r denotes in L_K a 1-ary operation symbol, to be interpreted inside $\mathcal{V}$ in the scalar multiplication by r. The other symbols in L_K are interpreted in the obvious way. The assumption on the cardinality of K has the only role of ensuring L_K countable. Moreover, what we have said so far easily extends to right or left modules over a (countable) ring R with identity; the corresponding language is obviously denoted by L_R.

As already said, we should distinguish symbols and interpretations, for instance, a binary relation symbol R and the relation $R^{\mathcal{A}}$ embodying it in a structure $\mathcal{A}$ (sometimes an order relation in a partially ordered set, but elsewhere possibly the adjacency relation in a graph). But, to avoid too many complications, we will often confuse (and actually we already confused) the language symbols and their "most natural" interpretations. For instance, in Example 1.1.2, 6, we denoted in the same way the addition symbol + of L_R and its obvious interpretation in a given R-module M, namely the addition in M.
We will be interested in several algebraic notions concerning structures. In particular embeddings play a crucial role in Model Theory. So let's recall their definition.

Definition 1.1.3 *Let $\mathcal{A}$ and $\mathcal{B}$ two structures in a language L. A* **homomorphism** *of $\mathcal{A}$ into $\mathcal{B}$ is a function f from A into B such that*

(i) *for every constant c of L, $f(c^{\mathcal{A}}) = c^{\mathcal{B}}$;*

(ii) *for every positive integer n, for every n-ary operation symbol F in L and for every sequence $\vec{a} = (a_1, \ldots, a_n)$ in A^n, $f(F^{\mathcal{A}}(\vec{a})) = F^{\mathcal{B}}(f(\vec{a}))$ (hereafter $f(\vec{a})$ abridges $(f(a_1), \cdots, f(a_n))$);*

(iii) *for every positive integer n, for every n-ary relation symbol R of L and for every sequence $\vec{a}$ in A^n, if $\vec{a} \in R^{\mathcal{A}}$, then $f(\vec{a}) \in R^{\mathcal{B}}$.*

f is called an **embedding** of $\mathcal{A}$ into $\mathcal{B}$ if f is injective and, in (iii), $f(\vec{a}) \in R^{\mathcal{B}}$ implies $\vec{a} \in R^{\mathcal{A}}$ for every $\vec{a}$ in A^n. When there is some embedding of $\mathcal{A}$

into $\mathcal{B}$, we write $\mathcal{A} \subseteq \mathcal{B}$. An **isomorphism** of $\mathcal{A}$ onto $\mathcal{B}$ is a surjective embedding. When there is some isomorphism of $\mathcal{A}$ onto $\mathcal{B}$, we say that $\mathcal{A}$ and $\mathcal{B}$ are *isomorphic* and we write $\mathcal{A} \simeq \mathcal{B}$. Finally, an *endomorphism* (*automorphism*) of $\mathcal{A}$ is a homomorphism (isomorphism) of $\mathcal{A}$ onto $\mathcal{A}$.

Definition 1.1.4 *Let $\mathcal{A}$ and $\mathcal{B}$ be two structures of L such that $A \subseteq B$. If the inclusion of A into B defines an embedding of $\mathcal{A}$ into $\mathcal{B}$, $\mathcal{A}$ is called a* **substructure** *of $\mathcal{B}$, and $\mathcal{B}$ an* **extension** *of $\mathcal{A}$.*

Now let $\mathcal{B}$ be a structure of L, and A be a non-empty subset of B. We wonder if A is the domain of a substructure of $\mathcal{B}$. One promptly realizes that this may be false. Indeed

(i) if c is a constant of L, it may happen that $c^{\mathcal{B}}$ is not in A;

(ii) if F is an n-ary operation symbol in L, it may happen that the restriction of $F^{\mathcal{B}}$ to A^n is not an n-ary operation in A, in other words that A is not closed under $F^{\mathcal{B}}$;

(iii) on the contrary, if R is an n-ary relation symbol in L, then $R^{\mathcal{B}} \cap A^n$ is an n-ary relation in A.

So A is not necessarily the domain of a substructure of $\mathcal{B}$. However the closure of $A \cup \{c^{\mathcal{B}} : c \text{ constant in } L\}$ with respect to the operations $F^{\mathcal{B}}$, when F ranges over the operation symbols in L, does form the domain of a substructure of L, usually denoted $\langle A \rangle$, and called the *substructure generated* by A: in this case A is said to be a set of generators of $\langle A \rangle$. Notice that these notions can be introduced even in the case $A = \emptyset$, provided that L contains at least a constant symbol. $\mathcal{B}$ is called *finitely generated* if there exists a finite subset A of B such that $\mathcal{B} = \langle A \rangle$.
Finally, let $L \subseteq L'$ be two languages, $\mathcal{A}$ be an L-structure, $\mathcal{A}'$ be an L'-structure such that $A = A'$ and the interpretations of the symbols in L are the same in $\mathcal{A}$ and in $\mathcal{A}'$. In this case, we say that $\mathcal{A}'$ *expands* $\mathcal{A}$, or also that $\mathcal{A}'$ is an *expansion* of $\mathcal{A}$ to L'; $\mathcal{A}$ is called a *restriction* of $\mathcal{A}'$ to L.

1.2 Sentences

Given a language L, after forming the structures of L, one builds, in a complementary way, the *formulas* of L, in particular the *sentences* of L, and one defines when a formula (a sentence) is *true* in a given structure. This is the realm of Logic rather than of Algebra.

Actually there are several possible ways of introducing formulas and truth, according to our tastes or our mathematical purposes. We will limit ourselves in this book to the *first order* framework. Let us sketch briefly how formulas and truth are usually introduced in the first order logic. For simplicity let us work in the particular setting of natural numbers (full general details and sharp definitions can be found in any handbook of basic Mathematical Logic, such as [153]).

Consider the natural numbers and the corresponding structure $(\mathbf{N}, 0, s)$, where s denotes the successor function. The corresponding language L includes a constant (for 0) and a 1-ary operation symbol (to be interpreted in s). As announced at the end of the previous section, we denote these symbols by still using 0 and s: this is not completely correct, but simplifies our life. In the first order setting, formulas can be built by using additional symbols

- countably many element variables $v_0, v_1, \ldots, v_n, \ldots$ (just to respect our countable framework; otherwise we can use as many variables as we need),
- the basic connectives $\wedge$ (*and*), $\vee$ (*or*), $\neg$ (*not*) (and even $\rightarrow$ (*if ..., then*), $\leftrightarrow$ (*if and only if*) if you like),
- the quantifiers $\forall$ (*for all*) and $\exists$ (*there exists*),
- parentheses (,)

and a symbol $=$ to be interpreted everywhere by the equality relation. At this point one forms the *terms* of L. Essentially they are polynomials; in our case they are built starting from the constant 0 and the variables v_n (n natural) and using the operation symbols (so s in our setting). The second step is to construct the *atomic formulas* of L: basically they are equations between terms, but, when the language includes a k-ary relation symbol R, we have to include every statement saying that a k-uple of terms satisfies R. At this point the formulas of L are built from the atomic ones inductively in the following way:

1. one can negate, or conjunct, or disjunct some given formulas α, β, ... and get new formulas $\neg\alpha$, $\alpha \wedge \beta$, $\alpha \vee \beta$;
2. one can take a formula α and a variable v_n and form new formulas $\forall v_n \alpha$, $\exists v_n \alpha$;
3. nothing else is a formula.

For α and β formulas, $\alpha \to \beta$, $\alpha \leftrightarrow \beta$ just abridge $\neg\alpha \vee \beta$, $(\alpha \to \beta) \wedge (\beta \to \alpha)$ respectively. Let us propose some examples in our framework of natural numbers. The injectivity of s can be expressed by the following formula in our language L

$$\forall v_0 \forall v_1 (s(v_0) = s(v_1) \to v_0 = v_1,$$

while the formula

$$\forall v_0 \neg(0 = s(v_0))$$

says that 0 is not in the image of s. Actually these formulas are *sentences* (each occurring variable is under the influence of a corresponding quantifier). In general, an occurrence of a variable v in a formula α is *bounded* if it is under the influence of a quantifier $\forall v$, $\exists v$, and *free* otherwise; α is called a *sentence* if, as already said, each occurrence of a variable in α is bounded. When writing $\alpha(\vec{v})$, we want to emphasize that the variables freely occurring in the formula α are in the tuple $\vec{v}$.
2. and 3. are very restrictive conditions, and are the distinctive peculiarity of first order logic. Actually, in Mathematics, one sometimes uses $\forall$ and $\exists$ on subsets (rather than on elements) of a structure. This is just what happens in our setting concerning (N, 0, s) with respect to the Induction Principle. In fact, Induction says

for *every subset* X of N, if X contains 0 and is closed under s, then $X = \text{N}$.

This statement uses $\forall$ on subsets, and this is not allowed in first order logic. Accordingly, the Induction Principle cannot be written (at least literally in the form proposed some lines ago) in the first order framework. This might look very disappointing: consequently, one may search more powerful and expressive ways of constructing formulas, for instance by allowing quantification on set variables (this it the so-called second order logic). But actually first order logic enjoys several important and reasonable technical theorems, that get lost and do not hold any more in these alternative worlds. We will discuss these results later, but it may be useful to quote already now a theorem of Lindström saying (very roughly speaking) that "first order logic is the best possible one" (see [11] for a detailed exposition of Lindström theorem).

However, formulas and sentences are not sufficient to form a logic. What we need now to accomplish a complete description of our setting is a notion a *truth*. We want to define when a sentence of a language L is true in a structure $\mathcal{A}$ of L, and, more generallly, when a sequence $\vec{a}$ in A makes

a formula $\alpha(\vec{v})$ true in $\mathcal{A}$. This can be done in a very natural way, saying exactly what one expects to hear. For instance the sentence $\exists v(v^2+1=0)$ is true in the complex field just because **C** contains some elements $\pm i$ satisfying the equation $v^2+1=0$; and in the ordered field of reals $\sqrt{2}$ makes the formula $v^2=2 \wedge v \geq 0$ true because satisfies both its conditions, while $-\sqrt{2}$, or 1, or other elements cannot satisfy the same formula. See again [153], or any handbook of Mathematical Logic for more details on the definition of first order truth. We omit them here.
Incidentally we note that, according this notion of truth, $\alpha \vee \beta$ just means $\neg(\neg\alpha \wedge \neg\beta)$, and $\forall v_n \alpha$ says the same thing as $\neg\exists v_n(\neg\alpha)$. So we could avoid the connective $\vee$ and the quantifier $\forall$ in our alphabet and, consequently, in our inductive definition of formula, and to introduce $\alpha \vee \beta$ and $\forall v_n \alpha$ as abbreviations, just as we did for $\alpha \to \beta$ and $\alpha \leftrightarrow \beta$. In this perspective, formulas are obtained from the atomic ones by using $\wedge$, $\neg$, $\exists$ and nothing else.
Moreover one can see that, according to this definition of truth, up to suitable manipulations, each formula $\varphi(\vec{w})$ can be written as

$$(\star) \quad Q_1 v_1 \ldots Q_n v_n \alpha(\vec{v}, \vec{w})$$

where $Q_1, \ldots, Q_n$ are quantifiers, $\vec{v}=(v_1, \ldots, v_n)$ and $\alpha(\vec{v}, \vec{w})$ is a quantifier free formula, and even a disjunction of conjunctions of atomic formulas and negations. $(\star)$ is called the *normal form* of a formula. When $\varphi(\vec{w})$ is in its normal form and every quantifier Q_i $(1 \leq i \leq n)$ is universal $\forall$ (existential $\exists$), we say that $\varphi(\vec{w})$ is **universal** (**existential**, respectively).

Before concluding this section, we would like to emphasize that the study of this truth relation between structures and sentences is just Model Theory, at least according to the feeling in the fifties. In fact, one says that a structure $\mathcal{A}$ is a *model* of a sentence α, or of a set T of sentences in the language L of $\mathcal{A}$, and one writes $\mathcal{A} \models \alpha$, $\mathcal{A} \models T$ respectively, whenever α, or every sentence in T, is true in $\mathcal{A}$. Model Theory is just the study of this relationship between structures and (sets of) sentences. Tarski provides an authoritative corroboration of this claim, when he writes in 1954 [158]

> *Whithin the last years, a new branch of metamathematics has been developing. It is called theory of models and can be regarded as a part of the semantics of formalized theories. The problems studied in the theory of models concern mutual relations between sentences of formalized theories and mathematical systems in which these sentences hold.*

It is notable that this Tarski quotation is likely to propose officially for the first time the expression *theory of models*. Accordingly, one might fix 1954 as the birthyear -or perhaps the baptism year- of Model Theory (if one likes this kind of matters). Actually, several themes related to the theory of models predate the fifties; but one can reasonably agree that just in that period Model Theory took its first steps as an autonomous subject in Mathematical Logic and in general mathematics.

1.3 Embeddings

We already defined in 1.1 embeddings and isomorphisms among structures of the same language L. We followed the usual algebraic approach. However there are alternative and equivalent ways, of more logical flavour, to introduce these notions. Let us recall them. First we consider embeddings.

Theorem 1.3.1 *Let $\mathcal{A}$ and $\mathcal{B}$ be structures of L, f be a function from A into B. Then the following propositions are equivalent :*

(i) *f is an embedding of $\mathcal{A}$ into $\mathcal{B}$;*

(ii) *for every quantifier free formula $\varphi(\vec{v})$ in L and for every sequence $\vec{a}$ in A, $\mathcal{A} \models \varphi(\vec{a})$ if and only if $\mathcal{B} \models \varphi(f(\vec{a}))$;*

(iii) *for every atomic formula $\varphi(\vec{v})$ in L and for every sequence $\vec{a}$ in A, $\mathcal{A} \models \varphi(\vec{a})$ if and only if $\mathcal{B} \models \varphi(f(\vec{a}))$.*

The proof is just a straightfoward check using the definitions of embedding, term and (atomic or quantifier free) formula. Referring to definitions is a winning and straightforward strategy also in showing the following characterizations of the notion of isomorphism.

Theorem 1.3.2 *Let $\mathcal{A}$ and $\mathcal{B}$ be structures of L, f be a surjective function from A onto B. Then the following propositions are equivalent:*

(i) *f is an isomorphism of $\mathcal{A}$ onto $\mathcal{B}$;*

(ii) *for every quantifier free formula $\varphi(\vec{v})$ in L and for every sequence $\vec{a}$ in A, $\mathcal{A} \models \varphi(\vec{a})$ if and only if $\mathcal{B} \models \varphi(f(\vec{a}))$;*

(iii) *for every atomic formula $\varphi(\vec{v})$ in L and for every sequence $\vec{a}$ in A, $\mathcal{A} \models \varphi(\vec{a})$ if and only if $\mathcal{B} \models \varphi(f(\vec{a}))$;*

(iv) *for every formula $\varphi(\vec{v})$ in L and for every sequence $\vec{a}$ in A, $\mathcal{A} \models \varphi(\vec{a})$ if and only if $\mathcal{B} \models \varphi(f(\vec{a}))$.*

It can be observed that, when f is any embedding of $\mathcal{A}$ into $\mathcal{B}$, for every quantifier free formula $\alpha(\vec{v}, w)$ in L and every sequence $\vec{a}$ in A,

$$\text{if } \mathcal{A} \models \exists \vec{w}\alpha(\vec{a}, \vec{w}), \text{ then } \mathcal{B} \models \exists \vec{w}\alpha(f(\vec{a}), \vec{w})$$

or also, equivalently,

$$\text{if } \mathcal{B} \models \forall \vec{w}\alpha(f(\vec{a}), \vec{w}), \text{ then } \mathcal{A} \models \forall \vec{w}\alpha(\vec{a}, \vec{w}).$$

Definition 1.3.3 *Two structures $\mathcal{A}$ and $\mathcal{B}$ of L are* **elementarily equivalent** *($\mathcal{A} \equiv \mathcal{B}$) if they satisfy the same sentences of L.*

As an easy corollary of Theorem 1.3.2, we have:

Theorem 1.3.4 *Isomorphic structures are elementarily equivalent.*

Conversely, it may happen that elementarily equivalent structures $\mathcal{A}$ and $\mathcal{B}$ are not isomorphic. We will see counterexamples below. However it is an easy exercise to show that, for finite structures, elementary equivalence and isomorphism are just the same thing.

Now let us introduce a related notion: partial isomorphism.

Definition 1.3.5 *Let $\mathcal{A}$ and $\mathcal{B}$ be structures of L. A* **partial isomorphism** *between $\mathcal{A}$ and $\mathcal{B}$ is an isomorphism between a substructure of $\mathcal{A}$ and a substructure of $\mathcal{B}$. $\mathcal{A}$ and $\mathcal{B}$ are said to be partially isomorphic $\mathcal{A} \simeq_p \mathcal{B}$ if there is a non empty set I of partial isomorphisms between $\mathcal{A}$ and $\mathcal{B}$ satisfying the* **back-and-forth** *property: for all $f \in I$,*

(i) *for every $a \in A$, there is some $g \in I$ such that $f \subseteq g$ and a is in the domain of g;*

(ii) *for every $b \in B$, there is some $g \in I$ such that $f \subseteq g$ and b is in the image of g.*

Example 1.3.6 Two dense linear orderings without endpoints $\mathcal{A} = (A, \leq)$ and $\mathcal{B} = (B, \leq)$ are partially isomorphic.
In fact, let I include all the possible isomorphisms between a finite substructure of $\mathcal{A}$ and a finite substructure of $\mathcal{B}$. I is not empty, because, for every $a \in A$ and $b \in B$, $a \mapsto b$ defines a partial isomorphism in I. Now take any $f \in I$; let $a_0 < a_1 < \ldots < a_n$ list the elements in the domain of f and $b_0 < b_1 < \ldots < b_n$ those in the image of f; so $f(a_i) = b_i$ for every $i \leq n$.

Pick $a \in A$, and notice that there exists some $b \in B$ such that, for every $i \leq n$,

$$a_i \leq a \quad \Leftrightarrow \quad b_i \leq b.$$

This is trivial when a is in the domain of f. Otherwise, one uses the facts that $\mathcal{B}$ has no minimum when $a < a_0$, that $\mathcal{B}$ has no maximum when $a > a_n$, and, finally, that the order of $\mathcal{B}$ is dense in the remaining cases. Define $g \in I$ by putting

$$Dom\, g = Dom\, f \cup \{a\}, \quad Im\, g = Im\, f \cup \{b\},$$

$$g \subseteq f, \quad g(a) = b.$$

Clearly g satisfies (i). (ii) is proved in the same way.

Remark 1.3.7 • If $\mathcal{A} \simeq \mathcal{B}$, then $\mathcal{A} \simeq_p \mathcal{B}$.
In fact, let f be an isomorphism of $\mathcal{A}$ onto $\mathcal{B}$. $I = \{f\}$ does satisfy (i) and (ii).

- Conversely, partially isomorphic structures may not be isomorphic.

Indeed one can find two structures that admit a different cardinality, and yet are partially isomorphic. For instance, this is the case of two dense linear orderings without endpoints. We have just seen that they are always partially isomorphic, indipendently of their cardinalities; in particular $(\mathbf{R}, \leq) \simeq_p (\mathbf{Q}, \leq)$.

But one can also find partially isomorphic non isomorphic structures with the same cardinality. For instance, still consider dense linear orderings without endpoints, and notice that $(\mathbf{R}, \leq) \simeq (\mathbf{R}+\mathbf{Q}, \leq)$ ($(\mathbf{R}+\mathbf{Q}, \leq)$ denotes here the disjoint union of a copy of $(\mathbf{R}, \leq)$ and a copy of $(\mathbf{Q}, \leq)$, where $(\mathbf{R}, \leq)$ precedes $(\mathbf{Q}, \leq)$). Both $(\mathbf{R}, \leq)$ and $(\mathbf{R}+\mathbf{Q}, \leq)$ have the continuum power. But they cannot be isomorphic, because $(\mathbf{R}+\mathbf{Q}, \leq)$, unlike $(\mathbf{R}, \leq)$, contains some countable intervals, and any order isomorphism maps countable intervals onto countable intervals.
However, within countable models, partially isomorphic structures are also isomorphic.

Theorem 1.3.8 *Let $\mathcal{A}$ and $\mathcal{B}$ be countable partially isomorphic structures. Then $\mathcal{A} \simeq \mathcal{B}$.*

The proof is obtained as follows. First one list A and B in some way

$$A = \{a_n : n \in \mathbf{N}\}, \quad B = \{b_n : n \in \mathbf{N}\}.$$

Let I be a set of partial isomorphisms between $\mathcal{A}$ and $\mathcal{B}$ ensuring $\mathcal{A} \simeq_p \mathcal{B}$. Due to (i) and (ii) one enlarges a given $f_\emptyset \in I$ by defining, for every natural n, a function $f_n \in I$ such that, for any n,

1. $f_n \subseteq f_{n+1}$,

2. a_n is in the domain of f_{2n},

3. b_n is in the image of f_{2n+1}.

Put $f = \cup_{n \in \mathbf{N}} f_n$. Owing to 1., f is a function; 2. implies that its domain is A, and 3. ensures that its image is B. In order to conclude that f is an isomorphism, we have to check that, for every atomic formula $\varphi(\vec{v})$ of L and every sequence $\vec{a}$ in A, $\mathcal{A} \models \varphi(\vec{a})$ if and only if $\mathcal{B} \models \varphi(f(\vec{a}))$. But this is easily done, as there is some n such that $\vec{a}$ is in the domain of f_n, and f_n restricts f and is an isomorphism between its domain and its image.

A noteworthy consequence of the theorem is

Corollary 1.3.9 (Cantor) *Two countable dense linear orderings without endpoints are isomorphic.*

Hence linearity, density and lack of endpoints characterize the order of rationals up to isomorphism. It should be underlined that Cantor's original proof used a different argument; but a subsequent approach of Hausdorff and Huntington inaugurated the back-and-forth method. In fact, what they did was just firstly to observe that two dense linear orders without endpoints are partially isomorphic (according to our modern terminology), and consequently to deduce that, if one adds the countability assumption, then isomorphism follows; the latter point can be easily generalized to arbitrary structures (and actually this is what Theorem 1.3.8 says). Now let us compare $\simeq_p$ and $\equiv$.

Theorem 1.3.10 *Partially isomorphic structures are elementarily equivalent.*

In fact let $\mathcal{A}$ and $\mathcal{B}$ be partially isomorphic structures in a language L, and let I be a set of partial isomorphisms between $\mathcal{A}$ and $\mathcal{B}$ witnessing $\mathcal{A} \simeq_p \mathcal{B}$.

Then one can show that, for every choice of a formula $\varphi(\vec{v})$ in L, a function $f \in I$ and a sequence $\vec{a}$ in the domain of f,

$$\mathcal{A} \models \varphi(\vec{a}) \quad \Leftrightarrow \quad \mathcal{B} \models \varphi(f(\vec{a})).$$

Note that, when φ ranges over the sentences of L, this implies $\mathcal{A} \equiv \mathcal{B}$. The proof proceeds by a straightforward induction on $\varphi(\vec{v})$: the conditions (i) and (ii) are useful in handling the quantifier step. In fact suppose that $\varphi(\vec{v})$ is of the form $\exists w \alpha(\vec{v}, w)$. If $\mathcal{A} \models \varphi(\vec{a})$, then there is some $b \in A$ satisfying $\mathcal{A} \models \alpha(\vec{a}, b)$. According to (i), there is some $g \in I$ enlarging f such that b is in the domain of g. By induction, $\mathcal{B} \models \alpha(g(\vec{a}), g(b))$. Therefore $\mathcal{B} \models \exists w \alpha(g(\vec{a}), w)$, $\mathcal{B} \models \exists w \alpha(f(\vec{a}), w)$ and, at last, $\mathcal{B} \models \varphi(f(\vec{a}))$. The converse is proved in a similar way, using (ii) instead of (i).

As a consequence of this theorem, one can deduce that there exist elementarily equivalent structures which are not isomorphic. For instance, among dense linear orders with no endpoints, $(\mathbf{Q}, \leq)$, $(\mathbf{R}, \leq)$ and $(\mathbf{R} + \mathbf{Q}, \leq)$ are partially isomorphic, hence elementarily equivalent. But they cannot be isomorphic, as already observed.
Notice that Theorem 1.3.10 provides also another proof that isomorphism implies elementary equivalence (via partial isomorphism). However, we will see later that elementarily equivalent structures may not be partially isomorphic (we will produce a counterexample). Let us also quote here the following result, characterizing elementary equivalence in terms of partial isomorphism.

Theorem 1.3.11 (Fraïssé) *Let $\mathcal{A}$ and $\mathcal{B}$ be structures in a finite language L. Then $\mathcal{A} \equiv \mathcal{B}$ if and only if there is a decreasing sequences $\{I_n : n \in \mathbf{N}\}$ of non empty sets I_n of partial isomorphisms between $\mathcal{A}$ and $\mathcal{B}$ such that, for every natural n and every $f \in I_{n+1}$,*

(a) *for every $a \in A$, there is $g \in I_n$ such that g extends f and the domain of g includes a,*

(b) *for every $b \in B$, there is $g \in I_n$ such that g extends f and the image of g includes b.*

Now let us deal again with arbitrary embeddings. The characterization of isomorphism given in Theorem 1.3.2 (iv) suggests the following notion.

Definition 1.3.12 *Let $\mathcal{A}$ and $\mathcal{B}$ be structures of L. An embedding f of $\mathcal{A}$ into $\mathcal{B}$ is called* **elementary** *if, for every formula $\varphi(\vec{v})$ in L and every sequence $\vec{a}$ in A, $\mathcal{A} \models \varphi(\vec{a})$ if and only if $\mathcal{B} \models \varphi(f(\vec{a}))$.*

We say that $\mathcal{A}$ is **elementarily embeddable** in $\mathcal{B}$, and we write $\mathcal{A} < \mathcal{B}$, when there is some elementary embedding of $\mathcal{A}$ in $\mathcal{B}$. $\mathcal{A}$ is called an **elementary substructure** of $\mathcal{B}$ if $A \subseteq B$ and the inclusion of A in B defines an elementary embedding of $\mathcal{A}$ in $\mathcal{B}$ (hence, for every formula $\varphi(\vec{v})$ in L and every sequence $\vec{a}$ in A, $\mathcal{A} \models \varphi(\vec{a})$ if and only if $\mathcal{B} \models \varphi(\vec{a})$). In this case, we say also that $\mathcal{B}$ is an **elementary extension** of $\mathcal{A}$.

Remark 1.3.13 1. Every isomorphism is, of course, an elementary embedding.

2. If $\mathcal{A}$ is elementarily embeddable in $\mathcal{B}$, then $\mathcal{A}$ and $\mathcal{B}$ are elementarily equivalent (just restrict the definition of elementary embedding to sentences φ in L).

Examples 1.3.14 (a) In the language $L = \{<\}$, consider the following structures

$$\mathcal{A} = (\mathbf{N} - \{0\}, <), \quad \mathcal{B} = (\mathbf{N}, <).$$

The inclusion of $\mathbf{N} - \{0\}$ in $\mathbf{N}$ defines an embedding of $\mathcal{A}$ in $\mathcal{B}$ which is not elementary because 1 is a minimal element in $\mathcal{A}$, but not in $\mathcal{B}$ (in other words, satisfies $\exists w(w < v)$ in $\mathcal{B}$, but not in $\mathcal{A}$). On the contrary, the successor function from $\mathbf{N}$ in $\mathbf{N} - \{0\}$ is an isomorphism between $\mathcal{B}$ and $\mathcal{A}$.

(b) The real field $\mathbf{R}$ is a subfield of the complex field $\mathbf{C}$, and hence is a substructure in the language $L = \{0, 1, +, \cdot, -\}$. However $\mathbf{R}$ is not an elementary substructure of $\mathbf{C}$. In fact, for every positive real a, a satisfies the formula $\exists w(w^2 + v = 0)$ in $\mathbf{C}$, but not in $\mathbf{R}$. On the other hand, we will see later that every embedding of algebraically closed fields (or real closed fields) is elementary.

Another remarkable class of embeddings concerns existential formulas.

Definition 1.3.15 *Let $\mathcal{A}$ and $\mathcal{B}$ be structures in a language L. An embedding f of $\mathcal{A}$ into $\mathcal{B}$ is called* **existential** *if and only if, for every existential formula $\varphi(\vec{v})$ in L and for every sequence $\vec{a}$ in A,*

$$\mathcal{A} \models \varphi(\vec{a}) \Leftrightarrow \mathcal{B} \models \varphi(f(\vec{a})).$$

In more detail, we require that, for every quantifier free formula $\alpha(\vec{a}, \vec{w})$ in L and for every sequence $\vec{a}$ in A,

$$(\star)\ \mathcal{A} \models \exists \vec{w} \alpha(\vec{a}, \vec{w}) \text{ if and only if } \mathcal{B} \models \exists \vec{w} \alpha(f(\vec{a}), \vec{w}).$$

When $A \subseteq B$ and f is the inclusion of $\mathcal{A}$ into $\mathcal{B}$, we say that $\mathcal{A}$ is an **existential substructure** of $\mathcal{B}$. When there is some existential embedding of $\mathcal{A}$ in $\mathcal{B}$, we say that $\mathcal{A}$ is **existentially embedded** in $\mathcal{B}$ and we write $\mathcal{A} <_1 \mathcal{B}$.

Let us discuss briefly existential embeddings f. First of all, it is clear that elementary embeddings are existential, too. Furthermore, notice that $(\star)$ can be weakened to require

$$\text{if } \mathcal{B} \models \exists \vec{w} \alpha(f(\vec{a}), \vec{w}), \text{ then } \mathcal{A} \models \exists \vec{w} \alpha(\vec{a}, \vec{w})$$

because the inverse implication

$$\text{if } \mathcal{A} \models \exists \vec{w} \alpha(\vec{a}, \vec{w}), \text{ then } \mathcal{B} \models \exists \vec{w} \alpha(f(\vec{a}), \vec{w})$$

is satisfied by every embedding. Now, a quantifier free formula $\alpha(\vec{v}, \vec{w})$ can be equivalently written (by standard propositional techniques) as a disjunction

$$\bigvee_{j \leq s} \alpha_j(\vec{v}, \vec{w})$$

where s is a natural number and, for every $j \leq s$, $\alpha_j(\vec{v}, \vec{w})$ is a conjunction of atomic formulas (equations) and negations. One easily deduce that $\exists \vec{w} \alpha(\vec{v}, \vec{w})$ can be equivalently written as

$$\bigvee_{j \leq s} \exists \vec{w} \alpha_j(\vec{v}, \vec{w}).$$

It follows that f is existential if and only if

$$\mathcal{B} \models \exists \vec{w} \alpha(f(\vec{a}), \vec{w}) \text{ implies } \mathcal{A} \models \exists \vec{w} \alpha(\vec{a}, \vec{w})$$

for every $\vec{a}$ in A and for every finite conjunction $\alpha(\vec{v}, \vec{w})$ of equations and negations.

Now let us deal again with arbitrary elementary embeddings. Let L be a language, $\mathcal{A}$ be a structure of L. In order to examine the structures elementarily equivalent to $\mathcal{A}$, L is just the language we need. But, within these models, one meets those where $\mathcal{A}$ is elementarily embeddable; and these structures actually require a richer language, emphasizing the fact that $\mathcal{A}$ embeds (elementarily) in each of them. This larger language is built by adding to L a constant symbol for every $a \in A$, and will be called $L(A)$; the new constant corresponding to a could be written c_a, in order to remind a but to avoid any confusion with a. But we will often denote it ambiguously by a (for simplicity's sake).

$\mathcal{A}$ itself becomes a structure of $L(A)$ provided we interpret quite naturally, for every $a \in A$, the constant corresponding to a in a. The resulting structure will be denoted $\mathcal{A}_A$.
Which are the structures elementarily equivalent to $\mathcal{A}_A$ in $L(A)$? They are obtained as follows. Take an L-structure $\mathcal{B}$ where $\mathcal{A}$ embeds elementarily, say by f; for every $a \in A$, let $f(a)$ interpret the constant of a. One gets in this way a structure $\mathcal{B}_{f(A)}$ of $L(A)$, and it is easy to check that $\mathcal{B}_{f(A)} \equiv \mathcal{A}_A$. Conversely, every structure elementarily equivalent to $\mathcal{A}_A$ can be obtained in this way.
More generally, for every structure $\mathcal{A}$ of L and for every subset X of A, one can introduce a new language $L(X)$ by adding to L a new constant for every element x in X. $\mathcal{A}_X$ is the $L(X)$-structure expanding $\mathcal{A}$ and interpreting, for every $x \in X$, the constant symbol corresponding to x in x itself. Of course, there do exist other structures elementarily equivalent to $\mathcal{A}_X$. Let us see how to construct them. Let $\mathcal{B}$ be a structure of L.

Definition 1.3.16 *A function f from X into B is called* **elementary** *if, for every formula $\varphi(\vec{v})$ in L and for every sequence $\vec{x}$ in X,*

$$\mathcal{A} \models \varphi(\vec{x}) \quad \Leftrightarrow \quad \mathcal{B} \models \varphi(f(\vec{x})).$$

Notice that, when $X = A$, an elementary function from X in B is just an elementary embedding of $\mathcal{A}$ in $\mathcal{B}$. Moreover, for any X, an elementary function f from X in B enjoys the following properties.

(i) f is $1-1$ (use $\varphi(v_1, v_2) : v_1 = v_2$).

(ii) f^{-1} itself is an elementary function (from $f(X)$ in A).

(iii) $\mathcal{A}$ and $\mathcal{B}$ are elementarily equivalent (apply the definition of elementary function to the sentences φ of L).

Now take an L-structure $\mathcal{B}$. Let f be an elementary function from X into B, and, for every $x \in X$, let $f(x)$ interpret the constant of x in $L(X)$. One gets in this way a structure $\mathcal{B}_{f(X)}$ of $L(X)$, and even a model elementarily equivalent to $\mathcal{A}_X$. Conversely, every structure $\equiv \mathcal{A}_X$ can be obtained in this way.

Remark 1.3.17 Let $\varphi(\vec{v}, \vec{w})$ be a formula of L, $\vec{a}$ be a sequence in A, $\vec{x}$ be a sequence in X. Fussy people will like to distinguish

(a) $\mathcal{A} \models \varphi(\vec{a}, \vec{x})$ (in the sense that $\mathcal{A}$ satisfies the L-formula $\varphi(\vec{v}, \vec{w})$ if $\vec{v}$, $\vec{w}$ are embodied by $\vec{a}$, $\vec{x}$ respectively);

(b) $\mathcal{A}_X \models \varphi(\vec{a}, \vec{x})$ (in the sense that $\mathcal{A}_X$ satisfies the $L(X)$-formula $\varphi(\vec{v}, \vec{x})$ if $\vec{v}$ is embodied by $\vec{a}$);

(c) $\mathcal{A}_A \models \varphi(\vec{a}, \vec{x})$ (in the sense that $\mathcal{A}_A$ satisfies the $L(A)$-sentence $\varphi(\vec{a}, \vec{x})$).

However one easily shows that (a), (b) and (c) are equivalent. So we will use the simplest notation (that of (a)) to mean any of these conditions.
We conclude this section by mentioning without proof two theorems on elementary embeddings. We will state them for simplicity in the case when the involved embeddings are just inclusions but they extend to arbitrary (elementary) embeddings.
The former theorem provides a criterion to check whether a given subset of a structure $\mathcal{B}$ is the domain of an elementary substructure of $\mathcal{B}$ (remember the discussion at the end of 1.1).

Theorem 1.3.18 (Tarski-Vaught) *Let $\mathcal{B}$ be a structure of L, A be a subset of B. Then the following propositions are equivalent:*

(i) *A is the domain of an elementary substructure $\mathcal{A}$ of $\mathcal{B}$;*

(ii) *for every formula $\alpha(\vec{v}, w)$ of L and for every sequence $\vec{a}$ in A, if $\mathcal{B} \models \exists w \alpha(\vec{a}, w)$, then there exists an element $b \in A$ such that $\mathcal{B} \models \alpha(\vec{a}, b)$.*

Now let us introduce the latter result. Take a set I totally ordered by a relation $\leq$. For every $i \in I$, let $\mathcal{A}_i$ be a structure of L. Suppose that, for every choice of $i \leq j$ in I, $\mathcal{A}_i$ is a substructure of $\mathcal{A}_j$: this means that $A_i \subseteq A_j$ and

- for every constant c of L, $c^{\mathcal{A}_i} = c^{\mathcal{A}_j}$,
- for every n-ary operation symbol F of L, $F^{\mathcal{A}_i}$ is the restriction of $F^{\mathcal{A}_j}$ to A_i^n,
- for every n-ary relation symbol R of L, $R^{\mathcal{A}_i} = R^{\mathcal{A}_j} \cap A_i^n$.

Therefore we can build a new structure $\mathcal{A}$ in L, having domain $A = \bigcup_{i \in I} A_i$ (and hence including A_i for all $i \in I$), and interpreting the symbols of L as follows:

- for every constant c of L, $c^{\mathcal{A}} = c^{\mathcal{A}_i}$ where i is any element in I;

- for every n-ary operation symbol F of L and for every choice of $a_1, \ldots, a_n$ in A,

$$F^{\mathcal{A}}(a_1, \ldots, a_n) = F^{\mathcal{A}_i}(a_1, \ldots, a_n)$$

  where $i \in I$ satisfies $a_1, \ldots, a_n \in A_i$;

- for every n-ary relation symbol R of L and for every choice of $a_1, \ldots, a_n$ in A,

$$(a_1, \ldots, a_n) \in R^{\mathcal{A}} \quad \Leftrightarrow \quad (a_1, \ldots, a_n) \in R^{\mathcal{A}_i}$$

  where $i \in I$ satisfies $a_1, \ldots, a_n \in A_i$.

It is clear that $\mathcal{A}$ is well defined and extends $\mathcal{A}_i$ for every $i \in I$. Straightfoward techniques show:

Theorem 1.3.19 (Elementary Chain Theorem) *Suppose that, for every choice of $i \leq j$ in I, $\mathcal{A}_i$ is an elementary substructure of $\mathcal{A}_j$. Then, for every $i \in I$, $\mathcal{A}_i$ is an elementary substructure of $\mathcal{A}$.*

1.4 The Compactness Theorem

The Compactness Theorem is the most powerful tool -and indeed a key feature- in classical Model Theory. It states

Theorem 1.4.1 *Let S be an infinite set of sentences in a language L. Suppose that every finite subset of S has a model. Then S has a model.*

Notice that the converse is obvious, because a model of S is a model of every (finite or infinite) subset of S. But the theorem ensures that, if every finite subset of S has its own model (hence different subsets may admit different models), then there is a global model satisfying all the sentences in S. In fact, there are several situations where the Compactness Theorem applies and guarantees satisfiability for sets S of sentences for which it is very difficult to imagine a general model directly, but it is quite simple to equip every finite subset with a suitable private model: we will see some of them later in 1.5. In fact this section and (implicitly) the next one will be devoted to discussing this fundamental theorem and, in detail:

- its proof;

- its name;

- its role in producing "nonstandard" and, in some sense, unexpected models, and, as the reverse side of the same medal, in bounding the expressiveness of first order logic and in excluding that some familiar principles, like induction on naturals, may be written in any way in the first order framework;
- in spite of this, its plausibility, supported by metamathematical considerations on the nature of mathematical proofs;
- finally, some words about the already mentioned theorem of Lindström saying that the Compactness Theorem, together with a related result (the downward Löwenheim Skolem Theorem) fully characterizes first order logic.

As said, let us begin by discussing the proof. There are several possible ways to show the Compactness Theorem. For instance, there is an approach based on the algebraic notion of ultraproduct and due to Keisler (see [39]). Another classical proof was proposed by Henkin. Let us outline very quickly its idea.
What we have at the beginning is a(n infinite) set of sentences S such that every finite subset has a model. Some technical -and non trivial- preliminary work shows that there is no loss of generality in assuming that S satisfies two further conditions:

1. S is *complete*, in other words, for every sentence φ in L, either φ or $\neg\varphi$ is in S;
2. S is *rich*: if S contains a sentence of the form $\exists v \alpha(v)$, then there is a constant symbol c in L such that $\alpha(c)$ is in S.

At this point a quite artificial construction produces the model we are looking for. Basically, the domain is just the set of terms without variables in L (the so-called *Herbrand universe* of L); this is non-empty owing to 2. The L-structure arises in a rather reasonable way. 1 and 2 play a key role in showing that what we build is a model of S.
It is worth emphasizing that the model we get in our proof is countable (for a countable L; when the language has a larger cardinality λ, it is easy to check that our argument still works and produces a model of power $\leq \lambda$). So, as a byproduct of the Henkin proof, we have that, when S has a model, then S has a countable model: this is the so called *Downward Löwenheim-Skolem Theorem*, and is a notable result. We shall discuss its relevance in 1.5.

Now let us treat the reasons of the theorem name. Actually compactness recalls topology. In fact, we will see later in Chapter 5 that the theorem has a topological content and implicitly says that a certain topological space is compact.
We shall see within a few lines in 1.5 that the Compactness Theorem produces some strong and severe expressiveness restrictions in first order logic. For instance, we will show that, just owing to Compactness, conditions like finiteness, or popular statements such as the Minimum Principle, cannot be expressed in a first order way. On the other hand, one should agree that what the Compactness Theorem says is a quite reasonable statement, especially if one considers the following corollary. Let S be a set of sentences of L and σ be a sentence of L; we say that σ is a *logical consequence* of S, and we write $S \models \sigma$, when σ is true in all the models of S.

Corollary 1.4.2 *If $S \models \sigma$, then there is a finite subset S_0 of S such that $S_0 \models \sigma$.*

Proof. Clearly $S \models \sigma$ if and only if $S \cup \{\neg\sigma\}$ has no models. But, owing to Compactness, this is equivalent to say that there is a finite subset of $S \cup \{\neg\sigma\}$ without any models. With no loss of generality we can assume that this finite set is of the form $S_0 \cup \{\neg\sigma\}$ where $S_0 \subseteq S$. But, again, stating that $S_0 \cup \{\neg\sigma\}$ has no models is equivalent to say that $S_0 \models \varphi$. ♣

Now, another fundamental result in first order logic, deeply related to compactness -the Completeness Theorem- says that one can explicitly provide a notion of *provability* accompanying and supporting this concept of consequence in such a way that the logical consequences of a given S are just what is proved by S at the end of a sequence of rigorous deductions. So what compactness in conclusion emphasizes is the finitary nature of mathematical proofs; this feature can be regarded as an authoritative witness in its favour and, trough it, as a support to first order logic.

1.5 Elementary classes and theories

When considering, for a given language L, structures, formulas and truth, two problems arise quite naturally:

(a) given a set T of L-sentences, "*classify*" the models of T (their class will be denoted $Mod(T)$);

(b) given a class **K** of L-structure, "*characterize*" the set of the L-sentences true in all the structures of **K** (this set will be called the *theory of* **K** and denoted $Th(\mathbf{K})$).

According to its declared intents, Model Theory should be mainly concerned with Problem (a). However, also (b) arises quite naturally in the model theoretic framework. For instance, consider a class **K** formed by a single structure, like the complex field, or the real field. We will see later that **K** cannot be represented as $Mod(T)$ for any T. But it may be quite interesting to realize in an explicit way which sentences are true in the only structure in **K**.
However we have to admit that the previous statements of (a) and (b) are somewhat vague and unprecise. First of all, what do *classifying* or *characterizing* mean? This is not a minor question; on the contrary, it is a very delicate and central matter. For instance, the classification problem for a class of structures touches and overlaps several basic open questions in Algebra. So we should be more detailed about this crucial point. Of course, one can reasonably agree that a classification should identify isomorphic structures. But this is still a partial and indefinite answer; we should fix more precisely which criteria, tools and invariants we want to use in our classification problem. We shall try to clarify these fundamental questions in the next chapters. Here we limit ourselves to discuss other points, mainly concerning (a). T is a set of sentences in a language L.

1. Let σ be a sentence of L, and suppose that every model of T is also a model of σ (so σ a logical consequence of T $T \models \sigma$). Hence $Mod(T) = Mod(T \cup \{\sigma\})$. Consequently we can assume with no loss of generality for our purposes that

   for every sentence σ of L, if $T \models \sigma$, then $\sigma \in T$.

   A set of sentences in L with this property is called a *theory* of L. It is a simple exercise to show that, given a set T of sentences of L, T is a theory if and only if there is a class **K** of structures of L such that $T = Th(\mathbf{K})$ (hint: ($\Leftarrow$) is clear; to show ($\Rightarrow$) use $\mathbf{K} = Mod(T)$).

2. A theory T is called *consistent* if and only if T satisfies one of the following (equivalent) conditions:

   (i) for every sentence σ of L, either $\sigma \notin T$ or $\neg\sigma \notin T$;

   (ii) there is some sentence of L which is not in T;

(iii) $Mod(T) \neq \emptyset$.

To show the equivalence among (i), (ii) and (iii) is an easy exercise. (i) says that the consistent theories are just those excluding any contradiction. (iii) ensures that these theories are exactly those admitting at least one model. It is clear that, within Problem (a), we are exclusively interested in consistent theories. So we can assume in (a) that T is a consistent theory; accordingly hereafter *theory* will always abbreviate *consistent theory*.
A rigid model theoretic perspective might limit the classification analysis to the classes of models of (consistent) theories. But open minds could prefer a more general study, providing an abstract treatment of the classification problem for arbitrary classes of structures. Hence it is worth underlining that there do exist classes **K** of L-structures which are not of the form $\mathbf{K} = Mod(T)$ for any theory T of L. We propose here some examples; the Compactness Theorem is a fundamental tool in this setting.

Definition 1.5.1 *A (non-empty) class* **K** *of structures of* L *is said to be* **elementary** *(or also* **axiomatizable***) if there is a set* T *of sentences of* L *(without loss of generality, a theory* T *of* L*) such that* $\mathbf{K} = Mod(T)$.

Now let us propose a series of examples, as promised. Part of them aim at pointing out that several classes of structures are explicitly elementary because their definitions can be naturally written in a first order way. But other cases are not elementary: it is here that the Compactness Theorem plays its role and first order logic shows its expressiveness bounds.

Examples 1.5.2 1. Let $L = \emptyset$ (so the structures of L are the non-empty sets), **K** be the class of infinite sets. "*Infinite*" means "**admitting at least** $n+1$ **elements for every natural** n". Given n, the property "*there are at least* $n+1$ *elements*" can be expressed in a first order way by the following sentence of L

$$\sigma_n \quad : \quad \exists v_0 \ldots \exists v_n \bigwedge_{i<j\leq n} \neg(v_i = v_j).$$

Hence $\mathbf{K} = Mod(T)$ where $T = \{\sigma_n : n \in \mathbf{N}\}$, and so **K** is elementary.

2. Let again $L = \emptyset$, but now let **K** be the class of finite (non-empty) sets. "*Finite*" means "*having at most* $n+1$ *elements for some natural* n". Given n, the proposition "*there are at most* $n+1$ *elements*" can

be expressed in a first order way by the sentence $\neg\sigma_{n+1}$. But now $Mod(\{\neg\sigma_{n+1} : n \in \mathbf{N}\})$ is not $\mathbf{K}$, indeed it equals the class of the sets having only one element. So the approach in 1 does not work any longer. However, assume that $\mathbf{K}$ is elementary, hence $\mathbf{K} = Mod(T)$ for a suitable set T of sentences of L. Put

$$T' = T \cup \{\sigma_n : n \in \mathbf{N}\}.$$

Let T'_0 be a finite subset of T'. For some natural N,

$$T'_0 \subseteq T \cup \{\sigma_n : n \in \mathbf{N},\, n \leq N\}.$$

Notice that $T \cup \{\sigma_n : n \in \mathbf{N},\, n \leq N\}$ (and hence T'_0) has a model: it suffices to take a finite set with at least $N+1$ elements. At this point, owing to the Compactness Theorem, we deduce that T' itself has a model. This is a set both finite (as a model of T') and infinite (as a model of σ_n for every natural n). We get in this way a contradiction. Hence $\mathbf{K}$ is not elementary.

Notice that this argument works as well for every class $\mathbf{K}$ of finite *arbitrarily large* structures (in the sense that, for every positive integer n, there is a structure in $\mathbf{K}$ whose size is larger than n). A class $\mathbf{K}$ of this kind cannot be elementary; in other words, the theory of $\mathbf{K}$ does admit infinite models, too; notice that this applies, for instance, to the class of finite groups, as well as to the class of finite fields. So one can wonder which are the infinite models of the theory of these finite structures. We will consider the particular case of fields later in Example 6.

Now let us deal with orders.

3. Let $L = \{\leq\}$ where $\leq$ is a binary relation symbol (which we confuse, for simplicity, with its interpretation below -an order relation-). In L we consider the class $\mathbf{K}$ of linear orders. It is easily seen that $\mathbf{K}$ is elementary, because the properties defining linear orders are first order sentences (and indeed universal first order sentences) of L. For instance, linearity can be expressed by

$$\forall v_0 \forall v_1 (v_0 \leq v_1 \vee v_1 \leq v_0).$$

The set of the logical consequences of these sentences is the *theory of linear orders*; it is formed by the sentences true in every linear order.

In the same way the class of *dense linear orders without endpoints* is elementary; in fact, density is stated by

$$\forall v_0 \forall v_1 \exists v_2 (v_0 < v_1 \to v_0 < v_2 \wedge v_2 < v_1),$$

and lackness of endpoints by

$$\forall v_0 \exists v_1 (v_1 < v_0),$$

$$\forall v_0 \exists v_1 (v_0 < v_1)$$

($v_0 < v_1$ abbreviates here $v_0 \leq v_1 \wedge \neg(v_0 = v_1)$).

The set of the logical consequences of the sentences quoted so far is the *theory of dense linear orders without endpoints*; we will denote it by DLO^-. It is formed by the sentences true in every dense linear order with no endpoints. Recall that $(\mathbf{Q}, \leq)$, $(\mathbf{R}, \leq)$ are dense linear orders without endoints, and consequently their theories include DLO^- (and one may wonder if they actually equal DLO^-).

The reader can check directly that the following classes of L-structures are elementary:

- dense linear orders with a least but no last element, or a last but no least element, or both a least and a last element,
- infinite discrete linear orders with or without endpoints (an order is discrete when every element, but the least one -if any-, has a predecessor and every element, but the last one -if any-, has a successor).

4. We still work in $L = \{\leq\}$ (where $\leq$ is a binary relation symbol), but this time we deal with the class $\mathbf{K}$ of well ordered sets (so ordered sets where every non-empty subset has a least element). Hence $(\mathbf{N}, \leq) \in \mathbf{K}$ owing to the Minimum Principle, while any dense linear order $(A, \leq)$, even with a minimum, does not lie in $\mathbf{K}$ (in fact, given $b > a$ in A, which is the least element $> a$ in A?). So the situation is, in some sense, opposite to the last (elementary) example 3.

   Suppose that $\mathbf{K}$ is elementary, so $\mathbf{K} = Mod(T)$ for a suitable set T of L-sentences. Put
   $$L' = L \cup \{c_n : n \in \mathbf{N}\}$$
   where, for every natural n, c_n is a constant symbol, and in L' look at the following set of sentences
   $$T' = T \cup \{c_{n+1} < c_n : n \in \mathbf{N}\}.$$

Let T'_0 be a finite subset of T', then there is some natural N such that

$$T'_0 \subseteq T \cup \{c_{n+1} < c_n : n \in \mathbf{N},\, n \leq N\}.$$

Then T'_0 has a model because $T \cup \{c_{n+1} < c_n : n \in \mathbf{N}, n \leq N\}$ has: it suffices to take the well ordered set $(\mathbf{N}, \leq)$, to interpret c_0, $c_1, \ldots, c_{N+1}$ in $N+1, N, \ldots, 0$ respectively, and any further constant c_n (with $n > N$) arbitrarily. By the Compactness Theorem, T' does admit a model

$$\mathcal{A}' = (A, \leq, (c_n^{\mathcal{A}'})_{n \in \mathbf{N}}).$$

Let $\mathcal{A} = (A, \leq)$, then $\mathcal{A}$ is a model of T, and hence is a well ordered set; however it contains the non-empty subset

$$X = \{c_n^{\mathcal{A}'} : n \in \mathbf{N}\}$$

admitting no minimum, because, for every natural n, $c_{n+1}^{\mathcal{A}'} < c_n^{\mathcal{A}'}$. So we get a contradiction. Consequently $\mathbf{K}$ is not elementary. In other words there are linearly ordered sets which are not well ordered but satisfy the same first order sentence as well ordered sets.

5. Let now $L = \{0, 1, +, -, \cdot\}$ be our language for fields. We consider in L the class $\mathbf{K}$ of fields. $\mathbf{K}$ is elementary. In fact the definition itself of field can be written as a series of first order sentences (in most cases, of universal first order sentences) in L. For instance

$$\forall v_0 \exists v_1 (\neg(v_0 = 0) \to v_0 \cdot v_1 = 1),$$

says that any non zero element has an inverse.

Also the class of algebraically closed fields is elementary, although the corresponding check is a little subtler. In fact what we have to say now is that, for every natural n, any (monic) polynomial of degree $n+1$ has at least one root. So the point is how to quantify over polynomials of degree $n+1$. However recall that such a polynomial is just an ordered sequence of length $n+2$ of elements in the field: the first is the coefficient of degree 0, the last is the coefficient of degree $n+1$ (and equals 1 if we deal with monic polynomials); so what we have to write is just, for every n,

$$\forall v_0 \forall v_1 \ldots \forall v_n \exists v (v_0 + v_1 \cdot v + \ldots + v_n \cdot v^n + v^{n+1} = 0)$$

(where v^i has the obvious meaning, for every $i \leq n+1$).

The logical consequences of the sentences listed so far form the *theory of algebraically closed fields*, usually denoted ACF. Now let p be a prime, or $p = 0$. Also the class of (algebraically closed) fields of characteristic p is elementary; for, it suffices to add to the previous sentences the one saying that the sum of p times 1 is 0 when p is a prime, or, when $p = 0$, the negations of all these sentences. In conclusion, for every p prime or equal to 0, we can introduce the theory ACF_p of *algebraically closed fields of characteristic* p. Among the algebraically closed fields in characteristic 0 recall the complex field $\mathbf{C}$, as well as the (countable) field $\mathbf{C}_0$ of complex algebraic numbers; their theories contain ACF_0, and one may wonder if they equal ACF_0. Recall also that every field $\mathcal{K}$ has a (minimal) algebraically closed extension $\overline{\mathcal{K}}$; in particular, for p prime, $\overline{\mathbf{Z}/p\mathbf{Z}}$ is an example of algebraically closed field in characteristic p.

Since we are treating fields, let us consider again finite fields, and, more exactly, the infinite models of their theory we met in example 2: the so called *pseudofinite fields*. As observed before, one can ask which is the structure of these fields. J. Ax equipped them with a very elegant axiomatization, explaining the essential nature of finite fields in the first order setting: in fact, pseudofinite fields are just the fields $\mathcal{K}$ such that:

- ⋆ $\mathcal{K}$ is perfect,
- ⋆ $\mathcal{K}$ has exactly one algebraic extension of every degree,
- ⋆ every absolutely irreducible variety over $\mathcal{K}$ has a point in K.

All these conditions can be written in a first order way, although this is not immediate to check.

6. A first order language for the class $\mathbf{K}$ of *ordered fields* is $L = \{0, 1, +, -, \cdot, \leq\}$. $\mathbf{K}$ is elementary in L because it equals $Mod(T)$ where T is the set of the following sentences in L:

   (i) the field axioms (see Example 5);

   (ii) those characterizing the linear orders (see Example 3);

   (iii) the sentence saying that sums and products of nonnegative elements are nonnegative

$$\forall v_0 \forall v_1 (0 \leq v_0 \wedge 0 \leq v_1 \rightarrow 0 \leq v_0 + v_1 \wedge 0 \leq v_0 \cdot v_1).$$

Also the class of real closed ordered fields (those satisfying the Intermediate Value Property for polynomials of degree ≥ 1) is elementary, it suffices to add the new sentences:

(iv) for every natural n,

$$\forall v_0 \forall v_1 \ldots \forall v_n \forall u \forall w \exists v (v_0 + v_1 \cdot u + \ldots + v_n \cdot u^n + u^{n+1} < 0 \wedge$$

$$\wedge\, 0 < v_0 + v_1 \cdot w + \ldots + v_n \cdot w^n + w^{n+1} \wedge u < w \rightarrow$$

$$\rightarrow u < v \wedge v < w \wedge v_0 + v_1 \cdot v + \ldots + v_n \cdot v^n + v^{n+1} = 0).$$

The logical consequences of (i), (ii), (iii), (iv) form the *theory of real closed ordered fields*, usually denoted RCF. Examples of real closed ordered fields are the ordered field of the real numbers $\mathbf{R}$, as well as the (countable) ordered field $\mathbf{R}_0$ of real algebraic numbers. Their theories include RCF, and one may wonder if actually they equal RCF.

7. Let $\mathcal{R}$ be a (countable) ring with identity. Consider the language $L_R = \{0, +, -, r \,(r \in R)\}$ of (left) $\mathcal{R}$-modules. The class of left $\mathcal{R}$-modules is elementary because it equals the class of models of the following sentences in L_R:

   (i) those axiomatizing the abelian groups in the language with 0, $+$ and $-$;

   (ii) for every $r, s \in R$, if $r+s$ and $r \cdot s$ denote the sum and the product (respectively) of r and s in R,

$$\forall v_0((r+s)v_0 = rv_0 + sv_0),$$

$$\forall v_0((r \cdot s)v_0 = r(sv_0)),$$

$$\forall v_0 \forall v_1 (r(v_0 + v_1) = rv_0 + rv_1),$$

   (iii) finally, if 1 denotes the identity element in $\mathcal{R}$,

$$\forall v_0(1v_0 = v_0).$$

The logical consequences of the previous sentences form the theory ${}_{\mathcal{R}}T$ of left $\mathcal{R}$-**modules**. Of course, there is no reason to prefer the left to the right, at least in this case; indeed, one can check that even the class of right $\mathcal{R}$-modules is elementary, and consequently one can introduce the theory $T_{\mathcal{R}}$ of right $\mathcal{R}$-modules.

Let us come back to our classification problem for elementary, or also non-elementary classes. The following fundamental theorem can suggest that, even in the elementary case, this problem is not simple, as the class of models of a theory T can include many pairwise non-isomorphic structures.

Theorem 1.5.3 (Löwenheim-Skolem) *Let T be a theory in a (countable) language L. Suppose that T has some infinite model. Then, for every infinite cardinal λ, T admits some model of power λ.*

The proof just uses Compactness in the extended framework of languages of arbitrary cardinalities. In fact one enlarges L by λ many new constant symbols c_i ($i \in I, |I| = \lambda$) and one gets in this way an extended language L'. In L' one considers the following set of sentences

$$T' = T \cup \{\neg(c_i = c_j) \, : \, i, \, j \in I, \, i \neq j\}.$$

Any finite portion T'_0 of T' has a model; in fact it turns out that, for some finite subset I_0 of I,

$$T'_0 \subseteq T \cup \{\neg(c_i = c_j) \, : \, i, \, j \in I_0, i \neq j\}.$$

so, in order to obtain a model of T'_0, it is sufficient to refer to an infinite model $\mathcal{A}$ di T, as ensured by the hypothesis, and to interpret the finitely many constants c_i ($i \in I_0$) in pairwise different elements of A. At this point, Compactness applies and gives a model of T' (hence of T) of power $\leq \lambda$. But this model has to include the λ many distint interpretations of the c_i's, and so its power is exactly λ.

Therefore, if a theory T of L has at least an infinite model then T has a model in each infinite power (and two models with different cardinalities cannot be isomorphic). Of course, one may wonder how strong is the assumption that T has some infinite model. Not so much, if one recalls that a theory T admitting finite models of arbitrarily large size must admit also some infinite models. Another reasonable question may concern how many models T admits in any fixed infinite cardinal λ. One can check that their number cannot exceed 2^λ, but this upper bound can be reached, for every λ, by some suitable T's. The opposite case, when T has just one model in power λ (up to isomorphism), will be of some interest in the next chapters; we fix it in the following definition.

Definition 1.5.4 *Let T be a theory with some infinite model, λ be an infinite cardinal. T is said to be* **λ-categorical** *if and only it any two models of T of power λ are isomorphic.*

We wish to devote some more lines to the Löwenheim-Skolem theorem. Among other things, it confirms that elementary equivalence is a weaker relation than isomorphism. In fact, take an infinite structure $\mathcal{A}$, and use the Löwenheim-Skolem to build a model $\mathcal{A}'$ satisfying the same first order sentences as $\mathcal{A}$ but having a different cardinality. It is easily checked that $\mathcal{A}, \mathcal{A}'$ are elementary equivalent; but, of course, they cannot be isomorphic. Now recall what we pointed out in 1.2 : the Induction Principle (in its usual form) cannot be written in the first order style in the language for $(\mathbf{N}, 0, s)$ because first order logic forbids quantification on set variables. However, as far as we know, one might find an equivalent statement that can be expressed in the first order setting; in this sense, Induction might become a first order statement. Well, the Löwenheim-Skolem theorem excludes this extreme possibility. For, the Induction Principle characterizes $(\mathbf{N}, 0, s)$ up to isomorphism, while the Löwenheim-Skolem theorem ensures us that any tentative first order equivalent translation (even involving infinitely many sentences) has some uncountable models. So this translation cannot exist.

The Löwenheim-Skolem theorem emphasizes other similar expressiveness restrictions in first order logic. For instance, it is well known that the ordered field of reals is, up to isomorphism, the only *complete* ordered field (here completeness means that every non-empty upperly, lowerly bounded set of reals has a least upper bound, a greatest lower bound respectively). So completeness cannot be expressed in a first order way, because any tentative first order translation should be true in some real closed field with a non-continuum power.

On the other side, we will see that the Löwenheim-Skolem theorem is a very useful and powerful technical tool in first order model theory (just as the Compactness Theorem). And actually the expressiveness restrictions remarked before are only the other side of the picture of these technical advantages. This is just the content of the Lindström theorem quoted before in 1.2. Indeed, what Lindström shows is that, if you have a logic (namely a reasonable system of formulas and truth) and you demand that your logic satisfies the Compactness Theorem and the weaker form of the Löwenheim-Skolem Theorem, called Downward Löwenheim-Skolem Theorem, introduced in 1.4 and requiring -for countable languages- that any set of sentences admitting a model does have a countable model, then your logic is the first order logic. In this sense the first order framework is (Leibnizianly) the best possible one.

1.6 Complete theories

Let us deal again now with one of the main themes in Model Theory, i.e. classifying structures in a given class **K**. Due to our first order setting, we limit our analysis to elementary classes **K** = Mod(T), where T is a first order theory. This choice is not so partial and narrow as it may appear. In fact, it certainly includes the cases when T is explicitly given and equips **K** with an effective list of first order axioms, as in the positive examples of the last section; but it is also concerned with other, and worse situations. For instance, think of the theory T of finite sets, or groups, or fields, or, in general, of a class of finite arbitrarily large structures, so that T has also infinite models. Alternatively, think of the theory T of a single infinite structure $\mathcal{A}$: due to the Löwenheim-Skolem Theorem, T has some models non-isomorphic to $\mathcal{A}$. In these cases, T is introduced by specifying some crucial models, but this does not determine in an explicit way a priori which first order sentences belong to T, and which are excluded; indeed we could just be interested in finding an effective axiomatization as in the previous examples, and we could aim both at describing T and also -as a related matter- at classifying its models.
These are the settings we wish to consider. Actually we should also admit that we have not clearly explained up to now which kind of classification we pursue; however we have agreed that this classification should identify isomorphic models but distinguish non isomorphic structures. Also, we have seen that isomorphic models satisfy the same order order sentences. So a preliminary classification is just up to elementary equivalence, and aims at distinguishing non elementarily equivalent structures. Once this is done, we could restrict our analysis to structures satisfying the same first order conditions; i.e. fix a structure $\mathcal{A}$ and classify up to isomorphism the models of its theory $T = Th(\{\mathcal{A}\})$ (by the way, let us abbreviate for simplicity $Th(\{\mathcal{A}\})$ by $Th(\mathcal{A})$).
Which is an intrinsic syntactical characterization of such a theory T? Basically it is "complete" according to the following definition.

Definition 1.6.1 *A (consistent) theory T of L is said to be* **complete** *if, for every sentence φ of L, either $\varphi \in T$ or $\neg\varphi \in T$.*

In fact, it is easily observed on the ground of the definition of truth in first order logic that, given a structure $\mathcal{A}$ in a language L, for every L-sentence φ, either φ is true in $\mathcal{A}$ or $\neg\varphi$ is; equivalently, either $\varphi \in Th(\mathcal{A})$ or $\neg\varphi \in Th(\mathcal{A})$.

On the other hand, every complete theory T can be represented in this way. In fact, fix any model $\mathcal{A}$ of T. Clearly $T \subseteq Th(\mathcal{A})$. Conversely, let φ be a sentence of $Th(\mathcal{A})$, then $\neg\varphi \notin Th(\mathcal{A})$ and so $\neg\varphi \notin T$; as T is complete, $\varphi \in T$.
Notice that the same argument shows that, if $T \subseteq T'$ are consistent theories and T is complete, then $T = T'$.
Now notice what follows.

Remark 1.6.2 Every (consistent) theory T of L can be enlarged in at least one way to a complete theory in L. In fact, it suffices to consider $Th(\mathcal{A})$ where $\mathcal{A}$ is any model of T. A complete theory extending T is called a *completion* of T.

So our classification project can be organized as follows.

- First, determine structures up to elementary equivalence, in other words find all the completions of a given theory T;

- then, classify up to isomorphism the models of a complete T.

We deal in this section with the former problem, hence with completions and, definitively, with complete theories. Incomplete theories are easy to meet.

Example 1.6.3 For instance, the theory of groups it is not complete (as there are both abelian and nonabelian groups, and commutativity can be written in a universal first order sentence). In the same way, the theory of fields is not complete (why?). Also the theory of linear orders is not complete (as there are both dense and non dense total orders, as well as orders with or without a minimum or a maximum).

On the other hand, the previous remark pointing out that a complete T is the theory of any model of T seems to provide a great deal of complete theories; but these examples are not satisfactory. In fact, as already said, what we reasonably expect is to have complete theories T equipped with an explicit list of basic axioms, ensuring that the sentences in T are just the consequences of these axioms. Now, when we look at $Th(\mathcal{A})$ for some structure $\mathcal{A}$ (the field of complex numbers, or the ordered field of reals, and so on), this list of axioms is lacking; indeed we could wish to obtain such a basic axiomatization in the mentioned cases. A possible strategy to solve these problems might be the following. Given $\mathcal{A}$, prepare a tentative explicit axiomatization and the corresponding theory T. Of course, $\mathcal{A}$ should

be a model of T. At this point, check if T is complete, by some suitable procedures. If yes, $T = Th(\mathcal{A})$.
Unfortunately, checking completeness for a theory T as before is not simple. We mention here a celebrated sufficient (but non-necessary) condition, founded on the notion of λ-categoricity.

Theorem 1.6.4 (Vaught) *Let T be a theory of L. Suppose that every model of T is infinite and T is λ-categorical for some infinite cardinal λ. Then T is complete.*

Proof. Suppose towards a contradiction that T is not complete; let φ be a sentence such that $\varphi \notin T$ and $\neg\varphi \notin T$. As $\varphi \notin T$, there exists a(n infinite) model $\mathcal{A}_0$ of T such that $\mathcal{A}_0 \models \neg\varphi$. In a similar way, there exists a(n infinite) model $\mathcal{A}_1$ of T such that $\mathcal{A}_1 \models \varphi$. For every $i \leq 1$, put $T_i = Th(\mathcal{A}_i)$; then $T_i \supseteq T$, and T_i has an infinite model. By the Löwenheim-Skolem theorem, T_i has a model $\mathcal{B}_i$ of power λ. Both $\mathcal{B}_0$ and $\mathcal{B}_1$ are models of T, hence they are isomorphic because T is λ-categorical. However $\mathcal{B}_0 \not\equiv \mathcal{B}_1$ because $\mathcal{B}_0 \models \neg\varphi$, while $\mathcal{B}_1 \models \varphi$. ♣

Here are some consequences of Vaught's Theorem.

Corollary 1.6.5 *Let $L = \emptyset$. Then the theory I_∞ of infinite sets is complete.*

Proof. Clearly T has no finite models. Moreover two (infinite) sets in the same power are isomorphic; in other words I_∞ is λ-categorical in any $\lambda \geq \aleph_0$. Consequently I_∞ is complete. ♣

Corollary 1.6.6 *The theory DLO^- of dense linear orders without endpoints is complete.*

Proof. DLO^- has no finite models; this is easy to check, by using density or the absence of endpoints. Moreover the famous theorem of Cantor on dense linear orders recalled in 1.3 ensures that DLO^- is $\aleph_0$-categorical: every countable linear order without endpoints is isomorphic to the order of rationals. Hence DLO^- is complete. ♣

Recall that both $(\mathbf{Q}, \leq)$ and $(\mathbf{R}, \leq)$ are models of DLO^-; it follows that $DLO^- = Th(\mathbf{Q}, \leq) = Th(\mathbf{R}, \leq)$.

Corollary 1.6.7 *For every $p = 0$ or prime, the theory ACF_p of algebraically closed fields of characteristic p is complete.*

Proof. An algebraically closed field $\mathcal{K}$ is always infinite; in fact, if $a_0, \ldots, a_n$ are distinct elements of $\mathcal{K}$, the polynomial $(x-a_0)\cdot \ldots \cdot (x-a_n)+1$ in $\mathcal{K}[x]$ has a root α in $\mathcal{K}$; α cannot equal $a_0, \ldots, a_n$, and so is a new element. At this point, in order to apply Vaught's Theorem, we have to prove λ-categoricity for some infinite λ. But this is just a consequence of Steinitz's analysis of algebraically closed fields. For, this analysis essentially implies (in our terminology) that, fixed $p=0$ or prime, the theory ACF_p of algebraically closed fields of characteristic p is λ-categorical for every uncountable cardinal λ (so Vaught's Theorem applies and yields completeness). Let us recall briefly why (we will provide an alternative, detailed proof of the completeness of ACF_p in Chapter 2). Any algebraically closed field in characteristic p can be obtained as

$$\mathcal{K} = \overline{\mathcal{K}_0(S)}$$

where $\mathcal{K}_0$ is the prime subfield of $\mathcal{K}$ (hence $\mathcal{K}_0$ is isomorphic to the rational field if $p=0$, or to the field with p elements if p is prime), S is a transcendence basis of $\mathcal{K}$ (namely a maximal algebraically independent subset), and $\overline{}$ denotes the algebraic closure in $\mathcal{K}$. Furthermore the isomorphism type of $\mathcal{K}$ is fully determined by the cardinality of S (the *transcendence degree* of $\mathcal{K}$). Accordingly, one can realize that ACF_p has

- $\aleph_0$ pairwise non isomorphic countable models (correspondingly to the transcendence degrees $0, 1, \ldots, \aleph_0$),
- for every uncountable cardinal λ, exactly one isomorphism class of models of power λ, because all these models share the same transcendence degree λ.

Hence ACF_p is λ-categorical in every cardinal $\lambda > \aleph_0$, and consequently complete. ♣

In particular, two algebraically closed fields $\mathcal{K}_1$ and $\mathcal{K}_2$ having the same characteristic p but different transcendence degrees $d_1 \neq d_2$ are elementarily equivalent, but cannot be isomorphic. Hence, when $\mathcal{K}_1$ and $\mathcal{K}_2$ are countable, they are not even partially isomorphic.
Notice also that the field of complex numbers is a model of ACF_0, and so ACF_0 equals its theory: we find in this way an explicit list of axioms (that of ACF_0) for the theory of the complex field.
Let us propose a further application of Vaught's Theorem to deduce completeness. Perhaps at this point someone may expect to meet RCF and the theory of the real field in our list of examples. But we have to delay this appointment. Indeed, RCF is complete (and hence equals the theory

of the ordered field of reals), but RCF is not λ-categorical for any infinite cardinal λ. So a different approach is necessary: we shall follow this new strategy in the next chapter. On the contrary, Vaught's Theorem applies to vectorspaces over a countable field. Let us see why.

Corollary 1.6.8 *Let $\mathcal{K}$ be a countable field. Then the theory ${}_{\mathcal{K}}T'$ of infinite vectorspaces over $\mathcal{K}$ is complete.*

Proof. Clearly ${}_{\mathcal{K}}T'$ has no finite models. Moreover we know that two (infinite) vectorspaces with the same dimension over $\mathcal{K}$ are isomorphic. Consequently, for every cardinal λ bigger than $\aleph_0$, there is a unique isomorphism class for all the $\mathcal{K}$-vectorspace of power λ (in fact, each of them has dimension λ). In other words, ${}_{\mathcal{K}}T'$ is λ-categorical for every cardinal $\lambda > \aleph_0$. By Vaught's Theorem, ${}_{\mathcal{K}}T'$ is complete. ♣

On the contrary, ${}_{\mathcal{K}}T'$ may not be categorical in $\aleph_0$. In fact, when $\mathcal{K}$ is infinite, $\mathcal{K}, \mathcal{K}^2, \ldots, \mathcal{K}^{(\aleph_0)}$ are countable $\mathcal{K}$-vectorspaces with distinct dimensions, and so cannot be isomorphic, hence they are not even partially isomorphic. So elementary equivalence cannot imply partial isomorphism (and isomorphism). The reader may check directly what happens when $\mathcal{K}$ is finite.

We conclude this section by introducing another notion related to completeness. It will be used in Chapter 3 to show that RCF is complete. Recall that a complete theory T equals $Th(\mathcal{A})$ for every model $\mathcal{A}$, and hence a theory T is complete if and only if any two models of T are elementarily equivalent.

Definition 1.6.9 *A theory T is model complete if every embedding of models of T is elementary.*

It is easy to exhibit theories which are not model complete. For instance, the previous examples 1.3.14 ensure that the theory of $(\mathbf{N}, <)$, as well as the theory of fields, are not model complete. On the contrary, it is not simple to give explicit examples of model complete theories. Chapter 3 will be devoted to this point, and to discussing the relevance of this notion within Model Theory.

1.7 Definable sets

Formulas include equations, and Algebra aims at finding solutions of equations. More generally, given a language L, a formula $\varphi(\vec{v})$ of L and a structure $\mathcal{A}$ of L, one could try to determine all the sequences $\vec{a}$ in A for which $\mathcal{A} \models \varphi(\vec{a})$. As we shall see in the next chapters, this is not just a minor, collateral exercise; on the contrary, in treating this framework, we are moving to the core of modern model theory. So let us give the corresponding definition.

Definition 1.7.1 *Let $\mathcal{A}$ be a structure of L, n be a positive integer. A subset D of A^n is called* **definable** *in $\mathcal{A}$ if there is a formula $\varphi(\vec{v})$ of $L(A)$ such that D equals the set of the elements $\vec{a}$ in A^n for which $\mathcal{A} \models \varphi(\vec{a})$ (n is the length of $\vec{v}$, of course).*

In this case one says that the $L(A)$-formula $\varphi(\vec{v})$ *defines* D, and one writes

$$D = \varphi(\mathcal{A}^n).$$

The elements of A occurring as constants in the formula $\varphi(\vec{v})$ are called a sequence of *parameters* defining D.
Let us propose a simple example.
Consider a polynomial $p(x_1, x_2) \in \mathbf{R}[x_1, x_2]$, and look at the algebraic curve of the solutions of $p(x_1, x_2)$ in $\mathbf{R}^2$. This is a definable set, because it is formed by the elements in $\mathbf{R}^2$ satisfying the formula $p(v_1, v_2) = 0$ (together with the coefficients of $p(x_1, x_2)$ -the parameters of the formula in $\mathbf{R}$-).
A function f having domain $\subseteq A^n$ and image $\subseteq A^s$ for some positive integer s is called definable when its graph (hence the set of sequences $(\vec{a}, \vec{b})$ in A^{n+s} such that $f(\vec{a}) = \vec{b}$) is.
If $X \subseteq A$ and the parameters $\vec{x}$ in a formula defining D are in X, then D is said to be X-*definable*. In particular D is $\emptyset$-*definable* if and only if there exists a formula $\varphi(\vec{v})$ in L such that $D = \varphi(\mathcal{A}^n)$ is the set of the sequences of A^n satisfying $\varphi(\vec{v})$ in $\mathcal{A}$.
For $D \subseteq A^n$,

(i) D is definable if and only if D is A-definable;

(ii) if $X \subseteq Y \subseteq A$ and D is X-definable, then D is Y-definable, too;

(iii) D is definable if and only if there exists a finite subset X of A such that D is X-definable (($\Leftarrow$) follows from (ii); in order to show ($\Rightarrow$), just let X be the set of the parameters in a defining $L(A)$-formula $\varphi(\vec{v})$).

An element $a \in A$ is X-definable if its singleton is. Of course, every a is A-definable (by $v = a$). But, when $a \notin X$, things are not so trivial.

Remark 1.7.2 Fix a structure $\mathcal{A}$ of L and a positive integer n.

1. Let X be a subset of A. The X-definable subset of A^n form a subalgebra of the Boolean algebra of all the subset of A^n.

   In other words, both A^n and $\emptyset$ are X-definable (by the formulas $v_1 = v_1$ and $\neg(v_1 = v_1)$ respectively), and, if D_0 and D_1 are two X-definable subsets of A^n, then even their union $D_0 \cup D_1$, their intersection $D_0 \cap D_1$ and the complement $A^n - D_0$ are X-definable (if $\varphi_0(\vec{v})$ and $\varphi_1(\vec{v})$ are two $L(X)$-formulas defining D_0 and D_1 respectively, just look at the formulas $\varphi_0(\vec{v}) \vee \varphi_1(\vec{v})$, $\varphi_0(\vec{v}) \wedge \varphi_1(\vec{v})$, $\neg\varphi_0(\vec{v})$).

   $\mathcal{B}_n(X, \mathcal{A})$ will denote below the Boolean algebra of the subsets of A^n X-definable in $\mathcal{A}$.

2. Definable sets are closed also under projections, in the following sense. Let D be a subset of A^n definable in $\mathcal{A}$. Let π be the projection of $\mathcal{A}$ onto some fixed $i \leq n$ coordinates. Then $\pi(D)$ (a subset of A^i) is still definable. For instance, if the $L(A)$-formula $\varphi(\vec{v})$ defines D, then $\exists v_2 \ldots \exists v_n \varphi(\vec{v})$ defines the image of D by its projection onto the first coordinate (of course, $\vec{v}$ abridges here $(v_1, v_2, \ldots, v_n)$).

3. Every finite subset of A^n is definable.

   In fact let D be a finite subset of A^n, and let $\vec{d_0}, \ldots, \vec{d_t}$ be its elements. For every $j \leq t$, put $\vec{d_j} = (d_{j1}, \ldots, d_{jn})$. Then

   $$\bigvee_{j \leq t} \bigwedge_{1 \leq i \leq n} (v_i = d_{ji})$$

   defines D in $\mathcal{A}$. In particular, in a finite structure $\mathcal{A}$ every subset of A^n is definable; moreover, owing to 1, in any structure $\mathcal{A}$ even the cofinite subsets of A^n are definable.

4. $\mathcal{A}$ is infinite, then there exist some subsets of A^n which are not definable in $\mathcal{A}$.

   This follows from a simple cardinal counting argument. In fact, we know that there are $2^{|A|}$ distinct subsets of A^n, while the subsets of A^n definable in $\mathcal{A}$ cannot exceed the $L(A)$-formulas defining them. Consequently the definable subsets of A^n are at most $|A|$ (for a countable L, of course).

An explicit example of an infinite structure with a non-definable subset is the following. Let $L = \emptyset$, so the structures of L are just the non-empty sets A. Take an infinite set A. We have seen that every finite or cofinite subset of A is definable. We claim that no other subset of A is definable. In fact, let D be a subset of A such that both D and its complement $A - D$ are infinite. Suppose towards a contradiction that D is definable, and so $D = \varphi(A, \vec{b})$ for a suitable L-formula $\varphi(v, \vec{w})$ and a sequence $\vec{b}$ of parameters from A. Take $d \in D$, $d' \in A - D$, d, d' out of $\vec{b}$. We can find a bijection f of A onto A (and hence an automorphism of A) fixing $\vec{b}$ pointwise and mapping d in d'. As $d \in D$, $A \models \varphi(d, \vec{b})$; as isomorphisms preserve satisfiability, $A \models \varphi(d', \vec{b})$; consequently $d' \in D$ -a contradiction-. In conclusion, D is not definable.

Let us concentrate our attention on infinite structures from now on. In fact, owing to Remark 1.7.2, 3, there is no point in exploring definability in the finite case. First let us give some more examples of definable sets, suggesting some intriguing connections between this part of Model Theory and other branches of Mathematics.

Examples 1.7.3 1. (**Definable sets and Complex Algebraic Geometry**) Let $L = \{0, 1, +, \cdot, -\}$ be the language of fields. We have seen at the beginning of this section that, over the real field, polynomials determine definable sets. Let us investigate this example more generally and closely. Accordingly consider an arbitrary field $\mathcal{K}$ and a positive integer n. Algebraic Geometry deals with *algebraic varieties* in $\mathcal{K}^n$. These are the zero sets in $\mathcal{K}^n$ of finite systems of polynomials

$$q_0(\vec{x}), \ldots, q_t(\vec{x}) \in K[\vec{x}].$$

Hence algebraic curves are examples of algebraic varieties. Moreover every algebraic variety as before is definable in $\mathcal{K}$, for instance by the (quantifier free) formula

$$\bigwedge_{j \leq t} q_j(\vec{v}) = 0.$$

So one may wonder how close and deep this connection between definable sets in $\mathcal{K}$ and algebraic varieties in $\mathcal{K}$ is. Of course, we cannot expect that any definable set is a variety (although this is certainly true when $\mathcal{K}$ is finite). In fact, owing to Remark 1.7.2,1 before, every finite Boolean combination of definable sets is also definable. But, with respect to this point, algebraic varieties behave in a different way.

- The union of two (and consequently of finitely many) algebraic varieties in an algebraic variety. This is a simple exercise of Algebra, essentially using the fact that, in a field $\mathcal{K}$, the product of two nonzero elements is different from 0.
- The intersection of two, or finitely many, or even infinitely many algebraic varieties is still an algebraic variety. This is a trivial exercise in the finite case, and a deep theorem in Algebra -known as Hilbert's Basis Theorem- otherwise.

Notice that these properties (together with the easy observation that K^n and $\emptyset$ are algebraic varieties -for, they are the zero sets of the zero polynomial, and of any nonzero constant polynomial in $\mathcal{K}[\vec{x}]$ respectively-) show that the algebraic varieties of $\mathcal{K}^n$ are the closed sets in a suitable topology of $\mathcal{K}^n$ (the *Zariski topology*). However

- the complement of an algebraic variety of $\mathcal{K}^n$ is not necessarily an algebraic variety of $\mathcal{K}^n$.

So there are definable sets of $\mathcal{K}^n$ which are not algebraic varieties. Indeed Algebraic Geometry introduces the notion of *constructible set* to define a finite Boolean combination of algebraic varieties of $\mathcal{K}^n$. Remark 1.7.2,1 before ensures that every constructible set is definable.

In certain fields $\mathcal{K}$ the converse is also true, and hence definable just means constructible. For instance, this is what happens when $\mathcal{K}$ is an algebraically closed field (and so, in particular, when $\mathcal{K}$ is the complex field). This is not a trivial result, but a deep theorem of Tarski and Chevalley, and will be discussed in the next Chapter.

2. **(Definable sets and Real Algebraic Geometry)** Let $L = \{0, 1, +, \cdot, - \leq\}$ be our language for ordered fields. Fix an ordered field $\mathcal{K}$, and a positive integer n. Algebraic Geometry studies the sets of the elements of K^n satisfying disequations like $q(\vec{x}) \geq 0$ where $q(\vec{x}) \in K[\vec{x}]$, and calls *semialgebraic set* any finite Boolean combination of them. It is clear that every semialgebraic set is definable in $\mathcal{K}$, and even by a quantifier free formula (a Boolean combination of atomic formulas $q(\vec{v}) \geq 0$). A theorem of Tarski and Seidenberg ensures that, when $\mathcal{K}$ is a real closed ordered field (in particular when $\mathcal{K}$ is the ordered field of reals), then the definable sets of $\mathcal{K}^n$ are exactly those semialgebraic. So a close connection arises between Model Theory and Algebraic Geometry also in this framework. The Tarski and Seidenberg theorem will be treated in detail in the next chapter.

Notice also that the order relation $\geq$ is definable in the real field $\mathbf{R}$ even within the language of fields $\{0, 1, +, \cdot, -\}$: in fact, it suffices to recall that the nonnegative reals are exactly the squares, and hence to define

$$v_1 \geq v_2$$

by the formula

$$\exists w(v_1 - v_2 = w^2).$$

Consequently every semialgebraic set D in $\mathbf{R}$ is definable in the real field even within the language of fields, just by replacing any formula

$$q(\vec{v}) \geq 0$$

(with $q(\vec{x}) \in \mathbf{R}[\vec{x}]$) by the equivalent formula

$$\exists w(q(\vec{v}) = w^2).$$

However, notice that the latter formula requires a quantifier.

3. (**Definable sets and recursive sets**) Now we consider the language $L = \{+, \cdot\}$ and, in L, the structure $(\mathbf{N}, +, \cdot)$. First notice that every natural n is $\emptyset$-definable in $(\mathbf{N}, +, \cdot)$. This can be easily shown by using an induction argument on n; if $n = 0$ or $n = 1$, just take the formulas

$$"v_1 = 0" : v_1 + v_1 = v_1, \quad "v_1 = 1" : v_1 \cdot v_1 = v_1 \wedge \neg("v_1 = 0")$$

respectively, while, for $n \geq 1$, $n + 1$ is $\emptyset$-definable by

$$"v_1 = n + 1" : \exists z_0 \exists z_1 ("z_0 = 1" \wedge "z_1 = n" \wedge v_1 = z_0 + z_1).$$

Consequently in $(\mathbf{N}, +, \cdot)$ definable sets just equal $\emptyset$-definable sets.

Definability in $(\mathbf{N}, +, \cdot)$ is deeply related to recursion theory. Let us see why. A basic aim in *recursion theory* is to provide a sharp definition of the notion of *algorithm*. According to the Church and Turing thesis,

algorithm means *Turing machine*,

in the sense that the problems with a solving algorithm are just those handled by a Turing machine. Actually the Church-Turing model of computation dates back to the thirties, and, from the practical point of view, is undoubtedly surpassed by the new advances in computer science. However, as an abstract and theoretic proposal, it is still valid and commonly agreed, at least in any discrete setting. This is clearly our framework when we are dealing with natural numbers. Incidentally, recall that every discrete context can be easily translated into the world of natural numbers, owing to the Gödel coding procedures, equipping effectively each element with its own natural label. So let's work with the structure $(\mathbf{N}, +, \cdot)$. On the ground of the Church and Turing thesis, one can define in a sharp way which are the subsets $D \subseteq \mathbf{N}^n$ (with n a positive integer) admitting a decision algorithm, namely a procedure running in finitely many steps and establishing, for every $\vec{a} \in \mathbf{N}^n$, if $\vec{a}$ is in D or not. These sets D are called *recursive*. A crucial result in Recursion Theory -indeed a key step within the proof of Gödel First Incompleteness Theorem- ensures that every recursive set is definable in $(\mathbf{N}, +, \cdot)$. More generally, any *recursively enumerable* $D \subseteq \mathbf{N}^n$ is definable in $(\mathbf{N}, +, \cdot)$. Recall that recursive enumerability is a weaker notion than recursiveness, and just requires that there is some algorithm effectively listing the elements of the involved set D. On the contrary, there do exist some subset of $\mathbf{N}^n$ which are not recursively enumerable (and hence are not even recursive), but are definable. These remarks witness that now definable sets are a very complicated class, because they inherit the complexity of the class of recursively enumerable sets with the corresponding intrinsic hierarchies, and possibly more. So, when dealing with $(\mathbf{N}, +, \cdot)$, definable sets are not so clean as in the complex, or in the real field.

4. **(Definable sets and decidable theories)** We again work in the language $L = \{+, \cdot\}$, but this time we examine the structure $(\mathbf{Z}, +, \cdot)$. Identify natural numbers and nonnegative integers. Then $\mathbf{N}$ becomes an $\emptyset$-definable subset of $\mathbf{Z}$; in fact, a celebrated theorem of Lagrange ensures that, among the integers, the nonnegative elements are just the sums of 4 squares. Hence the formula

$$\exists w_1 \exists w_2 \exists w_3 \exists w_4 (v = w_1^2 + w_2^2 + w_3^2 + w_4^2)$$

defines $\mathbf{N}$ inside $(\mathbf{Z}, +, \cdot)$. Also the sum and product operations in $\mathbf{N}$ are $\emptyset$-definable in $(\mathbf{Z}, +, \cdot)$, because they just restrict to $\mathbf{N}$ the addition and multiplication in $\mathbf{Z}$. So we can conclude that the whole structure $(\mathbf{N}, +, \cdot)$ is "$\emptyset$-definable" in $(\mathbf{Z}, +, \cdot)$.

Now it is well known that the theory of $(\mathbf{N}, +, \cdot)$ is not *decidable*, in the sense that the set of the natural codes of the sentences true in $(\mathbf{N}, +, \cdot)$ is not recursive. In other words, no general algorithm can decide, for every sentence φ di L, if φ is true in $(\mathbf{N}, +, \cdot)$ or not. On the other hand, the previous observations let us effectively translate every sentence φ of L in a sentence φ' of L such that

$$(\mathbf{N}, +, \cdot) \models \varphi \quad \Leftrightarrow \quad (\mathbf{Z}, +, \cdot) \models \varphi'.$$

Consequently even the theory of $(\mathbf{Z}, +, \cdot)$ is undecidable.

The method sketched here is often used to show undecidability for theories interpreting as described other theories whose undecidability is known, or also, specularly, to deduce decidability for the theories that can be interpreted in the previous way in some other decidable theory. Hence definability plays a crucial role also within the decision problem for theories.

5. (**Modules and *pp*-definable subgroups**) Let $\mathcal{R}$ a (countable) ring with identity. Consider the language $L_R = \{0, +, -, r\,(r \in R)\}$ of $\mathcal{R}$-modules, and in L_R the class $\mathcal{R} - Mod$ of (left) $\mathcal{R}$-modules. A formula $\varphi(\vec{v})$ of L_R is called a *positive primitive formula* (or also, more synthetically, a **pp-formula**) if $\varphi(\vec{v})$ is of the form

$$\exists \vec{w}(A \cdot {}^t\vec{v} = B \cdot {}^t\vec{w})$$

where A and B are matrices with coefficients in R and suitable sizes, $\cdot$ denotes the usual row-by-column multiplication for matrices, and ^{t} is the transpose operation. Equivalently, if one puts $\vec{v} = (v_1, \ldots, v_n)$, $\vec{w} = (w_1, \ldots, w_m)$, $A = (r_{ij})_{i,j}$ and $B = (s_{ih})_{i,h}$, where j ranges from 1 to n, h from 1 to m, and i from 1 to some suitable positive integer t, then $\varphi(\vec{v})$ can be written

$$\exists w_1 \ldots \exists w_m \bigwedge_{1 \leq i \leq t} \Big(\sum_{1 \leq j \leq n} r_{ij} v_j = \sum_{1 \leq h \leq m} s_{ih} w_h \Big).$$

Hence, in every $\mathcal{R}$-module $\mathcal{M}$, $\varphi(\mathcal{M}^n)$ is the set of the sequences $\vec{a} = (a_1, \ldots, a_n)$ in M^n for which the linear system

$$A \cdot {}^t\vec{a} = B \cdot {}^t\vec{w}$$

has some solution in M^m.

Let us examine some particular cases.

- Let $r \in R$, $\varphi(v) : rv = 0$. Then $\varphi(v)$ is a pp-formula (take an empty $\vec{w}$, or write $\varphi(v)$ in the equivalent form $\exists w(rv = 0w)$). For every $\mathcal{R}$-module $\mathcal{M}$,

$$\varphi(\mathcal{M}) = \{a \in M : ra = 0\}$$

is just the *annihilator* of r in $\mathcal{M}$, hence a(n additive) subgroup of $\mathcal{M}$, and even a submodule of $\mathcal{M}$ when $\mathcal{R}$ is commutative, or when, simplerly, r is in the centre of $\mathcal{R}$.

- Take again $r \in R$, and now consider $\varphi(v) : \exists w(v = rw)$. Then $\varphi(v)$ is a pp-formula (for which A e B?). If $\mathcal{M}$ is an $\mathcal{R}$-module, then

$$\varphi(\mathcal{M}) = rM$$

is a subgroup of $\mathcal{M}$, and even a submodule at least when $\mathcal{R}$ is commutative, or, simplerly, when r is in the centre of $\mathcal{R}$.

In general it is easy to check that, for every pp-formula $\varphi(\vec{v})$, $\varphi(\mathcal{M}^n)$ is an additive subgroup of $\mathcal{M}^n$; on the other hand, $\varphi(\mathcal{M}^n)$ is not always a submodule, although this is certainly true when $\mathcal{R}$ is commutative, as we saw in the previous examples. $\varphi(\mathcal{M}^n)$ is called a *pp-definable subgroup* of M^n. Every coset D of $\varphi(\mathcal{M}^n)$ in M^n is definable in $\mathcal{M}$: in fact, let $\vec{a}$ be any element in D, then the formula $\varphi(\vec{v} - \vec{a})$ defines the whole coset D.

A theorem of Baur and Monk ensures that, in every $\mathcal{R}$-module $\mathcal{M}$, any definable sets is a finite Boolean combination of cosets of pp-definable subgroups (see Chapter 2).

1.8 References

There exist several excellent handbooks providing the backgrounds of Mathematical Logic necessary in this chapter. Among them, let us mention once again Shoenfield [153], but also Malitz [103] or Ebbinghaus-Flum-Thomas [37]. The key reference for basic Model Theory is [18]. We also refer to Devlin [31] for Set Theory, to Odifreddi [121] for Recursion Theory and to Jacobson [65] for Algebra. [173] is the historical source quoted at the end of 1.2. Ax's analysis of pseudofinite fields is given in [3]

Chapter 2

Quantifier Elimination

2.1 Elimination sets

Let L be a language. It may happen that two different L-formulas $\varphi(\vec{v})$ and $\varphi'(\vec{v})$ admit the same meaning in a structure $\mathcal{A}$ of L, or in a class of L-structures, for instance among the models of a given L-theory T. For example, in the ordered field of reals (and even in every real closed field), the formula $\varphi(v)\ :\ v \geq 0$ (being nonnegative) is the same thing as $\varphi'(v)\ :\ \exists w(v = w^2)$ (being a square). Similarly, in the ordered domain of integers, $\varphi(v)\ :\ v \geq 0$ (being positive) has the same interpretation as $\varphi'(v)\ :\ \exists w_1 \exists w_2 \exists w_3 \exists w_4\, (v = \sum_{1 \leq i \leq 4} w_i^2)$ (being the sum of four squares): this is a celebrated theorem of Lagrange, already mentioned in the last chapter.

So, fix a consistent, possibly incomplete theory T in a countable L. We shall say that two L-formulas $\varphi(\vec{v})$ and $\varphi'(\vec{v})$ are equivalent with respect to T, and we shall write $\varphi(\vec{v}) \sim_T \varphi'(\vec{v})$, when

$$\forall \vec{v}(\varphi(\vec{v}) \leftrightarrow \varphi'(\vec{v})) \in T,$$

equivalently when

$$\varphi(\mathcal{A}^n) = \varphi'(\mathcal{A}^n)$$

for all models $\mathcal{A}$ of T.

The notion of elimination set arises quite naturally at this point. An elimination set for T is a set F of L-formulas such that every L-formula $\varphi(\vec{v})$ is T-equivalent to a suitable Boolean combination of formulas of F.

Clearly the set of all the L-formulas is an elimination set for T. But, of course, this is not an interesting case, and we reasonably expect simpler sets

F. In particular, when the set of atomic formulas in L is an elimination set for T, we say that T has the *quantifier elimination* in L. In detail

Definition 2.1.1 *Let T be a theory in a language L. T has the* **elimination of quantifiers** *(q.e.) in L if and only if every formula $\varphi(\vec{v})$ of L is equivalent in T to a quantifier free L-formula $\varphi'(\vec{v})$ (so to a finite Boolean combination of atomic formulas).*

One easily realizes that every T gets the elimination of quantifiers in a suitable language extending L. In fact, put $L = L_0$, $T = T_0$, and enlarge L_0 to a language L_1 containing an n-ary relation symbol R_φ for every formula $\varphi(\vec{v})$ of L_0 (n is the length of $\vec{v}$, of course); then add the following sentences to T_0

$$\forall\vec{v}(\varphi(\vec{v}) \leftrightarrow R_\varphi(\vec{v}))$$

for every $\varphi(\vec{v})$, and get a new theory T_1; it is clear that the atomic formulas of L_1 form an elimination set in T_1 for the formulas in L_0. By repeating this procedure countably many times, one eventually defines a language $L' \supseteq L$ and a theory T' of L' "naturally" extending T and having the elimination of quantifiers in L'.
Unfortunately this procedure has a quite artificial and abstract flavour. Indeed, what we would like to obtain, given a theory T in a language L, is showing that T has the quantifier elimination directly in L or, otherwise, determining a smallest extension $L' \supseteq L$, possibly suggested by the algebraic analysis of the models of T, where T (or, more exactly, its natural extension to L') has the elimination of quantifiers, or also a reasonably simple elimination set of formulas, in L'. In fact, there are good reasons to believe that such a language L' is, in some some, "the" proper language of T.

Which are the main advantages of an elimination set, in particular of quantifier elimination? They concern several applications.

1. The main one (at least from a historical point of view) is *decidability*. Actually the first and most celebrated quantifier elimination results are related to the decision theme. Let us explain why. Recall that a theory T is decidable if there is an algorithm checking in finitely many steps, for every sentence α in the language L of T, whether α is in T or not. Now suppose that F is an elimination set for T and that the following are available:

    - an *effective* procedure translating any L-sentence into a T-equivalent Boolean combination of sentences in F (or even an effective reduction of any L-formula into a T-equivalent Boolean combination of formulas in F);

- an algorithm to decide, for every Boolean combination α of sentences of F, whether α is or not in T.

Then, clearly, T is decidable, and actually we have got a decision algorithm (by successively applying the previous two procedures).

2. Another noteworthy application of quantifier elimination concerns *definability*. In fact, if F is an elimination set for T, then the definable sets of a model $\mathcal{A}$ of T reduce to

$$\varphi(\mathcal{A}^n, \vec{x})$$

where $\varphi(\vec{v}, \vec{w})$ is a finite Boolean combination of formulas of F and $\vec{x} \in A$; in particular, if T has the quantifier elimination in L, then the definable sets of $\mathcal{A}$ are just the ones of the form

$$\varphi(\mathcal{A}^n, \vec{x})$$

where $\varphi(\vec{v}, \vec{w})$ is a quantifier free formula and $\vec{x}$ in A.

3. A third application regards the classification of *completions* of T. Recall that T is possibly incomplete; but we know that T has some (non-unique!) complete extension in L. So we are led to consider the problem of finding all the complete extensions of T in L, in other words classifying the isomorphism classes of models of T up to elementary equivalence. Now, if $\mathcal{A}$ and $\mathcal{B}$ are two models and $\mathcal{A}$ is not elementarily equivalent to $\mathcal{B}$, then there is some sentence φ in L such that $\mathcal{A} \models \varphi$ and $\mathcal{B} \models \neg\varphi$. As F is an elimination set for T in L, we can assume that φ is a Boolean combination of sentences in F. Indeed, one easily realizes that one can choose φ directly in F.

   For instance, we will see in this chapter that the theory ACF of algebraically closed fields has the quantifier elimination in $L = \{+, \cdot, -, 0, 1\}$. Consequently the classification of algebraically closed fields up to elementary equivalence depends on the quantifier free sentences in L, which are of the form $m = n$, where m and n are integers (m abbreviates in the previous formula the addition of m summands equal to 1 if $m > 1$, and $-(-m)$ if $m < -1$; similarly for n). This implies that the complete extensions of ACF are fully determined by the characteristic of their models, and hence coincide with the theories ACF_p where p is 0, or a prime.

4. Finally, let us deal with *model completeness*. Assume that T has quantifier elimination in L. We claim that, in this case, every embedding between models of T is elementary, in other words T is model complete. In fact, let $\mathcal{A}$ and $\mathcal{B}$ be models of T, f be an embedding of $\mathcal{A}$ into $\mathcal{B}$. Given a formula $\varphi(\vec{v})$ in L, let $\varphi'(\vec{v})$ a quantifier free formula equivalent to $\varphi(\vec{v})$ in L. Take $\vec{a}$ in A. As f is an embedding,

$$\mathcal{A} \models \varphi'(\vec{a}) \quad \Leftrightarrow \quad \mathcal{B} \models \varphi'(f(\vec{a})).$$

As $\forall \vec{v}(\varphi(\vec{v}) \leftrightarrow \varphi'(\vec{v})) \in T$,

$$\mathcal{A} \models \varphi(\vec{a}) \quad \Leftrightarrow \quad \mathcal{B} \models \varphi(f(\vec{a})).$$

Hence f is elementary.

This chapter is devoted to illustrating several key examples of quantifier elimination, starting from the earliest (Langford's results on discrete or dense linear orders) to include those perhaps most classical and celebrated (Tarski's elimination procedures for the real and the complex fields). We shall treat other eliminations sets as well (most notably, Baur-Monk's *pp*-elimination theorem for modules over a given ring).
These examples will lead us to introduce two basic notions in Model Theory, strong minimality and o-minimality respectively. We shall discuss them at the end of the chapter. The final section will be devoted to some computational aspects of the quantifier elimination procedures.
It should be underlined that the interest in quantifier elimination arose several years before the official birth of Model Theory. In fact it was at the beginning of the twentieth century that Löwenheim and, later, Skolem provided some procedures translating formulas into a simpler form avoiding quantifiers (they are, more or less, the artificial method we sketched at the beginning of this section). Moreover, the earliest explicit examples of quantifier elimination in some specific algebraic structures treat discrete and dense linear orders and date back to the twenties (they were obtained by Langford in 1927). In these results, as well as in Tarski's theorems, the major emphasis seems to be on decidability: the elimination of quantifiers is a step towards decidability, just as described before. But over the years this emphasis on decidability reduced and was replaced by an increasing interest in definability. Actually, definability is the main theme where Model Theory and quantifier elimination meet.

2.2 Discrete linear orders

We begin here our analysis of quantifier eliminable theories. First we treat infinite linear orders. Accordingly our basic language is $L = \{\leq\}$. More precisely we deal with:

- theories of discrete linear orders (in this section),
- theories of dense linear orders (in the next one).

As already said, the quantifier elimination results in these cases were firstly shown by Langford in 1927; Tarski pursued the analysis to get decidability and to classify the complete theories of infinite discrete and dense total orders.
Recall that a(n infinite) linear order $\mathcal{A} = (A, \leq)$ is **discrete** if and only if

(i) $\forall a \in A$, if there is some $a' \in A$ such that $a < a'$, then there exists a least $b \in A$ for which $a < b$ (b is called the *successor* of a and is denoted $s(a)$);

(ii) $\forall a \in A$, if there is some $a' \in A$ such that $a' < a$, then there exists a maximal $b \in A$ for which $b < a$ (b is called the *predecessor* of a; obviously $a = s(b)$).

Accordingly we can distinguish 4 classes of (infinite) discrete linear orders:

1. the class of discrete linear orders with a least, but no last element (like $(\mathbf{N}, \leq)$);
2. the class of discrete linear orders with a last, but no least element (for instance, $\mathbf{N}$ with respect to the relation reversing its usual order);
3. the class of (infinite) discrete linear orders with both a least and a last element (like the disjoint union of two discrete linear orders $(A, \leq)$, $(B, \leq)$, the former in 1, the latter in 2, with $a < b$ for all $a \in A$ and $b \in B$);
4. the class of discrete linear orders without endpoints (like $(\mathbf{Z}, \leq)$).

Each of these classes is elementary. Moreover one can show that its theory has the elimination of quantifiers in a suitable language extending L, and is complete even in L. Here we limit ourselves to prove, for simplicity, these results in the case 1, in other words for discrete linear orders with a least but no last element.

Accordingly consider the language $L' = \{\leq, 0, s\}$ where 0 is a constant (to be interpreted in the least element) and s is a 1-ary operation symbol (to be interpreted in the function mapping any element into its successor).
It is easy to write down a first order set of axioms for our class in L'. Let dLO^+ denote the corresponding theory. By the way, notice that suitable formulas in the restricted language L define the minimal element and the successor function in every model of dLO^+. This implies that the axioms of dLO^+ can be rewritten also in L, at the cost of some more complications (and quantifiers). For instance, expressing the existence of a minimal element requires the L-sentence

$$\exists w \forall v (w \leq v)$$

instead of

$$\forall v (0 \leq v).$$

But we momentarily prefer to treat dLO^+ in L'. Observe that:

- $(\mathbf{N}, \leq, 0, s)$ is a model of dLO^+;
- if $\mathcal{A}$ is another model of dLO^+, then $\mathcal{A}$ contains a substructure $(\{s^n(0^{\mathcal{A}}) : n \in \mathbf{N}\}, \leq, 0^{\mathcal{A}}, s^{\mathcal{A}})$ isomorphic to a $(\mathbf{N}, \leq, 0, s)$, and moreover some further copies of $(\mathbf{Z}, \leq, s)$ (as $0^{\mathcal{A}}$ is the only element without any predecessor).

Theorem 2.2.1 *dLO^+ has the elimination of quantifiers in L'.*

Proof. Take a formula $\varphi(\vec{v})$ in our language L'; we look for an equivalent formula $\varphi'(\vec{v})$ without quantifiers.
Our first step is to show that we can assume that $\varphi(\vec{v})$ is of the form

$$\exists w \bigwedge_{i \leq r} \alpha_i(\vec{v}, w)$$

where each $\alpha_i(\vec{v}, w)$ is an atomic formula, or its negation, and w actually occurs in $\alpha_i(\vec{v}, w)$ for every $i \leq r$.
We wish to underline that this step is quite general, and does not depend on our particular language L'. Let us see why. First of all, we can tacitly assume that $\varphi(\vec{v})$ is of the form

$$Q_1 w_1 \ldots Q_m w_m \alpha(\vec{v}, \vec{w})$$

where the Q_j's $(1 \leq j \leq m)$ denote quantifiers $\forall$ or $\exists$, $\alpha(\vec{v}, \vec{w})$ is a quantifier free formula, and even a disjunction of conjunctions of atomic formulas and negations (and $\vec{w}$ abridges $(w_1, \ldots, w_m)$, of course). The strategy at this point is first to eliminate Q_m, and then to repeat the procedure and remove the quantifier string completely. We recall that $\forall$ is equivalent to $\neg\exists\neg$ and consequently agree that it is enough to deal with the case when Q_m is $\exists$, namely with

$$\exists w\, \alpha(\vec{v}, w)$$

where α is a disjunction of conjunctions of atomic formulas or negations, $w = w_m$ and $\vec{v}$ is possibly enlarged to include $w_1, \ldots, w_{m-1}$. As $\exists$ is distributive with respect to $\vee$, namely $\exists w(\alpha' \vee \alpha'')$ is equivalent to $(\exists w\alpha') \vee (\exists w\alpha'')$, there is no loss of generality for our purposes in assuming that α is just a conjunction of atomic formulas or negations. In conclusion we are dealing with

$$\exists w \bigwedge_{i \leq r} \alpha_i(\vec{v}, w)$$

where each $\alpha_i(\vec{v}, w)$ is an atomic formula, or its negation. We can also assume that w actually occurs in $\alpha_i(\vec{v}, w)$ for every $i \leq r$; otherwise let $j \leq r$ deny this condition and notice that our formula

$$\exists w \bigwedge_{i \leq r} \alpha_i(\vec{v}, w)$$

is equivalent to

$$\alpha_j(\vec{v}) \wedge \exists w \bigwedge_{i \neq j} \alpha_i(\vec{v}, w);$$

at this point it suffices to eliminate the quantifier $\exists$ in the latter part of the formula

$$\exists w \bigwedge_{i \neq j} \alpha_i(\vec{v}, w).$$

This completes our preliminary step. As already said, this does not depend on our particular framework.

Now let us work with our formula

$$\exists w \bigwedge_{i \leq r} \alpha_i(\vec{v}, w)$$

and our language L'. We wonder which is the form of any α_i. A look at L' shows that α_i is $t = t'$, or $t \leq t'$, or the negation of one of these formulas, where t and t' are terms in 0, w, $\vec{v}$ (and w actually occurs in t or t'). Recall that $\neg(t \leq t')$ means $t > t'$, and so on. Deduce that α_i is, with no loss of generality, either $t = t'$ or $t > t'$, with t and t' as before. Notice that t and

t' are of the form $s^p(u)$ where p is a nonnegative integer and u ranges over 0, w, $\vec{v}$. Now recall that s is injective and deduce that the formula under exam ensures that there is a solution w for a finite set of conditions saying that w or a successor $s^q(w)$ (q a nonnegative integer) is equal, or bigger, or smaller than a term $s^p(u)$ where p is, again, a nonnegative integer and u ranges over w, 0, $\vec{v}$: as s is injective, we can assume that, in each of these equations and inequations, s occurs only in one side (on the left, or on the right). Our aim is to translate this formula into an equivalent one avoiding w and simply stating quantifier free conditions on $\vec{v}$ (and 0).
To obtain this, proceed as follows. Trivialities like $w = w$ or $w < s^p(w)$ for a positive p can be ignored and deleted (they can be preliminarily listed and hence are easily recognized); if nothing else occurs, then replace the whole formula by $0 = 0$. On the contrary, when meeting a condition that cannot be satisfied by any w, like $w = s^p(w)$, or $0 = s^p(w)$ for a positive p (also these conditions can be preliminarily listed), then replace our formula with $\neg(0 = 0)$ (or with $\neg(v_1 = v_1)$ if you like and $\vec{v}$ is not empty).
Otherwise, as soon as you meet one equation like $w = s^p(v_i)$, delete w and $\exists$, and replace w with $s^p(v_i)$ throughout our formula. Proceed in the same way if an equation $w = s^p(0)$ occurs. Similarly, when meeting a condition $s^q(w) = v_i$, consider any further occurrence of w in the formula and, again using the injectivity of s, represent it as $s^{q'}(w)$ for a suitable nonnegative integer $q' \geq q$; finally delete w and $\exists$ and replace each occurrence $s^{q'}(w)$ by $s^{q-q'}(v_i)$.
At last, assume that only disequations occur. Again using the injectivity of s, we can suppose that all of them concern the same term $s^q(w)$ in w. So our formula states that $s^q(w)$ is smaller that certain terms $t_0, \ldots, t_h$ in 0 and $\vec{v}$, and larger that some other terms $t_{h+1}, \ldots, t_k$. We obtain an equivalent formula avoiding w and its quantifier in the following way. List (in a suitable disjunction) all the possible orderings of $t_0, \ldots, t_k$ in $\leq$ according to which $t_0, \ldots, t_h$ precede $t_{h+1}, \ldots, t_k$; for every ordering, let t, t' denote respectively the greatest element among $t_0, \ldots, t_h$ and the least among $t_{h+1}, \ldots, t_k$; add $s(t) < t'$ (in order to provide $s^q(w)$ with suitable room).
This concludes the elimination procedure. ♣

Corollary 2.2.2 *dLO^+ is model complete (in L') and complete (both in L' and L).*

Proof. Clearly dLO^+ is model complete in L'. Moreover $(\mathbf{N}, \leq, 0, s) \models dLO^+$, and $(\mathbf{N}, \leq, 0, s)$ is embeddable in every model of dLO^+. As dLO^+ is model complete, all the corresponding embeddings are elementary. Accordingly all the models of dLO^+ are elementarily equivalent to $(\mathbf{N}, \leq, 0, s)$ and hence to each other. This shows that dLO^+ is complete in L'. Now recall that both the minimal element and the successor function (so the interpretations of the symbols in $L' - L$) are $\emptyset$-definable by L-formulas in the models of dLO^+. Then it is an easy exercise to deduce that dLO^+ is complete in L, too. ♣

Corollary 2.2.3 *dLO^+ is decidable (in L' and in L).*

Proof. Reduce any sentence of L' into an equivalent quantifier free statement. This is a Boolean combination of formulas $s^m(0) \geq s^n(0)$ where m and n are non-negative integers, and dLO^+ can easily check its membership. This procedure works even for L-sentences. ♣

Corollary 2.2.4 *Let $\mathcal{A} \models dLO^+$. The subsets of A definable in $\mathcal{A}$ (in L or in L') are just the finite unions of (open or closed) intervals in $\mathcal{A}$ (possibly having $+\infty$ as a right endpoint).*

Proof. Let $\varphi(v, \vec{w})$ be a L'-formula. As dLO^+ has the elimination of quantifiers in L', we can assume thant $\varphi(v, \vec{w})$ is quantifier free; owing to Theorem 2.2.1 (and its proof), for every $\vec{a}$ in A, $\varphi(\mathcal{A}, \vec{a})$ is a union of intersections of intervals, and hence a union of intervals. ♣

This accomplishes our analysis of discrete linear orders with a last but no least elements. How to deal with the other three cases of infinite discrete linear orders listed before? They can be handled in a similar way to get quantifier elimination and consequently completeness. In particular it turns out that the four cases exhaust all the possible completions of the theory of infinite discrete linear orders; in other words, these completions are fully characterized by saying if the corresponding models admit or lack a least and a greatest element.
Actually the case without endpoints deserves some more comments. In fact, in this framework, the enlarged language L' needs no natural "constant" symbol (just because endpoints are lacking), and takes the only additional operation symbol s. Accordingly, properly speaking, the elimination of quantifiers fails in this extended language, because we have no constant to build atomic sentences. For instance the (true) sentence $\exists w(w = w)$ cannot be translated into an equivalent quantifier free sentence; the same happens for the (false) sentence $\exists w(s(w) = w)$. So the right statement here is as

follows: an elimination set for the theory of discrete linear orders without endpoints is the set of atomic formulas *plus* a unique sentence (such as "there is no least element", or "there is no last element"). We do not discuss the proof here. In fact, we shall treat this case in detail when considering dense linear orders in the next section.
Finally, notice that decidability can be shown (in L) in the 4 possible cases. Consequently the (incomplete) theory dLO of infinite discrete linear orders is decidable, too; in fact, a sentence φ of L belongs to dLO if and only if it is in each of its 4 completions.

2.3 Dense linear orders

Now we deal with dense linear orders. The plan here is exactly the same as in the discrete case. We use the language $L = \{\leq\}$ and we distinguish 4 possible cases:

1. there is a least element, but no last element (just as among non-negative rationals with respect to the usual order);
2. there is a last element, but no least element (now non-positive rationals provide an example);
3. there are both a least element and a last element (look at the rationals, or even at the reals, in the closed interval $[0, 1]$);
4. there are no endpoints (this is the case of $(\mathbf{Q}, \leq)$).

In 1, 2, 3 one shows elimination of quantifiers in a language with one or two additional constants to be intepreted into the endpoints; 4 deserves a more specific treatment, because quantifier free formulas need an auxiliary single L-sentence to form an elimination set (even in L): we provide full details below. In all these cases it is easy to deduce completeness in L. This implies that these 4 classes exhaust all the possible completions of the theory of dense linear orders.
As already said, here we limit our analysis to dense linear orders without endpoints. We just met their theory in Chapter 1; we called it DLO^- and we observed that it is $\aleph_0$-categorical, hence complete. We treat now quantifier elimination (in L), and in this way we provide an alternative and detailed proof of its completeness.

Theorem 2.3.1 *The quantifier free formulas of L together with a single sentence of DLO^- (such as $\exists v(v = v)$) are an elimination set of DLO^-.*

Proof. We follow the same approach as in the discrete case. But now the successor symbol does not make sense, our language is smaller and hence our setting is simpler: L-terms are just variables (no constant arises because there are no endpoints). Accordingly what we have to do is to eliminate the quantifier in a formula

$$\exists w\, \alpha(\vec{v}, w)$$

where $\alpha(\vec{v}, w)$ is a conjunction of conditions saying that w is equal, or smaller, or larger than some v in $\vec{v}$. To obtain this, proceed as follows. Again ignore trivialities like $w = w$ (they can be preliminarily listed and easily recognized); if nothing else occurs, just replace our formula with $\exists v(v = v)$. On the contrary, when meeting a condition that cannot be satisfied by any w, like $w < w$ (also these negative statements can be preliminarily listed), replace our formula with $\neg\exists v(v = v)$. Otherwise, as soon as you meet one equation $w = v_i$, delete w and $\exists$, and replace w with v_i throughout our formula. At last, if only disequations occur and hence our formula states that w is smaller than certain variables ($v_1, \ldots, v_h$ with no loss of generality) and larger that others ($v_{h+1}, \ldots, v_k$), then get the required quantifier free formula in the following way. List (in a suitable disjunction) all the possible orderings of $v_1, \ldots, v_k$ in $\leq$ according to which $v_1, \ldots, v_h$ precede $v_{h+1}, \ldots, v_k$; for every ordering, let v, v' denote respectively the maximal element among $v_1, \ldots, v_h$ and the least among $v_{h+1}, \ldots, v_k$; $v < v'$ and the density assumption are sufficient to ensure that an intermediate w exists. When $h = 0$ or $h = k$, one uses the lack of endpoints.
This concludes the elimination procedure. ♣

Corollary 2.3.2 *DLO^- is model complete and complete.*

Proof. Model completeness is a straightforward consequence. Completeness can be deduced as follows, using model completeness. First notice that any two dense linear orders with no endpoints $(A, \leq)$ and $(B, \leq)$ embed in a common extension (for instance, their sum $(A + B, \leq)$, where $A + B$ is the disjoint union of A and B, $\leq$ enlarges the orderings in A and B and, in addition, satisfies $a < b$ for every choice of $a \in A$ and $b \in B$). As DLO^- is model complete, each of these embeddings is elementary, in particular $(A, \leq)$ and $(B, \leq)$ are elementarily equivalent to their sum, and hence to each other. ♣

As already recalled, completeness was also observed in the previous chapter via $\aleph_0$-categoricity and the Vaught criterion. By the way, notice that the

Vaught Theorem provides a completeness proof even when endpoints arise. In fact, also the remaining classes of dense linear orders (with least and/or last element) have an $\aleph_0$-categorical theory.
The decidability of DLO^- can be easily shown. Indeed, by proceeding as in the discrete case, one sees that even the theory of arbitrary dense orders (with or without endpoints) is decidable.
Now let us deal with definability.

Corollary 2.3.3 *Let $\mathcal{A} \models DLO^-$. The subsets of A definable in $\mathcal{A}$ are just the finite unions of (open or closed) intervals, possibly with infinite endpoints.*

Proof. Proceed as for dLO^+. ♣

2.4 Algebraically closed fields (and Tarski)

Tarski obtained his celebrated quantifier elimination procedures for the complex field and the ordered field of reals in the thirties. Owing to the stop due to the World War, he published his results only in 1948. We consider here the complex case, and we delay the real one to the next section. We should underline that Tarski dealt with theories of single structures (the complex field, the ordered field of reals) rather than on axiomatizable classes (ACF, RCF). But a careful analysis of the proofs singles out which kind of algebraic conditions are necessary to ensure the quantifier elimination result: so one realizes that what makes the machinery work is just being algebraically closed in the complex case, and the intermediate value property for polynomials in the real case. This is a crucial result, specially towards the aim of finding a nice axiomatization for the theory of the complex field, or of the ordered field of reals.
As promised, here we consider the complex case, but we prefer an approach dealing with the whole class of algebraically closed fields.

Theorem 2.4.1 (Tarski) *The theory ACF of algebraically closed fields has the elimination of quantifiers in the language $L = \{+, \cdot, -, 0, 1\}$.*

Proof. Take a formula $\varphi(\vec{v})$ of L, we are looking for an equivalent quantifier free formula $\varphi'(\vec{v})$. As before, we can limit our analysis to the case when $\varphi(\vec{v})$ is of the form

$$\exists w\, \alpha(\vec{v}, w)$$

where $\alpha(\vec{v}, w)$ is a finite conjunction of atomic formulas and negations, all containing w. In our language, atomic formulas are just equalities of terms, hence equations. Using $-$, one can express each of them as

$$p(\vec{v}, w) = 0$$

where $p(\vec{y}, x)$ is a polynomial with integer coefficients. Accordingly $\varphi(\vec{v})$ is

$$\exists w \, (\bigwedge_{i \leq k} p_i(\vec{v}, w) = 0 \wedge \bigwedge_{j \leq h} \neg(q_j(\vec{v}, w) = 0))$$

where the p_i's and the q_j's are polynomials with integer coefficients, all having a positive degree, n_i and m_j respectively, in x, hence with respect to w.

Basic field theory tells us that a sequence of elements in a field excludes 0 if and only if its product is not 0. Accordingly, we can assume that at most one inequation occurs in $\varphi(\vec{v}, w)$, say

$$\neg(q(\vec{v}, w) = 0).$$

where $q(\vec{y}, x)$ is the product of the polynomials $q_j(\vec{y}, x)$ when $j \leq h$; let m denote the degree in x of $q(\vec{y}, x)$.

At this point one might wonder whether we can reduce the number of equations in our formula $\varphi(\vec{v}, w)$ to get at most a single equation. This is true, and can be shown by using again pure field theory (so without appealing to algebraic closure), but requires some more subtlety. The idea is that, for a given field K and a sequence $\vec{b}$ in K, the common roots of the polynomials $p_i(\vec{b}, x)$ are just the roots of their greatest common divisor, and that there is a quantifier free formula in $\vec{v}$, defining the coefficients (in x) of this greatest common divisor, and independent of K and $\vec{b}$. The former claim is clear. Let us explain the details of the latter.

Consider $p_i(\vec{y}, x)$ for $i \leq k$. For every i, write $p_i(\vec{y}, x)$ as a polynomial in x with coefficients in $\mathbf{Z}[\vec{y}]$

$$p_i(\vec{y}, x) = \sum_{r \leq n_i} p_{i,r}(\vec{y}) x^r.$$

Take two of these polynomials, for instance p_0 and p_1, and suppose for simplicity $n_0 \geq n_1$. We claim that there is quantifier free formula in $\vec{v}$ yielding, whenever $p_1(\vec{v}, x)$ is not the null polynomial (in x), the coefficients in x of the quotient and the remainder of the division of $p_0(\vec{v}, x)$ by $p_1(\vec{v}, x)$. To get this formula, just follow the usual division procedure for polynomials.

This is a tedious but straightforward exercise. For instance, the first step is to write that either

$$p_{1,n_1}(\vec{v}) = 0$$

or the coefficients of $p_0(\vec{v}, x)$ and $p_1(\vec{v}, x)$ satisfy

$$p_{1,n_1}(\vec{v})\, p_0(\vec{v}, x) = p_{0,n_0}(\vec{v}) p_1(\vec{v}, x) x^{n_0 - n_1} + P(\vec{v}, x),$$

where $P(\vec{v}, x)$ is a polynomial of degree $< n_0$ in x, and, in the latter case, the required quotient admits

$$p_{0,n_0}(\vec{v})(p_{1,n_1}(\vec{v}))^{-1} x^{n_0 - n_1}$$

as a coefficient of maximal degree.
At this point, recall the Euclidean algorithm of repeated divisions, yielding the greatest common divisor (in x) of our polynomials $p_i(\vec{y}, x)$ $(i \leq k)$ as the last nonzero remainder in a finite sequence of successive divisions. Again, a suitable quantifier free formula in $\vec{v}$ determines the coefficients in x of our greatest common divisor, whenever the $p_i(\vec{v}, x)$'s $(i \leq k)$ are not all zero.
In conclusion, we can assume that our formula $\varphi(\vec{v}, w)$ has one of the following three forms:

1. $\exists w (p(\vec{v}, w) = 0)$,
2. $\exists w \neg (q(\vec{v}, w) = 0)$,
3. $\exists w (p(\vec{v}, w) = 0 \wedge \neg (q(\vec{v}, w) = 0))$

where p and q are as before.
First consider 1. In any field, 1 is equivalent to say that, if $\vec{v}$ annihilates all the coefficients of $p(\vec{y}, x)$ in x of positive degree in x, then $\vec{v}$ assigns the value 0 also to the term of degree 0 in x; this can be written as a suitable quantifier free formula in $\vec{v}$.
Now consider 2. In any infinite field, 2 is equivalent to say that $\vec{v}$ does not annihilate the coefficients of the polynomial $p(\vec{y}, x)$ in x. Again, the latter statement can be expressed by a quantifier free formula in $\vec{v}$.
Finally let us deal with 3. We claim that, in any algebraically closed field, 3 is equivalent to the statement

$$(\star) \qquad p(\vec{v}, x) \text{ does not divide } q(\vec{v}, x)^n;$$

recall that n is the degree of $p(\vec{y}, x)$ with respect to x, and notice that $(\star)$ can be expressed as a quantifier free formula in $\vec{v}$ (just use the previous remarks

about divisibility, and write that the remainder of the division in ($\star$) is not 0). The direction from left to right is true in every field K: if, for a given sequence $\vec{b}$ in K, the annihilator of $p(\vec{b}, x)$ is not included in the annihilator of $q(\vec{b}, x)$, then $p(\vec{b}, x)$ cannot divide $q(\vec{b}, x)$ and $(q(\vec{b}, x))^n$. Conversely, take K and $\vec{b}$ as before; assume that every root of $p(\vec{b}, x)$ annihilates $q(\vec{b}, x)$, too; for K algebraically closed, this implies that every linear factor of $p(\vec{b}, x)$ divides $q(\vec{b}, x)$ and hence that $p(\vec{b}, x)$ divides $q(\vec{b}, x)^n$.
This accomplishes our proof. ♣

Now let us comment this quantifier elimination result, and propose some noteworthy consequences. First of all, we want to emphasize that the quantifier elimination property characterizes the algebraically closed fields among infinite fields. In fact, it is a profound result of Macintyre, McKenna e Van den Dries that an infinite field whose theory eliminates the quantifiers in the language $L = \{+, -, \cdot, 0, 1\}$ must be algebraically closed.
An obvious consequence of quantifier elimination is the following.

Corollary 2.4.2 *ACF is model complete.*

Clearly ACF is not complete. In fact, for every prime p, the sentence $p = 0$ is true in every algebraically closed field of characteristic p and false in every algebraically closed field of characteristic $\neq p$. However, as we already showed in Chapter 1,

Corollary 2.4.3 *For every $p = 0$ or prime, the theory ACF_p is complete.*

Proof. In Chapter 1 we provided a proof founded on Vaught's Theorem. An alternative approach, using quantifier elimination (indeed model completeness), is the following. Fix p. There is a minimal algebraically closed field $\mathcal{K}_p$ of characteristic p: this is the algebraic closure of the prime subfield. $\mathcal{K}_p$ is embeddable in every algebraically closed field of the same characteristic. Owing to the model completeness of ACF_p, all these embedding are elementary. In particular, all the algebraically closed fields of characteristic p are elementarily equivalent to $\mathcal{K}_p$, and consequently to each other. ♣

As we already observed in section 2.1, the theories ACF_p exhaust all the possible completions of ACF in L when p ranges over the primes and 0 (furthermore, each of them has the quantifier elimination in L, just because it extends ACF).
An application of the Compactness Theorem lets us say even more. In fact, we have seen that the theory of the complex field is just ACF_0, and so is axiomatized by ACF and, in addition, by the infinitely many sentences stating

$\neg(p = 0)$ for every prime p. Let σ be any sentence in ACF_0. Compactess tells us that σ is a consequence of ACF and finitely many sentences concerning the characteristic. Hence σ is true in every algebraically closed field of prime characteristic p for all but finitely many p's. So we have shown the following result.

Theorem 2.4.4 *Let σ be a sentence of the language L. σ is true in some (equivalently every) algebraically closed field of characteristic 0 if and only if σ is true in some (equivalently every) field of characteristic p for all but finitely many primes p.*

Hence what is true in the complex field (and in any algebraically closed field of characteristic 0) is satisfied by the algebraically closed fields of characteristic p for almost all primes p. We'll see later in this section a nice application of this model theoretic transfer principle to Algebra.
Now let us deal briefly with decision problems. As already said, decidability follows in a very simple way from quantifier elimination.

Corollary 2.4.5 *The theory ACF of algebraically closed fields is decidable.*

Proof. It suffices to decide if a given quantifier free sentence σ of L is in ACF or not. With no loss of generality, σ is a conjunction of disjunctions of atomic sentences and negations. As a conjunction is in a theory if and only if each conjunct is, we can write σ (up to equivalence, using $-$) as

$$(\bigvee_i m_i = 0) \vee (\bigvee_j \neg(n_j = 0)).$$

where the m_i's and the n_j's are positive integers. So our sentence just says that the characteristic divides $\prod_i m_i$ or is coprime with some n_j (or suitable variants when no equation, or inequation arises). This can be easily checked in the fixed framework. ♣

We shall add some more comments about the decidability of ACF in the last section of this chapter.
Now let us deal with definability. We have seen in Chapter 1 that, in any field $\mathcal{K}$, constructible sets (in particular algebraic varieties) are definable. Theorem 2.4.1 implies that, within algebraically closed fields, the converse is also true.

Corollary 2.4.6 *In an algebraically closed field $\mathcal{K}$, for every positive integer n, a subset of K^n is definable if and only if is constructible.*

Proof. It suffices to show that, if $X \subseteq K^n$ is definable, then X is constructible. Let $\varphi(\vec{v}, \vec{w})$ be a formula of L and $\vec{a}$ be a sequence in K satisfying

$$X = \varphi(\mathcal{K}^n, \vec{a})$$

where n is the length of $\vec{v}$. Using quantifier elimination, we can replace $\varphi(\vec{v}, \vec{w})$ by an equivalent formula which excludes quantifiers and consequently is a finite Boolean combination of equations

$$q(\vec{v}, \vec{w}) = 0$$

where $q(\vec{x}, \vec{y})$ is a polynomial with coefficients in the subring generated by 1. Hence $X = \varphi(\mathcal{K}^n, \vec{a})$ is the Boolean combination of the algebraic varieties defined by the formulas

$$q(\vec{v}, \vec{a}) = 0,$$

and so is a constructible set. ♣

Notice that in every field $\mathcal{K}$ the subsets of K^n definable by quantifier free formulas are constructible. Quantifier elimination ensures that, when $\mathcal{K}$ is algebraically closed, no further definable set arises.
The following proposition underlines the geometrical content of Tarski's Theorem.

Theorem 2.4.7 (Chevalley) *Let $\mathcal{K}$ be an algebraically closed field, n be a positive integer, $X \subseteq K^{n+1}$, X' be the projection of X onto the first n coordinates. If X is constructible, then X' is also constructible.*

Proof. If $\varphi(\vec{v}, w)$ defines X, then $\exists w \varphi(\vec{v}, w)$ defines X'. ♣

Now let us consider 1-ary definable sets in an algebraically closed field $\mathcal{K}$. In this restricted framework, the following proposition holds.

Corollary 2.4.8 *Let $\mathcal{K}$ be an algebraically closed field, $X \subseteq K$ be definable in $\mathcal{K}$. Then X is either finite or cofinite.*

Proof. For every $q(x, \vec{a}) \in K[x]$, $q(v, \vec{a}) = 0$ defines K if $q(x, \vec{a})$ is zero, and a finite set otherwise. A finite Boolean combination of finite or cofinite sets is still finite or cofinite. ♣

Actually we can say even more. Indeed, in any (possibly non-algebraically closed) field $\mathcal{K}$, a subset X of K definable by a quantifier free formula is either

finite or cofinite. Quantifier elimination extends this property to arbitrary 1-ary subsets when $\mathcal{K}$ is algebraically closed.
To conclude this section, we want to propose a nice application of Model Theory to Algebra within algebraically closed fields. This is the so called *injectivity-implies-surjectivity* Theorem, due to J. Ax [3]. Compactness, and the consequent remark that the sentences true in the complex field are just those satisfied by the algebraically closed fields of characteristic p for almost all primes p, are used to deduce

Theorem 2.4.9 *Any injective morphism f from an algebraic variety V over the complex field into V itself is surjective.*

Proof. We already noticed that any algebraic variety is a definable set, and is even defined by a finite conjunction of equations (possibly with parameters). In particular let the formula

$$\bigwedge_{j \leq t} p_j(\vec{v}, \vec{a}) = 0$$

give V in this way (the p_j's are polynomials with integer coefficients, and $\vec{a}$ denotes a sequence of complex parameters). Analogously, a morphism between varieties is a map defined by a finite conjunction of equations. Accordingly let

$$\bigwedge_{i \leq s} q_i(\vec{v}, \vec{w}, \vec{a}) = 0$$

yield f (the q_i's are again polynomials with integer coefficients; we can freely use here the same parameters $\vec{a}$ as before; if necessary, we extend $\vec{a}$ to include new complex numbers). At this point it is an easy exercise to write a first order sentence in the language L (without parameters) saying:

> *for all $\vec{z}$, if $\bigwedge_{i \leq s} q_i(\vec{v}, \vec{w}, \vec{z}) = 0$ defines a morphism from the variety given by $\bigwedge_{j \leq t} p_j(\vec{v}, \vec{z}) = 0$ into itself, and this morphism is injective, then it is also surjective.*

Let n denote, as usual, the length of $\vec{v}$. What we have to show is that the complex field is a model of all these sentences when the p_j's and the q_i's range over the polynomials with integer coefficients, equivalently that ACF_0 includes these statements. Using compactness, we can alternatively check what happens in ACF_p when p is a prime, and so if the previous sentences are true in every algebraically closed field of characteristic p; as ACF_p is complete, it suffices to look at the behaviour of a single model of ACF_p, for instance of the algebraic closure $\overline{\mathbf{F}_p}$ of the field $\mathbf{F}_p$ with p elements: a positive answer in $\overline{\mathbf{F}_p}$ for sufficiently many p implies a positive answer for the complex field. So take an algebraic variety V over $\overline{\mathbf{F}_p}$ and an injective morphism f from V to V over $\overline{\mathbf{F}_p}$. Use the algebraic fact that $\overline{\mathbf{F}_p}$ is locally finite and represent V as the union of its intersections with F^n where F ranges over the finite subfields of $\overline{\mathbf{F}_p}$ (containing the parameters defining V and f). Recall the trivial principle that any injective function from a finite set to itself is also surjective. Deduce that the restriction of f to $V \cap F^n$ is surjective for every F. Extend this result to f: f is surjective, as required. ♣

2.5 Tarski again: Real closed fields

In this section we deal with the quantifier elimination theorem for real closed fields. This is the main result of Tarski in this framework, not only because, as we shall see, the proof is deeper and more complicated than in the complex case, but also because the ordered field of reals is intrinsecally related to geometry. It is certainly needless to recall that, for instance, in the Euclidean plane equipped with some fixed Cartesian axes, every point is essentially an ordered pair of reals, every straight line is the variety given by a polynomial with degree 1 and 2 unknowns over the reals, and so on. Accordingly, statements about points, lines, ... can be easily translated into statements about reals, addition, multiplication (often in a first order way). In particular, a decision algorithm about the theory of the ordered field of reals (the *elementary algebra* according to Tarski's terminology) should work for (first order) Euclidean geometry as well.
Actually Tarski's quantifier elimination procedure dealt with the reals rather than with the theory RCF. But, just as in the complex case, one can realize that the basic ingredients of the proof concern arbitrary real closed fields. So we state (and show) the result in this (seemingly enlarged) setting; but we shall deduce quickly that RCF is complete and hence equals the theory of $(\mathbf{R}, +, -, \cdot, 0, 1, \leq)$. We follow the elegant approach of Cohen [27] rather than Tarski's original proof.

Theorem 2.5.1 *The theory RCF of reals closed fields has the elimination of quantifiers in the language $L = \{+, -, \cdot, 0, 1, \leq\}$.*

Proof. By proceeding as in the case of algebraically closed fields, one preliminarily realizes that the heart of the matter is to eliminate the quantifier

$\exists$ in a formula

$$\exists w \alpha(w, \vec{v})$$

where $\alpha(w, \vec{v})$ is the conjunction of at most one equation $p(w, \vec{v}) = 0$ and a finite (possibly empty) set of disequations $q_j(w, \vec{v}) > 0$ (with $j \leq m$), where $p(x, \vec{y})$ and $q_j(x, \vec{y})$ $(j \leq m)$ are polynomials with integer coefficients.
So let us open a (long) parenthesis and examine an arbitrary polynomial $f(x) = \sum_{i \leq t} f_i x^i$ with coefficients in a real closed field K. It is known that, if f_i is not 0 for all $i \leq t$, then $f(x)$ has at most t roots in the field. Fix t. Then it is easily seen that

1) the function calculating, for every polynomial $f(x)$ as before, equivalently for every non-zero sequence $(f_0, \ldots, f_t)$ in K^{t+1}, how many roots $f(x)$ admits

as well as, for every r and s with $1 \leq r \leq s \leq t$,

2) the set of non-zero sequences $(f_0, \ldots, f_t)$ in K^{t+1} such that $f(x)$ has exactly s roots,

3) the function mapping any nonzero $(f_0, \ldots, f_t)$ into the r-th root of $f(x)$

are definable in any ordered field K in a uniform way (independent of K). We claim that, within real closed fields, for every t, these objects are definable by quantifier free formulas, still in a uniform way (independent of the underlying field). To see this, one uses the Sturm theory of real root counting. We proceed by induction on t.
The case $t = 0$ is clear: the number of roots is 0 if $f_0 \neq 0$, and undefined otherwise; 2 and 3 are empty objects.
So assume $t > 0$ and suppose our claim true for every natural value $< t$, in order to extend it to t. The idea here is to relate the zeroes of $f(x)$ to the roots of its derivative and the sign of $f(x)$ in these roots. Hence build the formal derivative $f'(x)$ of $f(x)$ with respect to x

$$f'(x) = \sum_{0 < i \leq t} i f_i x^{i-1}.$$

Preliminarily, notice that $f'(x) = 0$ if and only if $(f_1, \ldots, f_t) = (0, \ldots, 0)$. Except this case, induction equips us with quantifier free formulas defining (with respect to $(f_0, \ldots, f_t)$ via $(f_1, 2f_2, \ldots, tf_t)$)

1) the function counting, for every sequence $(f_0, \ldots, f_t)$ with $(f_1, \ldots, f_t) \neq (0, \ldots, 0)$, how many roots $f'(x)$ admits,

and, for $1 \leq r \leq s < t$,

2) the set of the sequences $(f_0, \ldots, f_t)$ in K^{t+1} such that $(f_1, \ldots, f_t)$ is not zero and $f'(x)$ has exactly s roots,

3) the function mapping any (suitable) non-zero $(f_0, \ldots, f_t)$ into the r-th root of $f'(x)$.

Now order the roots of $f'(x)$

$$\rho_1 < \ldots < \rho_s.$$

The intermediate value property, holding in every real closed field, ensures that $f'(x)$ cannot change its sign between two successive roots. Can we deduce that $f(x)$ is monotone (increasing or decreasing according to the sign of $f'(x)$) in the same interval? Certainly yes in the case of the real field: this is a well known result in elementary real analysis. But a complete algebraic (although non trivial) proof can be done for polynomials by using only the axioms of RCF. Consequently, in every real closed field K, $f(x)$ is monotone (increasing or decreasing according to the sign -positive or negative- of its derivative) in each interval (ρ_i, ρ_{i+1}), $1 \leq i < s$. Now look at $f(\rho_i)$ and $f(\rho_{i+1})$.

(i) If they are not 0 and their sign is the same, then (ρ_i, ρ_{i+1}) does not contain any root of $f(x)$ because $f(x)$ is monotone in the interval (notice that the cases when exactly one between ρ_i and ρ_{i+1} annihilates $f(x)$ can be handled in a similar way).

(ii) If $f(\rho_i)$ and $f(\rho_{i+1})$ admit opposite signs, then (ρ_i, ρ_{i+1}) does contain a root of $f(x)$ by the intermediate value property. The uniqueness of this root might follow from Rolle's Theorem (two distinct roots of $f(x)$ $\rho_i < a < b < \rho_{i+1}$ determine a new intermediate root of $f'(x)$, and this is impossible). Elementary analysis ensures that Rolle's Theorem is certainly true for the reals; but, again, one can give an alternative algebraic and non trivial proof (for polynomials) holding in every real closed field.

(iii) Assume at last $f(\rho_i) = f(\rho_{i+1}) = 0$. The argument in 2 again excludes any additional intermediate root of $f(x)$.

This machinery lets us count the roots in the interval $[\rho_1, \rho_s]$. But what can we say in $(-\infty, \rho_1)$ and $(\rho_s, +\infty)$? The same arguments as before ensure that $f(x)$ is monotone, and at least one root occurs in each of these half-lines. But our setting changes when we examine the existence of this root. For, every interval (ρ_i, ρ_{i+1}) $(1 \leq i \leq s)$ is bounded, while our half-lines are not. However the following algebraic fact helps us.

> Let $f(x) = \sum_{i \leq t} f_i x^i$ as before. Then $f(x)$ has no roots out of the interval $[-a, a]$ where $a = 3t\, max\{(|f_{t-i} f_t^{-1}| : 0 < i \leq t\} + 1$.

(The proof only uses the axioms of ordered fields). So a possible root less than ρ_1 should lie in $[-a, \rho_1)$, and a possible root greater than ρ_s should belong to $(\rho_s, a]$; moreover the absolute value function $|\ \ |$ can be defined in a quantifier free way, because, for every $b \in K$, $|b|$ is b when $b \geq 0$ and $-b$ otherwise. Hence we are led to a bounded framework, and we can proceed as in the previous cases.
In conclusion, we have provided a uniform procedure counting, for every nonzero $(f_0, \ldots, f_t)$ in K, how many roots $f(x)$ admits. The function calculating their number s is defined by a quantifier free formula (essentially checking the sign of $f(x)$ in the roots of its derivative and in $\pm a$). Similarly the set of non-zero sequences $(f_0, \ldots, f_t)$ in K for which $f(x)$ has exactly s roots can be defined by checking these sign relations and forming a suitable first order disjunction to list the cases when s occurs. Finally, the function producing, for every non-zero $(f_0, \ldots, f_t)$ and $1 \leq r \leq s$, the r-th root of $f(x)$ is easily defined on the same basis.
This accomplishes the proof of the claim and ends our parenthesis. Now we come back to quantifier elimination. Recall that we are considering a formula

$$(a) \qquad \exists w (p(w, \vec{v}) = 0 \wedge \bigwedge_{j \leq m} q_j(w, \vec{v}) > 0)$$

or

$$(b) \qquad \exists w \bigwedge_{j \leq m} q_j(w, \vec{v}) > 0$$

where $p(x, \vec{y})$ and $q_j(x, \vec{y})$ $(j \leq m)$ are polynomials with integer coefficients. Each of them can be written as a polynomial with coefficients in $\mathbf{Z}[\vec{y}]$ in the following way

$$p(x, \vec{y}) = \sum_{i \leq t} p_i(\vec{y}) x^i,$$

$$q_j(x, \vec{y}) = \sum_{i \leq t_j} q_{j,i}(\vec{y}) x^i.$$

(a) is quickly reduced to (b) because its formula is equivalent to

$$(\bigwedge_{i\leq t} p_i(\vec{v}) = 0 \wedge \exists w \bigwedge_{j\leq m} q_j(w, \vec{v}) > 0) \vee$$

$$\bigvee_{1\leq r\leq s\leq t} (\text{"}p(x, \vec{v}) \text{ has } s \text{ roots"} \wedge$$

$$\wedge \text{ "the } r\text{-th root } \rho_r(\vec{v}) \text{ satisfies } \bigwedge_{j\leq m} q_j(\rho_r(\vec{v}), \vec{v}) > 0\text{"}),$$

where the latter disjunct can be expressed by a quantifier free first order formula. So look at (b). For every $j \leq m$ and for every $s_j \leq t_j$, there are quantifier free formulas defining, for every real closed field K, the set of the sequences $\vec{b}$ such that $q(x, \vec{b})$ has s_j roots in x, and listing these roots

$$\rho_{j,1}(\vec{b}) < \ldots < \rho_{j,s_j}(\vec{b}).$$

One can compute the sign of $q_j(x, \vec{b})$ in the intervals

$$(-\infty, \rho_{j,1}(\vec{b})),$$

$$(\rho_{j,i}(\vec{b}), \rho_{j,i+1}(\vec{b})) \quad (1 \leq i < s_j),$$

$$(\rho_{j,s_j}(\vec{b}), +\infty)$$

in a uniform way (independent of K and $\vec{b}$) by looking at the (sign) value of

$$q_j(\rho_{j,1}(\vec{b}) - 1, \vec{b})$$

$$q_j(\frac{\rho_{j,i}(\vec{b}) + (\rho_{j,i+1}(\vec{b})}{2}, \vec{b})$$

$$q_j(\rho_{j,s_j}(\vec{b}) + 1, \vec{b})$$

respectively. List all the possible orderings of the roots (in x) of the $q_j(x, \vec{b})$'s when j ranges over the natural numbers $\leq m$, and divide in every case K into finitely many intervals such that the $q_j(x, \vec{b})$'s have a constant sign (with respect to x) in each of them; check these signs (in the way suggested before) and form a suitable disjunction picking the intervals where all these signs are positive. This procedure is independent of K and $\vec{b}$ and provides the required quantifier free formula. ♣

Corollary 2.5.2 *RCF is model complete.*

Corollary 2.5.3 *RCF is complete; in particular, RCF is the theory of the ordered field of reals (as well as of every real closed field).*

Proof. There is a minimal ordered real closed field, embedded in any model of RCF. This is the ordered field $\mathbf{R}_0$ of real algebraic numbers. The model completeness of RCF ensures that every real closed field is an elementary extension of $\mathbf{R}_0$. In particular all the real closed fields are elementarily equivalent to $\mathbf{R}_0$ and, consequently, to each other. ♣

This is the first completeness proof we give about RCF; in fact Vaught's criterion does not apply because RCF is not categorical in any infinite power.
We have seen that real closed fields eliminate quantifiers in their language $L = \{+, -, \cdot, 0, 1, \leq\}$. Notably, they are fully characterized by this property: for, Macintyre, McKenna and Van den Dries showed that an ordered field, whose theory has the quantifier elimination in L, must be real closed. We also notice that RCF does not preserve quantifier elimination in the restricted language $L' = \{+, -, \cdot, 0, 1\}$ without order. Actually one can remember that, even in checking solvability of the popular equation $ax^2 + bx + c = 0$ with degree 2 and 1 unknown over the reals (or over any real closed field), one needs a disequation $b^2 - 4ac \geq 0$ to ensure roots, and hence to eliminate $\exists$ in the formula $\exists w(v_2 w^2 + v_1 w + v_0 = 0)$. More formally, recall that, with respect to the theory of the real field, the formulas

$$\varphi(v) : \ v \geq 0,$$

$$\varphi'(v) : \ \exists w(v = w^2)$$

are equivalent. As RCF is complete and hence equals the theory of the real field, the same holds in every real closed field. Consequently the L'-formula (with the quantifier $\exists$)

$$\varphi'(v) : \ \exists w(v = w^2)$$

defines the set of non-negative elements in every real closed field. However $\varphi(v)$ cannot be equivalent in RCF to any quantifier free L'-formula $\varphi''(v)$. In fact $\varphi(\mathbf{R})$ is the half-line $[0, +\infty)$ of $\mathbf{R}$, and so is both infinite and coinfinite, while $\varphi''(\mathcal{K})$ is either finite or cofinite for every field $\mathcal{K}$: see the proof of Corollary 2.4.8.
Now we discuss decidability: as already said, this was the main consequence of elimination of quantifiers, according to the general feeling in the fourties.

Corollary 2.5.4 *RCF is decidable.*

Proof. Owing to quantifier elimination, every L-sentence σ is equivalent in RCF to a Boolean combination of sentences $m = n$ or $m < n$ where m and

n are integers. This quantifier free statement can be easily checked in our framework. ♣

We shall comment this result later in 2.9. Now we examine another remarkable consequence of quantifier elimination, namely definability. Recall that, in an ordered field $\mathcal{K}$, every semialgebraic set (in other words, every finite Boolean combination of sets of solutions of disequations

$$q(\vec{x}) \geq 0$$

with $q(\vec{x}) \in \mathcal{K}[\vec{x}]$) is definable.

Corollary 2.5.5 *In a real closed ordered field $\mathcal{K}$, the definable sets are exactly the semialgebraic ones.*

Proof. Let n be a positive integer, $X \subseteq K^n$ be a set definable in $\mathcal{K}$. So there are a formula $\varphi(\vec{v}, \vec{w})$ of L and a sequence $\vec{a} \in K$ such that

$$X = \varphi(K^n, \vec{a}).$$

Owing to Tarski's quantifier elimination theorem, we can assume that $\varphi(\vec{v}, \vec{w})$ has no quantifier and hence is a finite Boolean combination of disequations

$$q(\vec{v}, \vec{w}) \geq 0$$

with $q(\vec{x}, \vec{y}) \in \mathbf{Z}[\vec{x}, \vec{y}]$. Consequently X is a finite Boolean combination of sets of solutions of disequations

$$q(\vec{v}, \vec{a}) \geq 0,$$

and so is a semialgebraic set. ♣

Here is a geometric restatement of Tarski's Theorem.

Theorem 2.5.6 (Tarski-Seidenberg) *Let $\mathcal{K}$ be a real closed ordered field, n be a positive integer, $X \subseteq K^{n+1}$, X' be the projection of X onto the first n coordinates. If X is semialgebraic, then X' is semialgebraic, too.*

This formulation is due to Thom, who also coined the name *semialgebraic set*. It has a more geometric flavour. Some mathematicians might appreciate this alternative terminology, for instance because it allows to state several results in a (seemingly) more agreable way (avoiding logic). Nevertheless, Tarski's original approach (via quantifier elimination and formulas) often provides quicker proofs, even in this geometric framework. Let us quote the following example from [168].

Example 2.5.7 Consider the following statement: *For a semialgebraic function f from $\mathbf{R}^{n+1}$ to $\mathbf{R}$, the set A of the sequences $\vec{x}$ in $\mathbf{R}^{n+1}$ such that*

$$lim_{y \to \infty} f(\vec{x}, y) \text{ is in } \mathbf{R}$$

is semialgebraic.

If one replaces everywhere *semialgebraic* by *definable*, this proposition may lose part of its (mathematical) glamour. But, using logic, one obtains a short proof: A is definable via the formula

$$\varphi(\vec{v}) \quad : \quad \exists w \forall \epsilon (\epsilon > 0 \longrightarrow \exists r \forall y (y > r \longrightarrow$$

$$\longrightarrow \exists z (f(\vec{v}, y) = z \wedge |z - w| < \epsilon)))$$

(recall that both f and the absolute value are definable). A direct approach via semialgebraic sets and projections is longer.
In a real closed field $\mathcal{K}$, the definable subsets $X \subseteq K$ have a very simple form.

Corollary 2.5.8 *Let $\mathcal{K}$ be a real closed field, X be a definable subset of $\mathcal{K}$. Then X is a finite union of intervals (closed or open, possibly with infinite endpoints).*

Proof. Let $q(x) \in K[x]$. We know that $q(v) = 0$ defines K if $q(x) = 0$ and a finite set (that is, the set of the roots $a_0 < \ldots < a_s$ of $q(x)$ in $\mathcal{K}$) otherwise. On the other hand, $q(v) > 0$ defines $\emptyset$ if $q(x) = 0$; otherwise $q(v) > 0$ defines the union of some intervals among $]-\infty, a_0[$, $]a_0, a_1[$, ..., $]a_s, +\infty[$ (recall that $\mathcal{K}$ satisfies the intermediate value property for polynomials). Hence any definable (equivalently, semialgebraic) set $X \subseteq K$ is a finite Boolean combination of intervals, and so a finite union of intervals. ♣

2.6 pp-elimination of quantifiers and modules

In this section we deal with (left) modules over a (countable) ring $\mathcal{R}$ with identity. In Chapter 1, we introduced a suitable language $L_R = \{0, +, -, r$ $(r \in R)\}$ for these structures, and we saw how to axiomatize their class by first order sentences in L_R. Let ${}_{\mathcal{R}}T$ denote the corresponding theory. A quick look at the axioms of ${}_{\mathcal{R}}T$ shows that each of them is a universal sentence $\forall \vec{v} \alpha(\vec{v})$ where $\alpha(\vec{v})$ is an atomic formula of L_R; this confirms that the class of the models of ${}_{\mathcal{R}}T$ (namely of the $\mathcal{R}$-modules) is closed under substructures. Now we wonder whether ${}_{\mathcal{R}}T$ has quantifier elimination in L_R. A trivial example shows that this is false even in the simple case when $\mathcal{R}$ is the ring $\mathbf{Z}$ of integers. In fact, just consider $\mathbf{Z}$ as a module over itself. In $\mathbf{Z}$ the formula

$$\varphi(v) : \exists w(v = 2w)$$

defines the set $2\mathbf{Z}$ of even integers. On the other side, every atomic formula $\varphi'(v)$ in $L_{\mathbf{Z}}$ is equivalent within ${}_{\mathbf{Z}}T$, and hence in the theory of the $\mathbf{Z}$-module $\mathbf{Z}$, to

$$rv = 0$$

for some non-negative integer r. This formula defines in $\mathbf{Z}$ $\{0\}$ if $r \neq 0$ and $\mathbf{Z}$ otherwise. No Boolean combination of these sets can equal $2\mathbf{Z}$. Therefore no quantifier free formula $\varphi'(v)$ of $L_{\mathbf{Z}}$ is equivalent to $\varphi(v)$ in $Th(\mathbf{Z})$, and so in ${}_{\mathbf{Z}}T$. It follows that ${}_{\mathbf{Z}}T$ does not eliminate the quantifiers in $L_{\mathbf{Z}}$.
However notice that $\varphi(v)$ is a typical pp-formula in $L_{\mathbf{Z}}$. Indeed we will see that, for any R, the pp-formulas of L_R are just the only obstruction to the elimination of quantifiers of ${}_{\mathcal{R}}T$ in L_R. Let us see why.
Take any (countable) ring $\mathcal{R}$ with identity. Recall that a pp-formula of L_R is an existential formula of the form

$$\varphi(\vec{v}) : \exists \vec{w}(A \cdot \vec{v} = B \cdot \vec{w})$$

where A and B are matrices with coefficients in R with suitable sizes, $\cdot$ denotes the usual row-by-column product between matrices, and $\vec{v}$, $\vec{w}$ should be viewed as column vectors (with suitably many rows). So, when $\vec{w} = \emptyset$, pp-formulas include the atomic formulas of L_R.
In Chapter 1 we pointed out that, for every pp-formula $\varphi(\vec{v})$ of L_R and every $\mathcal{R}$-module $\mathcal{M}$, $\varphi(\mathcal{M}^n)$ is a subgroup of $\mathcal{M}^n$ (called a pp-definable subgroup), but is not in general a submodule. Let us add here some more remarks about pp-formulas.

Remark 2.6.1 1. If $\varphi(\vec{v})$, $\psi(\vec{v})$ are pp-formulas of L_R, then also $\varphi(\vec{v}) \wedge \psi(\vec{v})$ is (equivalent in ${}_{\mathcal{R}}T$ to) a pp-formula.

2. Let $\varphi(\vec{v}, \vec{z})$ be a pp-formula of L_R, $\varphi(\vec{v}, \vec{z}) : \exists \vec{w}\, (A^t(\vec{v}, \vec{z}) = B\, \vec{w})$. Then $\varphi(\vec{v}, \vec{0})$ is a pp-formula, and hence, for every $\mathcal{R}$-module $\mathcal{M}$, $\varphi(\mathcal{M}^n, \vec{0})$ is a pp-definable subgroup of $\mathcal{M}^n$. Furthermore, for every $\vec{a} \in M$, $\varphi(\mathcal{M}^n, \vec{a}) = \emptyset$ or $\varphi(\mathcal{M}^n, \vec{a})$ is a coset of $\varphi(\mathcal{M}^n, 0)$ in M (in fact, given $\vec{b} \in \varphi(\mathcal{M}^n, \vec{a})$, it is easy to check $\varphi(\mathcal{M}^n, \vec{a}) = \varphi(\mathcal{M}^n, 0) + \vec{b}$).

3. Let $\varphi(\vec{v})$, $\psi(\vec{v})$ be pp-formulas of L_R with n free variables, and let k be a positive integer. It is simple to write a sentence in L_R ensuring that, in a given $\mathcal{R}$-module $\mathcal{M}$, the index of $\varphi(\mathcal{M}^n) \cap \psi(\mathcal{M}^n)$ in $\varphi(\mathcal{M}^n)$ is $\geq k$; in detail this sentence says

$$\exists \vec{v}_0 \ldots \exists \vec{v}_{k-1} \left(\bigwedge_{i<k} \varphi(\vec{v}_i) \wedge \bigwedge_{i<j<k} \neg\psi(\vec{v}_i - \vec{v}_j) \right).$$

We will denote it by $(\varphi : \psi) \geq k$. Any sentence of this form is called an *invariant statement* (we will see later the reason why). Notice that the finite Boolean combinations of invariant statements include the sentences saying:

- the index of $\varphi(\mathcal{M}^n) \cap \psi(\mathcal{M}^n)$ in $\varphi(\mathcal{M}^n)$ is k ("$= k$" means "$\geq k$" but "$\not\geq k+1$"); we will denote this formula by $(\varphi : \psi) = k$;
- the index of $\varphi(\mathcal{M}^n) \cap \psi(\mathcal{M}^n)$ in $\varphi(\mathcal{M}^n)$ is $\leq k$ ("$\leq k$" means "$\not\geq k+1$"); we shall denote this formula by $(\varphi : \psi) \leq k$.

At this point we can state and show the following fundamental theorem (of *pp-elimination of quantifiers* for modules).

Theorem 2.6.2 (Baur - Monk) *Let $\mathcal{R}$ be a (countable) ring with identity. Then the pp-formulas of L_R together with the invariant statements form an elimination set for ${}_RT$ in L_R. More precisely: for every formula $\alpha(\vec{v})$ of L_R, there are a Boolean combination β of invariant statements and a Boolean combination $\gamma(\vec{v})$ of pp-formulas such that*

$$\forall \vec{v}(\alpha(\vec{v}) \leftrightarrow \beta \wedge \gamma(\vec{v})) \in {}_RT.$$

We shall use in our proof the following result of group theory.

Lemma 2.6.3 (B. H. Neumann) *Let $\mathcal{G}$ be a group, a, $a_i \in G$, $\mathcal{H}$, $\mathcal{H}_i$ be subgroups of $\mathcal{G}$ (where i ranges among the naturals less than some fixed N), $aH \subseteq \bigcup_{i<N} a_i H_i$. Let I be the set of the naturals $i < N$ for which $|H : H \cap H_i| \leq N!$. Then $aH \subseteq \bigcup_{i \in I} a_i H_i$.*

Now let us begin the proof of the theorem.
Proof. We proceed by induction on $\alpha(\vec{v})$. If $\alpha(\vec{v})$ is an atomic formula, then $\alpha(\vec{v})$ is directly a pp-formula. The cases $\neg$ and $\wedge$ are easy to handle. So suppose that $\alpha(\vec{v})$ is of the form $\forall w \alpha'(w, \vec{v})$, where the induction hypothesis

ensures that there exist an invariant statement β' and a Boolean combination $\gamma'(w, \vec{v})$ of pp-formulas such that

$$\forall w \forall \vec{v}(\alpha'(w, \vec{v}) \leftrightarrow \beta' \wedge \gamma'(w, \vec{v})) \in {}_{R}T.$$

1st reduction: without loss of generality, $\alpha'(w, \vec{v})$ is a disjunction of pp-formulas or negations. In fact, $\forall \vec{v}(\alpha(\vec{v}) \leftrightarrow \beta' \wedge \forall w \gamma'(w, \vec{v})) \in {}_{\mathcal{R}}T$. Accordingly we can replace $\alpha'(w, \vec{v})$ by $\gamma'(w, \vec{v})$, which is a Boolean combination of pp-formulas, and hence is equivalent to a conjunction of disjunctions of pp-formulas or negations. Correspondingly put $\alpha'(w, \vec{v}) : \bigwedge_{j \leq s} \alpha'_j(w, \vec{v})$ where, for every $j \leq s$, $\alpha'_j(w, \vec{v})$ is a disjunction of pp-formulas or negations. $\forall w \alpha'(w, \vec{v})$ is equivalent in ${}_{R}T$ to $\bigwedge_{j \leq s} \forall w \alpha'_j(w, \vec{v})$. Then we can handle $\alpha'_j(w, \vec{v})$ (with $j \leq s$) instead of $\alpha'(w, \vec{v})$.
2nd reduction: $\alpha'(w, \vec{v})$ is of the form $\theta(w, \vec{v}) \to \bigvee_{i<N} \theta_i(w, \vec{v})$ where N is a positive integer, $\theta(w, \vec{v})$ and $\theta_i(w, \vec{v})$ (with $i < N$) are pp-formulas. In fact $\alpha'(w, \vec{v})$ is a single disjunction of pp-formulas and negations. But we know that any conjunction of pp-formulas is (equivalent in ${}_{R}T$ to) a pp-formula, and hence any disjunction of negations of pp-formulas is the negation of a single pp-formula. This clearly implies our claim.
Let us summarize the situation. We want to find β and $\gamma(\vec{v})$ such that, for every $\mathcal{R}$-module $\mathcal{M}$ and every sequence $\vec{a}$ in M,

$$(1) \qquad \theta(\mathcal{M}, \vec{a}) \subseteq \bigcup_{i<N} \theta_i(\mathcal{M}, \vec{a}).$$

if and only if $\mathcal{M} \models \beta \wedge \gamma(\vec{a})$. We know that, given $\mathcal{M}$ and $\vec{a}$, either $\theta(\mathcal{M}, \vec{a}) = \emptyset$ or $\theta(\mathcal{M}, \vec{a})$ is a coset of the pp-definable subgroup $\theta(\mathcal{M}, \vec{0})$. The same can be said about $\theta_i(\mathcal{M}, \vec{a})$ for every $i < N$. By the way, notice that $\exists w \theta(w, \vec{v})$, $\exists w \theta_i(w, \vec{v})$ (with $i < N$) are pp-formulas. (1) is certainly true when $\vec{a}$ satisfies $\neg \exists w \theta(w, \vec{v})$ (the negation of a pp-formula) in $\mathcal{M}$, and certainly false when $\vec{a}$ satisfies

$$\exists w \theta_i(w, \vec{v}) \wedge \bigvee_{i<N} \neg \exists w \theta_i(w, \vec{v})$$

in $\mathcal{M}$. So there is no loss of generality for our purposes in assuming $\theta(\mathcal{M}, \vec{a}) \neq \emptyset$ and $\theta_i(\mathcal{M}, \vec{a}) \neq \emptyset$ for every $i < N$. Let S be the set of the indices $i < N$ satisfying

$$|\theta(\mathcal{M}, \vec{0}) : \theta(\mathcal{M}, \vec{0}) \cap \theta_i(\mathcal{M}, \vec{0})| \leq N!\,.$$

Notice that S depends on $\mathcal{M}$ (and on $\vec{a}$, of course). However there are only finitely many possible ways of choosing S, and each of them is described by a suitable invariant statement. Let us assume, with no loss of generality, that S is just the set of the positive integers $\leq m$ for some $m \leq N$. We can apply B. H. Neumann's Lemma and deduce that (1) is equivalent to

$$(2) \qquad \theta(\mathcal{M}, \vec{a}) \subseteq \bigcup_{i<m} \theta_i(\mathcal{M}, \vec{a}).$$

Put $K = \theta(\mathcal{M}, \vec{0}) \cap \bigcap_{i<m} \theta_i(\mathcal{M}, \vec{0})$. As $\theta(\mathcal{M}, \vec{a})$ and $\theta_i(\mathcal{M}, \vec{a})$ for $i < m$ are union of cosets of K in M, (2) can be equivalently written

$$(3) \qquad \theta(\mathcal{M}, \vec{a})/K \subseteq \bigcup_{i<m} \theta_i(\mathcal{M}, \vec{a})/K.$$

As $\theta(\mathcal{M}, \vec{a})/K$ is finite, we can use some (hopefully) well known combinatorial arguments and restate (3) in the equivalent form

$$(4) \quad \sum_X (-1)^{|X|} |(\theta(\mathcal{M}, \vec{a}) \cap \bigcap_{i \in X} \theta_i(\mathcal{M}, \vec{a}))/K| = 0$$

where X ranges over the subsets of $\{0, 1, \ldots, m-1\}$. For every X, put

$$k(X) = |\theta(\mathcal{M}, \vec{0}) \cap \bigcap_{i \in X} \theta_i(\mathcal{M}, \vec{0}) : K|;$$

notice that, when $\theta(\mathcal{M}, \vec{a}) \cap \bigcap_{i \in X} \theta_i(\mathcal{M}, \vec{a}) \neq \emptyset$,

$$k(X) = |(\theta(\mathcal{M}, \vec{a}) \cap \bigcap_{i \in X} \theta_i(\mathcal{M}, \vec{a}))/K|.$$

Moreover $k(X) \leq N!^N$. Hence we have shown that $\mathcal{M}$ satisfies $\alpha(\vec{a})$ if and only if $\sum (-1)^{|X|} k(X) = 0$, where the sum concerns all the subsets X of $\{0, 1, \ldots, m-1\}$ such that $\mathcal{M} \models \exists w(\theta(w, \vec{a}) \wedge \bigwedge_{i \in X} \theta_i(w, \vec{a}))$, and hence if and only if $\mathcal{M}$ satisfies a convenient disjunction of conjunctions of invariant statements and pp-formulas. This is what happens for a given S. As there are only finitely many possible S's, one can find some suitable β and $\gamma(\vec{v})$, valid for every $\mathcal{R}$-module $\mathcal{M}$. ♣

Remark 2.6.4 1. Notice that the procedure given in the proof of Theorem 2.6.3 is effective, and provides explicitly for every $\alpha(\vec{v})$ the required formulas β and $\gamma(\vec{v})$. Furthermore β is actually a finite Boolean combination of invariant statements concerning pp-formulas $\varphi(v)$, $\psi(v)$ (with at most one free variable).

2. In particular, when α is a sentence of L_R, what the previous procedure produces is just a Boolean combination β of invariant statements (concerning pp-formulas $\varphi(v)$, $\psi(v)$ with at most one free variable) such that $\alpha \leftrightarrow \beta \in {}_{\mathcal{R}}T$.

3. Now fix an $\mathcal{R}$-module $\mathcal{M}$. Then, for every formula $\alpha(\vec{v})$ of L_R, there exists a Boolean combination $\gamma(\vec{v})$ of pp-formulas such that $\mathcal{M} \models \forall \vec{v}\,(\alpha(\vec{v}) \leftrightarrow \gamma(\vec{v}))$ (in fact, we know that $\alpha(\vec{v})$ is equivalent to $\beta \wedge \gamma(\vec{v})$ for some Boolean combination $\gamma(\vec{v})$ of pp-formulas and some sentence β; so, if $\mathcal{M} \models \beta$, then $\alpha(\vec{v})$ is equivalent to $\gamma(\vec{v})$, while, if $\mathcal{M} \models \neg\beta$, then $\alpha(\mathcal{M}^n)$ is empty and consequently $\alpha(\vec{v})$ is equivalent to $\delta(\vec{v}) \wedge \neg\delta(\vec{v})$, where $\delta(\vec{v})$ is an arbitrary pp-formula).

With respect to definable sets in modules, this is what Theorem 2.6.3 implies.

Corollary 2.6.5 *Let $\mathcal{M}$ be an $\mathcal{R}$-module, n be a positive integer. Then every set $X \subseteq M^n$ definable in $\mathcal{M}$ is a finite Boolean combination of cosets of pp-definable subgroups.*

Proof. There exist an L_R-formula $\alpha(\vec{v}, \vec{w})$ and a sequence $\vec{a}$ in M such that $X = \alpha(\mathcal{M}^n, \vec{a})$. We can assume that $\alpha(\vec{v}, \vec{w})$ is a Boolean combination of pp-formulas, and we know that, for every pp-formula $\varphi(\vec{v}, \vec{w})$, $\varphi(\mathcal{M}^n, \vec{a})$, when it is not empty, is a coset of the pp-definable subgroup $\varphi(\mathcal{M}^n, \vec{0})$. ♣

We can also characterize the complete extensions of ${}_{\mathcal{R}}T$ (and hence the $\equiv$-classes of $\mathcal{R}$-modules).

Corollary 2.6.6 *Let $\mathcal{M}, \mathcal{M}'$ be two R-modules. Then $\mathcal{M} \equiv \mathcal{M}'$ if and only if, for every choice of two pp-formulas $\varphi(v)$, $\psi(v)$, the indices of $\varphi(\mathcal{M}) \cap \psi(\mathcal{M})$ in $\varphi(\mathcal{M})$ and $\varphi(\mathcal{M}') \cap \psi(\mathcal{M}')$ in $\varphi(\mathcal{M}')$ are either finite and equal, or both infinite.*

Proof. It is clear that, if $\mathcal{M}$ and $\mathcal{M}'$ are elementarily equivalent, then, for every $\varphi(v), \psi(v)$ as before, and for every positive integer k,

$$\mathcal{M} \models (\varphi : \psi) \geq k \quad \Leftrightarrow \quad \mathcal{M}' \models (\varphi : \psi) \geq k.$$

The inverse implication follows from Remark 2.6.4,2. ♣

The previous result explains why "*invariant statements*" are called in this way: actually these sentences fully characterize any $\mathcal{R}$-module $\mathcal{M}$ up to elementary equivalence.

Now let us discuss the content of the previous results in some particular case. We deal with a principal ideal domain $\mathcal{R}$ (this setting includes the ring $\mathbf{Z}$ of integers, as well as any field). First let us examine a generic pp-formula of L_R

$$\alpha(\vec{v}) : \exists \vec{w}\,(A\vec{v} = B\vec{w}).$$

A and B can obtain a simpler form when $\mathcal{R}$ is a principal ideal domain. For, it is a fact of Algebra that, in this framework, there are two invertible matrices X, Y with coefficients in R such that the product $B' = XBY$ is diagonal. So $\alpha(\vec{v})$ is equivalent to

$$\exists \vec{w}\,(XA\vec{v} = B'Y^{-1}\vec{w})$$

and, unless replacing $\vec{w}$ by $Y^{-1}\vec{w}$, A by XA and B by B', one can suppose B diagonal in $\alpha(\vec{v})$. Consequently $\alpha(\vec{v})$ becomes of the form

$$\exists w_1 \ldots \exists w_m \,(\bigwedge_{1 \leq i \leq m} (\sum_{j=1}^{n} a_{ij} v_j = b_{ii} w_i)).$$

Now let us momentarily restrict our analysis to a smaller setting.
Case 1: $\mathcal{R} = \mathcal{K}$ is a field (so we are dealing with vectorspaces over $\mathcal{K}$). Assume that, for some i with $1 \leq i \leq m$, $b_{ii} \neq 0$. Then we can divide the i-th equation in $\alpha(\vec{v})$ by b_{ii} and consequently assume $b_{ii} = 1$. At this point it is easy to show that $\alpha(\vec{v})$ is equivalent to

$$\bigwedge_{1 \leq i \leq m, b_{ii} = 0} (\sum_{j=1}^{n} a_{ij} v_j = 0),$$

which is a conjunction of atomic formulas.
Combine this observation and Baur-Monk's Theorem, and deduce that every L_K-formula is equivalent in ${}_{\mathcal{K}}T$ to a conjunction of a quantifier free formula and a Boolean combination of invariant statements.
Moreover the pp-formulas with a unique free variable v reduce to $rv = 0$ for some $r \in K$, and hence to either $v = 0$ when $r \neq 0$ or to $v = v$ when $r = 0$. Consequently the only pp-definable subgroups of a vectorspace $\mathcal{V}$ are $\{0\}$ and V. Owing to Corollary 2.6.5, the subsets of V definable in $\mathcal{V}$ are just the finite Boolean combinations of the cosets of these subgroups, and so reduce to the finite or cofinite subsets.
Now let us examine invariant statements. In particular we direct our attention on the sentences of the form

$$(\star) \qquad (v = v : v = 0) \geq k$$

where k is a positive integer. In any given vectorspace $\mathcal{V}$ over $\mathcal{K}$, they witness if the size $|V|$ di V is finite or not, and, in the positive case, its value. We claim that they can even determine the $\equiv$-type of the vectorspace. Let us see why.
First assume K infinite. We know that, in this case, all the nonzero vectorspaces over $\mathcal{K}$ are elementarily equivalent (for, their theory is complete). In other words, when K is infinite, there are only two $\equiv$-classes of $\mathcal{K}$-vectorspaces: the former contains all the nonzero vectorspaces, and the latter reduces to the zero space. But a vectorspace $\mathcal{V}$ is $\{0\}$ if and only if $\mathcal{V} \models (v = v : v = 0) = 1$. In particular, the statements $(\star)$ determine the $\equiv$-type of any vectorspace.
Now assume K finite (say of size q). Now we meet infinitely many $\equiv$-classes of $\mathcal{K}$-vectorspaces. In fact, there is a class for every natural n, consisting of the (pairwise isomorphic) vectorspaces of dimension n over $\mathcal{K}$ (hence size q^n), while infinite vectorspaces form again a unique class. The sentences $(v = v : v = 0) \geq k$ can obviously distinguish these classes.
One can deduce that every invariant statement in L_K is a Boolean combination of sentences $(v = v : v = 0) \geq k$ where k ranges over the positive integers.
In conclusion, given a field $\mathcal{K}$, the atomic formulas of L_K together with the invariant statements $(v = v : v = 0) \geq k$, with k a positive integer, form an elimination set for ${}_{\mathcal{K}}T$. In particular this yields quantifier elimination in L_K for the theory of infinite $\mathcal{K}$-vectorspaces.

Case 2: Now let us enlarge our setting to arbitrary principal ideal domains $\mathcal{R}$. First let us examine a pp-formula $\alpha(\vec{v})$. We cannot expect any longer that, whenever $b_{ii} \neq 0$ in the i-th equation of $\alpha(\vec{v})$, one can divide the whole equation by b_{ii} and so obtain $b_{ii} = 1$. However $\alpha(\vec{v})$ is equivalent in ${}_{\mathcal{R}}T$ to a conjunction of formulas

(1) $\sum_{j=i}^{n} a_{ij} v_j = 0$ (for $b_{ii} = 0$),

(2) $\exists w (\sum_{j=i}^{n} a_{ij} v_j = qw)$ (where $q = b_{ii} \neq 0$).

The latter ones are divisibility conditions: we can abbreviate each of them by

(2) $q \mid \sum_{j=i}^{n} a_{ij} v_j$.

Of course, when q is a unit in $\mathcal{R}$, this is a trivial condition and can be forgotten. In the remaining cases, recall that q decomposes (uniquely) as a product of powers of pairwise distinct primes in $\mathcal{R}$, and q divides an element

$r \in R$ if and only if all these prime powers divide r. So there is no loss of generality in assuming that, in (2), q is a prime power.
The situation becomes clearer if we restrict our analysis to formulas having only one free variables. In fact, in this case, our pp-formula $\alpha(v)$ gets equivalent in ${}_{\mathcal{R}}T$ to a conjunction of formulas

(1)' $rv = 0$ (a torsion condition),

(2)' $p^l \mid sv$ (a divisibility condition);

here p, r, $s \in R$, p is a prime and l is a non-negative integer. Again, simple algebraic facts about principal ideal domains let us assume that s itself is a power of p, $s = p^h$ for some non-negative integer $h < l$. Every pp-formula in at most one free variable is a conjunction of torsion and divisibility conditions as before.
This result helps also the analysis of invariant statements $(\varphi(v) : \psi(v)) \geq k$. We avoid here too many details. However we wish to mention the following sentences (r, p are elements of R, p is prime, n, k are positive integers):

(3) $(pv = 0 \wedge p^{n-1}|v : pv = 0 \wedge p^n|v) \geq k$,

(4) $(pv = 0 \wedge p^n|v : v = 0) \geq k$,

(5) $(p^{n-1}|v : p^n|v) \geq k$,

(6) $(v = v : rv = 0) \geq k$.

The reader may check their truth (at least when $\mathcal{R}$ is the ring of integers) in some familiar abelian groups, like $\mathbf{Z}/q^h\mathbf{Z}$, the Prüfer groups $\mathbf{Z}/q^\infty\mathbf{Z}$, the localizations of $\mathbf{Z}$ at q (when q ranges over the primes, and h over the natural numbers), and the additive group of rationals, and realize in this way their meaning.
Indeed Wanda Szmielew (a student of Tarski's) showed that, for every $\mathcal{R}$-module $\mathcal{M}$, the $\equiv$-type of $\mathcal{M}$ is fully determined by the invariant statements (3)-(6) satisfied by $\mathcal{M}$.
In conclusion, owing to Baur-Monk's theorem, the formulas (1)-(6) are an elimination set for the theory ${}_{\mathcal{R}}T$ when $\mathcal{R}$ is a principal ideal domain.

2.7 Strongly minimal theories

We saw in 2.4 that the only (1-ary) definable sets in an algebraically closed field are the finite and cofinite ones. 2.6 told us that the same happens, for

instance, in every (infinite) vectorspace over a fixed countable field. One can also check that even pure sets (in a language with no symbols besides equality $=$) enjoy this feature; again, every definable set is either finite or cofinite (this was implicitly shown when we treated definable sets in 1.7, in particular when we provided an example of an infinite coinfinite non definable set).
So we find some non-trivial algebraic structures $\mathcal{A}$ whose definable 1-ary subsets reduce to the ones definable in the pure set A (with the equality relation $=$) by quantifier free formulas. Let us name these structures in the following way.

Definition 2.7.1 *An infinite structure $\mathcal{A}$ is said to be* **minimal** *if and only if the only definable subsets of A are those finite or cofinite. A complete theory T is said to be* **strongly minimal** *if and only if every model $\mathcal{A}$ of T is minimal.*

Hence any algebraically closed field is a minimal structure, and any theory ACF_p (with $p = 0$ or prime) is strongly minimal. The same is true for infinite vectorspaces, or pure sets.
It should be underlined that the minimality of a structure is not preserved by elementary equivalence. In other words there are minimal structures $\mathcal{A}$ such that the theory of $\mathcal{A}$ is not strongly minimal, and so admits some non-minimal models. Here is an example.

Example 2.7.2 Consider the theory dLO^+ of discrete orders with a least element 0 but no last element. We know that dLO^+ is complete, and has quantifier elimination in a language L with a constant (for 0) and a 1-ary operation symbol (for the successor function s) in addition to the relation symbol $\leq$. Consequently every definable subset of a model of T is a finite Boolean combination of intervals (possibly with an infinite right endpoint). Therefore $(\mathbf{N}, \leq, 0, s)$ is a minimal model of T, because every interval in $(\mathbf{N}, \leq, 0, s)$ is either finite or cofinite. However no other model $\mathcal{A} = (A, \leq, 0^{\mathcal{A}}, s^{\mathcal{A}})$ of T is minimal. In fact, let $a \in A$ satisfy $a \neq (s^{\mathcal{A}})^n(0^{\mathcal{A}})$ for any natural n. Then both $[0^{\mathcal{A}}, a[$ and $[a, +\infty[$ are infinite intervals in $\mathcal{A}$.

We shall examine again strongly minimal theories in the next chapters. In particular, with respect to algebraically closed fields, we will prove a theorem of Macintyre showing (among other things) that the only integral domains with indentity having a strongly minimal complete theory are just the algebraically closed fields.

2.8 o-minimal theories

Turning now to linearly ordered structures $\mathcal{A} = (A, \leq, \ldots)$, we met in the previous sections some examples where the definable subsets of A reduce to the finite unions of intervals (possibly with infinite endpoints) in $(A, \leq)$. This is what happens in real closed fields (as observed in 2.5), but also in dense or discrete (infinite) linear orders (see the sections 2.2 and 2.3). This suggests the following definition.

Definition 2.8.1 *An infinite linearly ordered structure $\mathcal{A} = (A, \leq, \ldots)$ is called* **o-minimal** *if and only if every subset of A definable in $\mathcal{A}$ is a finite union of intervals (closed or open, possibly with infinite endpoints). A complete theory T of infinite linearly ordered structures is called* **o-minimal** *if and only if every model of T is o-minimal.*

"o" abridges "order", of course. This o-minimal setting clearly reminds minimality. In fact the minimal structures (and the strongly minimal theories) are the ones where every definable (1-ary) set is already defined by a quantifier free formula involving the only (language) symbol $=$. Similarly the o-minimal structures (and theories) are just those admitting a total order relation $\leq$ such that every definable (1-ary) is already defined by a quantifier free formula involving the only (language) symbol $\leq$.
In this sense the o-minimal structures and theories are the simplest ones in the presence of a total order relation. Nevertheless they include several non-trivial algebraic examples. We will study in more detail these structures and theories in the last chapter of this book.
But it is worth emphasizing since now that, in spite of the similarities underlined above between minimality and o-minimality, a relevant difference arises. In fact, we noticed that the theory of a minimal structure may admit some non-minimal models, and so fail to be strongly minimal. This does not happen in the o-minimal setting. In fact the following theorem hold.

Theorem 2.8.2 (Knight - Pillay - Steinhorn) *If T is the theory of a linearly ordered o-minimal structure, then every model of T is o-minimal.*

Accordingly, we can spare the adverb "strongly" in defining a theory with o-minimal models.
Coming back to real closed fields, we would like to mention here a result quite similar to the one recalled at the end of the previous section on algebraically closed fields. In fact, it was shown by Pillay and Steinhorn that the only

ordered rings (with identity) having an o-minimal theory are the real closed fields.
The proofs of these theorems will be provided in Chapter 9.

2.9 Computational aspects of q. e.

In this section we shortly discuss the quantifier elimination procedures with respect to effectiveness and fastness. Actually these criteria did not correspond to the spirit of the forties (and some decades later), when the main quantifier elimination results were proved. For, those times lived the influence of Gödel incompleteness and undecidability phenomena; so, according to that feeling, any decision algorithm (such as Tarski's method for real elementary algebra), or even a decidability theorem simply ensuring the existence of such a procedure without explicitly exhibiting it, were exactly the best answer one might expect. But later, in the seventies, the birth of modern computers and the beginning of their science changed this setting and inspired a prevalent interest in quickly running algorithms. Hence complexity theory introduced

- ⋆ the class P of the problems having a fast procedure to find solutions,
- ⋆ the class NP of the problems having a fast procedure to verify solutions (namely to check that a solution works).

We agree that a problem has a solving procedure when there is a Turing machine handling it and that an algorithm is fast when it runs in a polynomial time with respect to the length of the input. To realize the difference between finding or verifying solutions, look at the problem of factoring integers. To decompose a natural number ≥ 2 into its prime factors -more precisely, to find these factors- can be significantly slower (at least with respect to the currently available algorithms) than to check this decomposition when done. Just to quote a famous historical example, F. Cole announced during an AMS meeting in 1903 that the Mersenne number $2^{67} - 1$ is not prime. Factoring $2^{67} - 1$ is not easy (and certainly it was not in 1903, when computers were not available). But Cole's proof is quite short to write and needs only one line

$$2^{67} - 1 = 193797721 \times 761838257287$$

and can be checked very quickly.

Coming back to P and NP, it is trivial to observe that $P \subseteq NP$, because a procedure yielding solutions implicitly confirms these solutions. A fundamental problem in complexity theory (and, more generally, in the area linking computer science and mathematics) asks whether $P = NP$, hence whether, whenever a problem has a fast procedure verifying solutions, then it admits a (possibly slower but still) fast (=polynomial) procedure finding solutions.
According to the new spirit, what is primary even in quantifier elimination, mainly towards decidability, is to get fast methods. Let us discuss this feature for the elimination of quantifiers of real closed fields: this is perhaps the most interesting case, owing to its connection with elementary geometry. However a devastating result of Fischer and Rabin shows

Theorem 2.9.1 *Any algorithm deciding, or even verifying membership to RCF for sentences in the language of ordered fields requires a running time at least exponential with respect to the length of the input sentence.*

Algorithm still means Turing machine. Notice that Fischer-Rabin's Theorem only refers to the additive structure of the reals $(\mathbf{R}, +, -)$; recall that, in this restricted language, the reals inherit in an obvious way a structure of $\mathbf{Q}$-vectorspace, because the scalar multiplication by any rational can be defined using $+$, and so form a structure elementarily equivalent to $(\mathbf{C}, +, -)$. In this sense, the theorem applies also to the complex field, yielding the same negative lower bound for the decision procedures concerning algebraically closed fields.
By the way, it is not known any decidable theory with infinite models and a decision procedure running in polynomial time. Indeed, if such a theory exists, then $P = NP$.
In particular, Tarski's original elimination procedure is very inefficient and slow. However, recently faster and more powerful elimination methods for real closed fields have been introduced. We wish to quote the Collins procedure, called *cylindrical algebraic decomposition* CAD, working in the worst cases in a doubly exponential time with respect to the number of variables in the input formula. Implementations of CAD, and other real quantifier elimination methods are discussed, for instance, in [33].

But now we want to treat briefly another intriguing relationship between complexity theory and quantifier elimination (for arbitrary theories and structures). A few lines ago we have said *algorithm = Turing machine*, in accordance with the Church-Turing Thesis. But the Turing model of computation has an intrinsic discrete character, so that its applications to a continous framework (like **R** or **C**) seem laborious and unnatural (however, see [121] for a discussion of this point). Consequently, new models of computation, including real and complex numbers as possible inputs, and even working in arbitrary structures, have been introduced. We quote the Blum-Shub-Smale BSS model ([14], [13]), or also the Poizat approach [133]. These new perspectives extend the Turing point of view: the classical computability is just the computability over the field $\mathbf{F}_2$ with 2 elements; but now computability over arbitrary structures is allowed. In particular, for every structure $\mathcal{A}$, one can define in a suitable sense the classes P and NP (over $\mathcal{A}$), one can compare these classes and check if $P = NP$ over $\mathcal{A}$. To introduce these matters in detail would require a long time, so we refer the interested reader to the bibliography quoted at the end of the chapter. Remarkably, quantifier elimination arises in this setting. In fact, Poizat observed

Theorem 2.9.2 *If $P = NP$ over $\mathcal{A}$, then the theory of $\mathcal{A}$ eliminates the quantifiers.*

It is comparatively easy to realize why. Let us refer for simplicity to the Cole example quoted before. To prove that $2^{67} - 1$ is composite requires to find a non trivial divisor, and hence to obtain some witnesses of the (existential) sentence $\exists u \exists v (2^{67} - 1 = uv)$. But, after Cole, we have simply to check the quantifier free sentence

$$2^{67} - 1 = 193797721 \times 761838257287.$$

In other words, what $P = NP$ asks here is a procedure (indeed a quick procedure) of elimination of quantifiers. Theorem 2.9.2 provides several examples of structures for which $P \neq NP$. For instance, recall the Macintyre-McKenna-Van den Dries theorems characterizing the infinite fields whose theory eliminates the quantifiers in the language for fields (they are the algebraically closed fields), or the ordered fields whose theory eliminates the quantifiers in the language for ordered fields (the real closed fields). One easily deduces

Corollary 2.9.3 1. *$P \neq NP$ over the field of rationals, or over the field of reals (without the order relation).*

2. *$P \neq NP$ over the ordered field of rationals.*

On the other side, quantifier elimination is only a necessary condition towards $P = NP$ over a given structure. There do exist quantifier eliminable structures $\mathcal{A}$ such that $P \neq NP$ over $\mathcal{A}$: this is the content of the following nice result of Meer.

Theorem 2.9.4 *$P \neq NP$ over* $(\mathbf{R}, +, -)$.

In fact the theory of $(\mathbf{R}, +, -)$ is essentially the theory of nonzero vectorspaces over the rational field, and so admits the elimination of quantifiers in the language $L_{\mathbf{Q}}$ and even in $\{+, -\}$ because the action of any rational is easily defined by the additive structure without using quantifiers.
What can we say about the complex field, or the ordered field of reals? Their theories eliminate the quantifiers in the corresponding language. Nevertheless $P = NP$ is still an open question over these structures. Notably, in the complex case, a key (NP-complete) problem towards a definitive answer is related to the celebrated Hilbert Nullstellensatz (a classical algebraic result closely related to Model Theory, as we will see in the next chapter): in fact, it asks a quick procedure checking the solvability of a given finite system of polynomials over $\mathbf{C}$ (in arbitrarily many variables). One shows that $P = NP$ over the complex field if and only if this fast procedure exists. NP-complete problems over the ordered field of reals are discussed in the references quoted below.

2.10 References

Van den Dries [168] and Doner-Hodges [34] are two excellent and enjoyable expository papers, explaining Tarski's work on the quantifier elimination and, more generally, the history of this matter. They also include a rich list of references. Here let us mention [87] and [154] on the pioneeristic contributions of Löwenheim and Skolem to the elimination of quantifiers. Langford's elimination methods for dense or discrete orders are in [81] and in [82], while Tarski's subsequent contributions in this setting are in [160]. Tarski's elimination procedures for the real field and the complex field are given in [157], while Cohen's method in the real case is in [27]. [98] contains the Macintyre-McKenna-Van den Dries theorems saying that algebraically closed fields are the only infinite fields whose first order theory eliminates the quantifiers, and real closed fields are the only ordered fields with the same feature. Let us mention that R. Thom coined the word "semialgebraic" in [162].
Some more details about pp-elimination in modules (and a proof of Neumann's Lemma 2.6.3) can be found in M. Prest's book [136]; the Eklof-Fisher

paper [40] deals with the particular case of abelian groups (and modules over Dedekind domains).
The computational aspects of the quantifier elimination for the real field are discussed in [33], while the cylindrical algebraic decomposition algorithm CAD is in the Collins paper [29]. The Fisher-Rabin theorem ensuring that no decision algorithm for the real field runs in polynomial time is in [47].
[13], [14] describe the new Blum-Shub-Smale model of computation; [133] provides Poizat's approach to this theme. K. Meer's theorem that $P \neq NP$ over $(\mathcal{R}, +, -)$ (although the corresponding theory eliminates the quantifiers) is in [112].

Chapter 3

Model Completeness

3.1 An introduction

We already defined model completeness in Chapter 1: a theory T is called *model complete* if every embedding between models of T is elementary. We dealt with this notion also in Chapter 2, where we considered its connection with quantifier elimination and completeness. But now we wish to examine model completeness in a closer and more direct way, to discuss its genesis and motivations, as well as its importance and applications.

Model completeness deals with embeddings between structures. This perspective might look slightly oblique with respect to the fundamental purpose in model theory, namely to connect sentences and structures via truth; under this point of view, the most genuine relation among structures is elementary equivalence (that is, to satisfy the same sentences). Nevertheless some basic theorems in model theory, such as the Löwenheim-Skolem theorems, involve pretty naturally extensions, substructures, embeddings, and so draw attention to this subject. Furthermore, as we will see in Section 3.2, there are other possible ways of introducing model completeness. The first one still deals with embeddings and says that a theory T is model complete when each embedding between models of T preserves existential formulas. But another characterization is quite syntactical and resembles the way we defined quantifier elimination; it says that a theory T is model complete exactly when any formula $\varphi(\vec{v})$ in the language of T is equivalent in T to an appropriate existential formula $\varphi'(\vec{v})$.

The main motivations leading to model completeness come from algebra. For instance, consider field theory. Given a field $\mathcal{K}$, one looks at the irreducible polynomials $f(x) \in K[x]$. Algebra builds richer and richer extensions of

$\mathcal{K}$, equipping these polynomials with a single root, or all the possible roots. Eventually, one reaches the *algebraic closure* $\overline{\mathcal{K}}$ of $\mathcal{K}$: a minimal extension where every nonconstant polynomial $f(x)$ in $K[x]$, and even in $\overline{K}[x]$ itself, splits into linear factors, and so gets its own roots. Notice that, from a logical point of view, adding a root of a polynomial $f(x)$ means to satisfy the sentence $\exists w(f(w) = 0)$ with parameters from K (the coefficients of $f(x)$). $\mathcal{K}$ is algebraically closed when it equals $\overline{\mathcal{K}}$ and hence when it is able to satisfy all these sentences when $f(x)$ ranges over the nonconstant polynomials over K itself. Pursuing this logical approach, one can generalize and look at arbitrary L-structures $\mathcal{A}$ instead of pure fields, towards two possible objects:

- ⋆ to enlarge $\mathcal{A}$ to a richer $\overline{\mathcal{A}}$ satisfying every existential sentence $\exists \vec{w} \alpha(\vec{w})$ (with a quantifier free $\alpha(\vec{w})$), or even every sentence in $L(A)$, or (why not?) in $L(\overline{A})$, too;
- ⋆ to examine closely the structures $\overline{\mathcal{A}}$.

This program recalls A. Weil's notion of *universal domains* in [176]. Weil's idea (for the class of fields) was to determine large and rich structures, embedding every field under consideration. Of course, in the case of fields, universal domains are just algebraically closed fields of infinite transcendence degree. This strategy has now fallen into disuse within Algebraic Geometry, but it is still alive in Model Theory (and certainly it was in the sixties). Model completeness arises quite naturally in this framework: for, in a model complete theory T, for every model $\mathcal{A}$ and for every $L(A)$-sentence φ, whenever φ is true in some model extending $\mathcal{A}$, then $\mathcal{A}$ itself satisfies φ; so, it is worth devoting some specific pages to this matter. This is what we will do in this chapter. First we will give an abstract analysis of model completeness. Then we will emphasize its strong connection with Algebra. In fact, Algebra inspires the notion of model completeness, and several related concepts; but, conversely, we will see that some developments in Model Theory concerning model completeness do produce a significant progress in Algebra; indeed some alternative elegant proofs of the celebrated Hilbert Nullstellensatz, or of the Hilbert Seventeenth Problem, and, more notably, the solution of Artin's Conjecture on p-adic fields witness these fruitful contributions. Actually, this was the dream of Abraham Robinson (the father of model completeness): to quote his own words in his address to the 1950 ICM,

> *"Symbolic Logic can produce useful tools for the developments of actual mathematics, more particularly of Algebra and, it would*

> *appear, of Algebraic Geometry. This is the realization of an ambition... expressed by Leibniz in a letter to Huyghens as long ago as 1679".*

This point of view is developed one year later in [140]. The algebraic theorems recalled before do corroborate this program. Other deep confirmations (also concerning Geometry) will be provided in the next chapters.
Let us conclude this section by recalling some connections between model completeness, elimination of quantifiers and completeness.
First of all, remember that elimination of quantifiers implies model completeness. The converse is not true. For instance, we saw in the last chapter that the theory of real closed fields RCF loses the quantifier elimination property if one removes the relation symbol for $\leq$ from its language L: actually the order $\leq$ is definable in the restricted language $L_0 = \{+, -, \cdot, 0, 1\}$, as $v \geq 0$ is equivalent in RCF to

$$\exists w(w^2 = v),$$

but any possible definition needs quantifiers. However to forget $\leq$ does not affect model completeness: in fact every embedding $f : \mathcal{A} \to \mathcal{B}$ of real closed fields in the restricted language L_0 enlarges naturally and involves $\leq$ (because the nonnegative elements must equal the squares), so is elementary in both L and L_0.
On the other side, model completeness can yield completeness under some suitable additional hypotheses. We saw that this happens, for instance, for real closed fields (or also for discrete linear orders). The reason was that RCF has a *"minimal"* model, embeddable in every real closed field: the ordered field of real algebraic numbers (hence the real closure of the rationals). To extend this example towards a general setting, we need the following

Definition 3.1.1 *Let T be a theory. A model of T is prime if it is embeddable in every model of T.*

Examples 3.1.2

1. The (complex) algebraic numbers are a prime model of ACF_0.
2. The real algebraic numbers are a prime model among real closed ordered fields.
3. $(\mathbb{N}, \leq, 0, s)$ is a prime model of dLO^+.

Proposition 3.1.3 *Let T a model complete theory. If T has a prime model $\mathcal{A}$, then T is complete.*

Proof. Just adapt the argument of RCF. For every model $\mathcal{B}$ of T, there is an embedding of $\mathcal{A}$ into $\mathcal{B}$. Owing to model completeness, this embedding is elementary. In particular $\mathcal{B}$ is elementarily equivalent to $\mathcal{A}$. Hence all the models of T are elementarily equivalent to each other. Consequently T is complete. ♣

Another useful criterion deducing completeness from model completeness is the following one (we used it when dealing with dense linear orders in the last chapter).

Proposition 3.1.4 *Let T a model complete theory. Assume that any two models $\mathcal{A}$ and $\mathcal{B}$ of T admit a common extension $\mathcal{C}$ in $Mod(T)$. Then T is complete.*

Proof. Given $\mathcal{A}$ and $\mathcal{B}$, form a common extension $\mathcal{C}$. Owing to model completeness, the embeddings of both $\mathcal{A}$ and $\mathcal{B}$ in $\mathcal{C}$ are elementary. So $\mathcal{A}$ and $\mathcal{B}$ are elementarily equivalent to $\mathcal{C}$, and consequently to each other. ♣

However, be careful: model completeness does not imply completeness in general. Just to avoid any temptation about this point, recall algebraically closed fields. ACF is model complete, but it is not complete: one needs to specify the characteristic to get a prime model and hence to deduce, even via model completeness, that ACF_p is complete for every $p = 0$ or prime.

3.2 Abraham Robinson's test

Let T be a theory in a language L. We know that T eliminates the quantifiers if and only if every L-formula is equivalent in T to a suitable quantifier free formula (with the same free variables). Model completeness can be characterized in a similar way. Indeed one can show that T is model complete if and only every L-formula is equivalent in T to an existential formula, or also, if you like $\forall$ rather than $\exists$, to a universal formula (in fact, assume that every L-formula $\varphi(\vec{v})$ admits an existential L-formula equivalent in T; apply this property to $\neg\varphi(\vec{v})$ and yield the corresponding existential formula $\varphi'(\vec{v})$; conclude that $\varphi(\vec{v})$ is equivalent in T to $\neg\varphi'(\vec{v})$, which in its turn is obviously equivalent to a universal formula; the converse can be shown in the same way).
There is another remarkable related characterization of model completeness via embeddings. Recall that T model complete just means that every embedding between two models of T is elementary. But, notably, this is also

equivalent to require that **every** embedding between two models of T is existential (at a first sight, a weaker condition). This is the content of the so called Abraham Robinson Test for model completeness. We will apply this criterion to some algebraic settings in the next section. Now we want to show Robinson's Test and, at the same time, the previous characterizations of model completeness in terms of existential, or universal formulas.

Theorem 3.2.1 (A. Robinson) *Let T be a theory of L. The following propositions are equivalent:*

(i) *T is model complete;*

(ii) *every embedding from $\mathcal{A}$ into $\mathcal{B}$, where $\mathcal{A}$ and $\mathcal{B}$ are models of T, is existential;*

(iii) *for every L-formula $\varphi(\vec{v})$, there is an existential formula $\varphi'(\vec{v})$ equivalent to $\varphi(\vec{v})$ in T;*

(iv) *for every L-formula $\varphi(\vec{v})$, there is a universal formula $\varphi''(\vec{v})$ equivalent to $\varphi(\vec{v})$ in T.*

Proof. (i)$\Rightarrow$(ii) is clear, and (iii)$\Leftrightarrow$(iv) was already established.
Let us consider now (ii)$\Rightarrow$(iii),(iv).
We preliminarily show that, if (ii) holds, then every existential formula $\varphi(\vec{v})$ admits an equivalent universal formula, and conversely. So assume $\varphi(\vec{v})$ existential. Let n be the length of $\vec{v}$. Look at the set S of the universal formulas $\sigma(\vec{v})$ in L satisfying

$$\forall \vec{v}(\varphi(\vec{v}) \to \sigma(\vec{v})) \in T.$$

Notice that S is closed under conjunctions $\wedge$ (up to straightforward manipulations). What we need is a formula $\sigma(\vec{v})$ in S satisfying the further condition

$$\forall \vec{v}(\sigma(\vec{v}) \to \varphi(\vec{v})) \in T;$$

in fact, in this case, $\sigma(\vec{v})$ is a universal formula equivalent to $\varphi(\vec{v})$ in T, and we are done. Suppose towards a contradiction that no $\sigma(\vec{v}) \in S$ works. We can express this assumption in the following way: in a language L' extending L by a sequence $\vec{c}$ of n new constant symbols,

$$\text{for every } \sigma(\vec{v}) \in S,\ T \cup \{\sigma(\vec{c}), \neg\varphi(\vec{c})\} \text{ has a model}$$

(otherwise, for every model $\mathcal{M}$ of T and every $\vec{m}$ in $\sigma(\mathcal{M}^n)$, $\vec{m} \in \varphi(\mathcal{M}^n)$, and so $\forall \vec{v}(\sigma(\vec{v}) \to \varphi(\vec{v}))$ is in T). Recall that S is closed under conjunctions and use compactness to deduce that

$$T \cup \{\sigma(\vec{c}) : \sigma(\vec{v}) \in S\} \cup \{\neg\varphi(\vec{c})\}$$

has a model. In other words, there are a model $\mathcal{A}$ of T and a sequence $\vec{a}$ in A^n (the interpretation of $\vec{c}$ in $\mathcal{A}$) such that

$$\vec{a} \in \cap_{\sigma(\vec{v}) \in S}\, \sigma(\mathcal{A}^n)$$

but $\vec{a} \notin \varphi(\mathcal{A}^n)$. Put $\vec{a} = (a_1, \ldots, a_n)$. Now take any quantifier free formula $\theta(\vec{v}, \vec{w})$ in L and suppose that there is a sequence $\vec{b}$ in A for which $\mathcal{A} \models \theta(\vec{a}, \vec{b})$. Let h denote the length of $\vec{w}$ (and of $\vec{b}$); in particular put $\vec{b} = (b_1, \ldots, b_h)$. Then $\forall \vec{w} \neg\theta(\vec{v}, \vec{w})$ is a universal L-formula, but $\vec{a}$ cannot satisfy it. Hence $\forall \vec{w} \neg\theta(\vec{v}, \vec{w}) \notin S$. It follows that, in some model $\mathcal{A}^*$ of T, a suitable sequence $\vec{a^*}$ in $\varphi(\mathcal{A}^{*n})$ does not satisfy $\forall \vec{w} \neg\theta(\vec{v}, \vec{w})$; hence, for some $\vec{b^*}$, $\mathcal{A}^* \models \theta(\vec{a^*}, \vec{b^*})$. We can express this fact in the following way. Consider the language $L(A)$; to avoid any danger of confusion, distinguish the elements of A and their names in $L(A)$, and denote by c_a the constant symbol corresponding to a, for every $a \in A$. What we have just shown is that

$$T \cup \{\varphi(c_{a_1}, \ldots, c_{a_n})\} \cup \{\theta(c_{a_1}, \ldots, c_{a_n}, c_{b_1}, \ldots, c_{b_h})\}$$

has a model. Again using compactness and the fact that the quantifier free formulas of L are closed under conjunction, we see that the union of $T \cup \{\varphi(c_{a_1}, \ldots, c_{a_n})\}$ and of the whole collection of the formulas $\theta(c_{a_1}, \ldots, c_{a_n}, c_{b_1}, \ldots, c_{b_h})$ where h ranges over nonnegative integers, $\vec{b} = (b_1, \ldots, b_h)$ over A^h, $\theta(\vec{v}, \vec{w})$ over quantifier free L-formulas and $\mathcal{A} \models \theta(\vec{a}, \vec{b})$, has a model. Let $\mathcal{A}'$ denote the restriction of this structure to L -hence a model of T-, and let $a' = c_a^{\mathcal{A}'}$ for every $a \in A$. Therefore, for every quantifier free L-formula $\theta(\vec{v}, \vec{w})$ and every $\vec{b}$ in A, $\mathcal{A}' \models \theta(\vec{a'}, \vec{b'})$ if and only if $\mathcal{A} \models \theta(\vec{a}, \vec{b})$. So $a \mapsto a'$ for every $a \in A$ embeds $\mathcal{A}$ into $\mathcal{A}'$. (ii) implies that this embedding is existential. Consequently, as $\mathcal{A}' \models \varphi(\vec{a'})$, one deduces $\mathcal{A} \models \varphi(\vec{a})$ -a contradiction-. So a universal $\sigma(\vec{v})$ equivalent to $\varphi(\vec{v})$ exists.
The converse reduction (from universal to existential formulas) can be handled by passing to negations.
At this point, the proof is straightforward. Take an arbitrary L-formula $\varphi(\vec{v})$. We are looking for an equivalent existential, or universal formula. If $\varphi(\vec{v})$ has no quantifier, then there is nothing to prove, and we are done. Otherwise write $\varphi(\vec{v})$ as

$$Q_1 z_1 \ldots Q_k z_k \alpha(\vec{v}, z_1, \ldots, z_k)$$

where k is a positive integer, the Q_j's ($1 \leq j \leq k$) are quantifiers ($\forall$ or $\exists$) and $\alpha(\vec{v}, z_1, \ldots, z_k)$ is quantifier free. Proceed by induction on k, using what was preliminarily shown: the details are an easy exercise.
At last, let us show that (iv) $\Rightarrow$(i). This is quite simple. We know that, if $f : \mathcal{A} \to \mathcal{B}$ is any embedding between models $\mathcal{A}$, $\mathcal{B}$ of T, then, for every universal L-formula $\varphi(\vec{v})$ and every sequence $\vec{a}$ in A, $\mathcal{B} \models \varphi(f(\vec{a}))$ implies $\mathcal{A} \models \varphi(\vec{a})$. But (iv) ensures that any L-formula is equivalent in T to a suitable universal formula, and so enlarges the previous statement to arbitrary formulas. ♣

As an immediate consequence of (i)$\Rightarrow$(iii), let us point out the following noteworthy fact.

Corollary 3.2.2 *In a model complete theory, the definable sets are just the projections of sets definable by quantifier free formulas.*

3.3 Model completeness and Algebra

Model completeness of ACF and RCF was shown in the last chapter as a consequence of elimination of quantifiers. Robinson's Test provides a direct proof (in these and in other relevant cases). Here we wish to illustrate this new approach. However, the main object in this section is to emphasize the role of model completeness towards some noteworthy applications to Algebra. We underlined in 3.1 A. Robinson's program, and his hope that a progress in model theory could supply Algebra with new important and fruitful tools and techniques. Model completeness really exemplifies this project. In fact, we will see that, just using the model completeness of ACF and RCF, A. Robinson found neat and elegant proofs of classical results, such as the Hilbert Nullstellensatz and the solution of Hilbert's Seventeenth Problem (a theorem of Artin). But we want to underline that the model theoretic approach can provide not only alternative proofs of previously known theorems, but also, and more notably, new and original answers to some formerly open famous algebraic problems, for instance Artin's Conjecture on p-adic fields (this will be treated in 3.4).
But now let us show the model completeness of ACF and RCF via the A. Robinson Test, as promised.

Theorem 3.3.1 (A. Robinson) *ACF is model complete.*

Proof. Suppose not. Owing to the Robinson Test, there is some embedding between algebraically closed fields which is not existential. Let $f : \mathcal{K} \to \mathcal{H}$ be a counterexample. With no loss of generality, we can assume that $\mathcal{K}$ is a subfield of $\mathcal{H}$ and f is just the inclusion (otherwise, replace $\mathcal{K}$ by its isomorphic copy inside $\mathcal{H}$). There are a quantifier free formula $\alpha(\vec{v}, \vec{w})$ in the language $L = \{+, -, \cdot, 0, 1\}$ and a sequence $\vec{a}$ in K such that $\exists\vec{w}\alpha(\vec{v}, \vec{w})$ is true in $\mathcal{H}$ and false in $\mathcal{K}$. Incidentally, recall that $\alpha(\vec{v}, \vec{w})$ is (equivalent to) a disjunction of conjunctions of equations and inequations; however, as $\exists$ is distributive with respect to $\vee$, we can assume that $\alpha(\vec{v}, \vec{w})$ is directly a single conjunction of equations and inequations. Let $\vec{b}$ satisfy $\alpha(\vec{a}, \vec{b})$ in $\mathcal{H}$. Form the extension $\mathcal{K}(\vec{b})$ of $\mathcal{K}$ by $\vec{b}$; its algebraic closure $\overline{\mathcal{K}(\vec{b})}$ embeds itself in $\mathcal{H}$ and satisfies $\exists\vec{w}\alpha(\vec{a}, \vec{w})$ because it includes $\vec{b}$. So there is no loss of generality in replacing $\mathcal{H}$ by $\overline{\mathcal{K}(\vec{b})}$ and hence in assuming that $\mathcal{H} = \overline{\mathcal{K}(\vec{b})}$ has a finite transcendence degree (> 0) over $\mathcal{K}$. Now take a transcendence basis $t_1, \ldots, t_s$ of $\mathcal{H}$ over $\mathcal{K}$, and split the embedding $\mathcal{K} \subseteq \overline{\mathcal{K}(t_1, \ldots, t_s)}$ by

$$\mathcal{K} = \mathcal{K}_0 \subseteq \mathcal{K}_1 \subseteq \ldots \subseteq \mathcal{K}_s = \mathcal{H}$$

where $\mathcal{K}_i = \overline{\mathcal{K}(t_1, \ldots, t_i)}$ for every $i = 1, \ldots, s$. There is some $i = 1, \ldots, s$ such that $\exists\vec{w}\alpha(\vec{a}, \vec{w})$ is true in $\mathcal{K}_i$ and false in $\mathcal{K}_{i-1}$. We can replace $\mathcal{K} \subseteq \mathcal{H}$ by $\mathcal{K}_{i-1} \subseteq \mathcal{K}_i$ and to assume

$$\mathcal{H} = \overline{\mathcal{K}(t)}$$

for a single transcendental element t over K. So $\exists\vec{w}\alpha(\vec{a}, \vec{w})$ is true in $\overline{\mathcal{K}(t)}$ and false in $\mathcal{K}$.

Now take any algebraically closed extension $\mathcal{K}'$ of $\mathcal{K}$ having transcendence degree ≥ 1. Hence $\mathcal{K}'$ enlarges $\overline{\mathcal{K}(u)}$ for some transcendental element u over K. Steinitz's analysis of algebraically closed fields tells us that $\overline{\mathcal{K}(t)}$ and $\overline{\mathcal{K}(u)}$ are isomorphic via a function enlarging the identity of K and mapping t into u. Then $\overline{\mathcal{K}(u)}$ and, consequently, $\mathcal{K}'$ satisfy $\exists\vec{w}\alpha(\vec{a}, \vec{w})$: our sentence is true in every algebraically closed extension $\mathcal{K}'$ of $\mathcal{K}$ with transcendence degree ≥ 1. Equivalently, in the language extending $L(K)$ by a new constant c, $\exists\vec{w}\alpha(\vec{a}, \vec{w})$ is a consequence of $Th(\mathcal{K}_K)$ plus the infinitely many sentences ensuring that c is transcendental over K, i. e. does not solve any nonzero polynomial with 1 unknown and coefficients from K. Use compactness and deduce that finitely many sentences suffice (in addition to $Th(\mathcal{K}_K)$) to imply $\exists\vec{w}\alpha(\vec{a}, \vec{w})$; in particular, c can be interpreted by a suitable element of K (out of the roots of a finite system of polynomials in $K[x]$). In conclusion, $\mathcal{K}$ satisfies $\exists\vec{w}\alpha(\vec{a}, \vec{w})$: a contradiction. Hence ACF is model complete. ♣

Which are the main algebraic ingredients of this Robinson proof? Basically, Steinitz's analysis of algebraically closed fields. More specifically, two key points should be underlined:

1. every field has an algebraic closure, and this is unique up to isomorphism enlarging the identity in the ground field;
2. if $\mathcal{K}$ is an algebraically closed field and t is transcendental over K, then the isomorphism class of $\mathcal{K}(t)$ over $\mathcal{K}$ is uniquely determined (in the sense that two extensions of this kind are isomorphic via a map extending the identity of K).

One can realize that real closed fields satisfy similar properties:

1. every ordered field has a *real closure* (a minimal real closed extension), and this is algebraic over the ground field, and unique up to isomorphism enlarging the identity in the ground field;
2. if $\mathcal{K}$ is a real closed ordered field and t is transcendental over K, then the isomorphism class of $\mathcal{K}(t)$ over $\mathcal{K}$ is fully characterized by the cut t determines over K.

So, when dealing with model completeness for RCF, one can reproduce the proof of the algebraically closed case (with some complications due to the order) and deduce

Theorem 3.3.2 (A. Robinson) *RCF is model complete.*

Robinson's Test can also be used to prove the model completeness of several theories we met in the previous chapter: discrete linear orders, dense linear orders and so on. The reader may check this, as an exercise. But here we prefer to discuss some very noteworthy applications of the model completeness of ACF and RCF to Algebra. They provide new elegant proofs of known algebraic facts.
First let us deal with algebraically closed fields and Hilbert's Nullstellensatz.

Theorem 3.3.3 (Hilbert Nullstellensatz) *Let $\mathcal{K}$ be an algebraically closed field, I be an ideal of the ring $\mathcal{K}[\vec{x}]$ (where $\vec{x}$ abridges, as usual, the sequence of unknowns $(x_1, \ldots, x_n)$). Then, for every polynomial $f(\vec{x}) \in K[\vec{x}]$,*

$$f(\vec{a}) = 0 \text{ for every } \vec{a} \in K^n \text{ such that } g(\vec{a}) = 0 \text{ for all } g(\vec{x}) \in I$$

if and only if

for some positive integer m $f^m(\vec{x}) \in I$.

Proof. The direction from right to left is clear. Conversely, suppose towards a contradiction that there exists some polynomial $f(\vec{x})$ in $K[\vec{x}]$ such that

$$f(\vec{a}) = 0 \text{ for every } \vec{a} \in K^n \text{ such that } g(\vec{a}) = 0 \text{ for all } g(\vec{x}) \in I$$

but

$$f^m(\vec{x}) \notin I \text{ for every positive integer } m.$$

Let J be an ideal of $K[\vec{x}]$ such that $J \supseteq I$, no power $f^m(\vec{x})$ of $f(\vec{x})$ (with m a positive integer) lies in J, and J is maximal with respect to these conditions (Zorn's Lemma ensures that such a J exists). We claim that J is prime. In fact, take two polynomials $g_0(\vec{x})$, $g_1(\vec{x})$ in $K[\vec{x}] - J$; then the ideals

J_0 generated by J and $g_0(\vec{x})$,

J_1 generated by J and $g_1(\vec{x})$

strictly include J; accordingly there are two positive integers m_0, m_1 such that $f^{m_0}(\vec{x}) \in J_0$, $f^{m_1}(\vec{x}) \in J_1$. So there exist two polynomials $q_0(\vec{x})$, $q_1(\vec{x}) \in K[\vec{x}]$ such that

$$f^{m_i}(\vec{x}) - g_i(\vec{x})q_i(\vec{x}) \in J_i, \ \forall i = 0, 1.$$

Consequently

$$f^{m_0+m_1}(\vec{x})$$

is in the ideal generated by J and $g_0(\vec{x}) \cdot g_1(\vec{x})$, and so $g_0(\vec{x}) \cdot g_1(\vec{x}) \notin J$. Hence J is prime, and $\mathcal{R} = \mathcal{K}[\vec{x}]/J$ is an integral domain extending $\mathcal{K}$ by the function mapping any $a \in K$ to $a + J$. Then $\mathcal{K}$ embeds into the algebraic closure $\mathcal{F}$ of the field of quotients of $\mathcal{R}$. As ACF is model complete, this embedding is elementary. Now take any (finite) set of generators $f_0(\vec{x}), \ldots, f_s(\vec{x})$ of I (I is finitely generated because $\mathcal{K}[\vec{x}]$ is Noetherian), and notice that the $L(K)$-sentence

$$\exists \vec{v} \ (\bigwedge_{i \leq s} f_i(\vec{v}) = 0 \wedge \neg(f(\vec{v}) = 0))$$

is true in $\mathcal{F}$ (owing to the sequence $(x_1 + J, \ldots, x_n + J)$). Consequently this sentence is true also in $\mathcal{K}$. So there exists some $\vec{a}$ in K^n such that $\vec{a}$ satisfies $f_0(\vec{x}), \ldots, f_s(\vec{x})$ and consequently all the polynomials in I, but $\vec{a}$ does not annihilate $f(\vec{x})$ -a contradiction-. ♣

Now we deal with RCF, and with Hilbert's Seventeenth Problem. This was solved by Artin in 1927. Indeed Artin himself and Schreier developed the algebraic notion of real closed field just to answer Hilbert's question. Later A. Robinson proposed a very nice and simple proof, founded on the model completeness of RCF. Here we want to report A. Robinson's approach. First let us introduce Hilbert's problem in detail.
Indeed the seventeenth question in the celebrated Hilbert 1900 list just concerns ordered fields (more properly, the ordered field of reals). Recall that, in any ordered field $\mathcal{K}$, a rational function $f(\vec{x}) \in K(\vec{x})$ is said to be *semidefinite positive* if and only if, for every sequence $\vec{a}$ in K (such that $f(\vec{a})$ is defined), $f(\vec{a}) \geq 0$. Of course, the sums of squares in $K(\vec{x})$ are semidefinite positive. Hilbert's Seventeenth Problem conjectures that the converse is also true when $\mathcal{K}$ is the ordered field of real numbers. As already said, Artin solved positively this question; indeed he extended the result to arbitrary real closed ordered fields $\mathcal{K}$. Now we provide A. Robinson's proof of this theorem.

Theorem 3.3.4 (Artin) *Let $\mathcal{K}$ be a real closed ordered field, $f(\vec{x})$ a semidefinite positive rational function in $\mathcal{K}(\vec{x})$. Then $f(\vec{x})$ can be expressed as a sum of squares in $\mathcal{K}(\vec{x})$.*

Proof. We need the following algebraic fact.

Fact 3.3.5 *Let $\mathcal{K}$ be a field, and assume that, for every natural t and for every choice of $a_0, \ldots, a_t \in K$, if $\sum_{i \leq t} a_i^2 = 0$, then $a_0 = \ldots = a_t = 0$ (such a field $\mathcal{K}$ is usually called formally real). Let a be an element of K such that a cannot be represented as a sum of squares in K. Then there exists a total order $\leq$ in $\mathcal{K}$ making $\mathcal{K}$ an ordered field and $a < 0$.*

Now let us begin our proof. Take a semidefinite positive $f(\vec{x}) \in K(\vec{x})$. Suppose towards a contradiction that $f(\vec{x})$ cannot be expressed as a sum of squares. Clearly $\mathcal{K}(\vec{x})$ is a formally real field. So, owing to Fact 3.3.5, there is some total order relation $\leq$ in $\mathcal{K}(\vec{x})$ with respect to which $\mathcal{K}(\vec{x})$ becomes an ordered field and $f(\vec{x}) < 0$. Notice that this order relation $\leq$ in $\mathcal{K}(\vec{x})$, when restricted to $\mathcal{K}$, does equal the primitive order of $\mathcal{K}$: in fact, in both these relations, the non-negative elements of $\mathcal{K}$ are just the squares in $\mathcal{K}$. In other words, $\mathcal{K}$ is a substructure of $\mathcal{K}(\vec{x})$ in our language for ordered fields. Recall that every ordered field admits a minimal real closed extension (its *real closure*), and accordingly embed $\mathcal{K}(\vec{x})$ into the real closure $\mathcal{R}$. Altogether we obtain an embedding of $\mathcal{K}$ into $\mathcal{R}$. As RCF is model complete, this embedding is elementary. But $\mathcal{K}(\vec{x})$, and consequently $\mathcal{R}$, satisfy the $L(K)$-sentence

$$\exists \vec{v}(f(\vec{v}) < 0)$$

(owing to the sequence $\vec{x}$). Accordingly also $\mathcal{K}$ satisfies $\exists \vec{v}(f(\vec{v}) < 0)$. But this contradicts our assumption that $f(\vec{x})$ is semidefinite positive. ♣

3.4 p-adic fields and Artin's Conjecture

Real numbers complete the rational field with respect to the usual metric topology. For every prime p, p-adic numbers complete the rationals with respect to an alternative topology, called p-adic. Let us shortly remind this topology. First write (uniquely) any nonzero rational a as

$$a = p^h \frac{r}{s}$$

where h, r and s are integers, s is positive and h, r, s are pairwise coprime; then define a function v_p from the multiplicative group $\mathbf{Q}^*$ of nonzero rationals into the ordered additive group of integers by putting, for a as before,

$$v_p(a) = h.$$

One gets in this way a group homomorphism from $\mathbf{Q}^*$ into the integers, satisfying the additional condition

$$v_p(a+b) \geq min\{v_p(a), v_p(b)\} \quad \forall a, b \in \mathbf{Q}^*$$

(this is straightforward to check). v_p can be formally extended to 0 in some artificial way; putting $v_p(0) = \infty$ is a reasonable choice, as p^h divides 0 among the integers for every natural number h. Now put, for every positive integer h,

$$O_h = \{a \in \mathbf{Q} : v_p(a) \geq h\}$$

and take the O_h's as basic open neighbourhoods of 0, hence their translations $O_h + b$ with $b \in \mathbf{Q}$ as basic open sets. One gets a new topology of the rationals: the p-adic topology. As already said, the set $\mathbf{Q}_p$ of p-adic numbers is the completion of $\mathbf{Q}$ with respect to this topology.
In order to realize as well as possible what a p-adic number is, and so to introduce $\mathbf{Q}_p$ in a more detailed way, one can follow several equivalent ways: see [36] for a general outline of this point. Here we limit ourselves to sketch

some basic ideas. A possible approach to p-adic numbers might use the well known fact that every positive integer a has a unique p-adic decomposition

$$a = a_o + a_1 p + \ldots + a_n p^n$$

where n and the a_i's are natural numbers, $a_n \neq 0$ and $a_i < p$ for all $i \leq n$. Under this perspective, $v_p(a)$ is just the least $i \leq n$ for which $a_i \neq 0$. Hence, for every nonzero natural $a_{n+1} < p$, it is trivial to realize

$$v_p((a + a_{n+1}p^{n+1}) - a) = v_p(a_{n+1}p^{n+1}) = n + 1.$$

So, when considering Cauchy sequences of integers with respect to the p-adic topology and equipping such a sequence with a limit, one naturally builds infinite sums

$$\sum_{i=0}^{\infty} a_i p^i$$

where the a_i's are nonnegative integers $< p$.
Enlarging the analysis from positive integers to arbitrary nonzero rationals leads to consider general infinite sums

$$(\star) \qquad \sum_{i=N}^{\infty} a_i p^i$$

where the a_i's are, as above, natural numbers $< p$, $a_N \neq 0$ but N is now any fixed -even negative- integer: in this sense $(\star)$ exhibits a typical p-adic number. 0 can be easily recovered in this framework as the infinite sum whose coefficients are constantly 0.
Hence, for every prime p, $\mathbf{Q}_p$ is the set of these infinite sums. One defines in a suitable way addition $+$ and multiplication $\cdot$ in $\mathbf{Q}_p$, extending the usual operations in $\mathbf{Q}$. But here we have to be careful: sum and product are not computed componentwise, but as suggested by the algebraic framework. For instance, the trivial identity $1 + (p-1) = p$ must be read

$$(1 + 0 \cdot p) + ((p-1) + 0 \cdot p) = 0 + 1 \cdot p.$$

However these operations equip $\mathbf{Q}_p$ with a field structure, extending the rational field; v_p can be enlarged to $\mathbf{Q}_p$ in very simple way, as, for $a = \sum_{i=N}^{\infty} a_i p^i$ as before (and $a_N \neq 0$), $v_p(a)$ is just N. One sees that

$$\mathbf{Z}_p = \{a \in \mathbf{Q}_p \, : \, v_p(a) \geq 0\}$$

is a local subring (the *ring of p-adic integers*), and

$$I_p = \{a \in \mathbf{Q}_p : v_p(a) > 0\}$$

is its maximal ideal. One easily checks that the quotient field $\mathbf{Z}_p/I_p$ is isomorphic to the field $\mathbf{F}_p$ with p elements.
We want to underline two further basic properties of $\mathbf{Q}_p$.

(a) The first one is quite trivial, and simply points out that $v_p(p) = 1$.

(b) The other is more substantial and concerns the so called *Hensel's Lemma*. This is a key result in locating roots of polynomials in $\mathbf{Q}_p$, in fact it states that,

> *if $f(x)$ is a monic polynomial in $\mathbf{Z}_p[x]$, then any decomposition of $f(x)$ modulo I_p in $\mathbf{F}_p[x]$ as a product of two relatively coprime monic polynomials lifts to a decomposition of $f(x)$ as a product of two monic polynomials in $\mathbf{Z}_p[x]$.*

This concludes our short summary about the algebraic structure of $\mathbf{Q}_p$ for every prime p. What we have sketched suggests some similarities with the reals. Actually both $\mathbf{Q}_p$ and $\mathbf{R}$ have a common topological genesis from the rationals (and, under this point of view, a common topological structure of locally compact field); moreover there do admit some reasonable criteria to locate roots of polynomials (the Sign Change property for the reals, Hensel's Lemma in the p-adic case).

Now we want to discuss another example, closely recalling $\mathbf{Q}_p$. We take any field $\mathcal{K}$ (but below we will be primarily interested in the field $\mathbf{F}_p$ with p elements for any prime p). We look at the formal Laurent series

$$a(t) = \sum_{i=N}^{\infty} a_i t^i$$

where $a_i \in K$ for every $i \geq N$ and N is a given integer. The corresponding set $K((t))$ inherits a field structure extending $\mathcal{K}$, provided that we define the addition $+$ *componentwise* and the multiplication $\cdot$ in the obvious way enlarging the product in $\mathcal{K}$. Again the set $K[[t]]$ of formal power series

$$\sum_{i=0}^{\infty} a_i t^i$$

is a local subring, whose maximal ideal just contains the power series with $a_0 = 0$. One easily deduce that the quotient ring is (isomorphic to) $\mathcal{K}$.
The function v_K from the multiplicative group $\mathcal{K}((t))^*$ (where *, as before, means to exclude 0) into the (ordered) additive group of integers sending any nonzero Laurent series $a(t)$ as before (with $a_N \neq 0$) to N again yields a group homomorphism sharing with v_p and $\mathbf{Q}_p$ the following property

$$v_K(a(t) + b(t)) \geq min\{v_K(a(t)), v_K(b(t))\}$$

for all $a(t)$ and $b(t)$ in $K((t))^*$.
Now assume $\mathcal{K} = \mathbf{F}_p$ for a given prime p. In this restricted framework

(a) $v_{\mathbf{F}_p}(p) \neq 1$. Notice that this distinguishes $\mathbf{F}_p((t))$ and $\mathbf{Q}_p$.

(b) However $\mathbf{F}_p((t))$, just as $\mathbf{Q}_p$, satisfies Hensel's Lemma.

An abstract notion including p-adics as well as formal Laurent series and other related examples towards a common general treatment is the concept of *valued field*: this is a structure $(\mathcal{K}, \mathcal{G}, v)$ where $\mathcal{K}$ is a field, $\mathcal{G}$ is an ordered abelian group, and v is a group homomorphism from the multiplicative group $\mathcal{K}^*$ in $\mathcal{G}$ satisfying the further assumption

$$v(a+b) \geq min\{v(a), v(b)\} \quad \forall a, b \in K^*.$$

The function v is called the *valuation map*. A general algebraic analysis promptly confirms some basic properties observed in the previous examples: in particular, for every valued field $(\mathcal{K}, \mathcal{G}, v)$, $\{a \in K : v(a) \geq 0\}$ is a local subring (the *valuation ring*), and its maximal ideal is just $\{a \in K : v(a) > 0\}$; the corresponding quotient field is called the *residue field* of $(\mathcal{K}, \mathcal{G}, v)$, and will be denoted by $\overline{\mathcal{K}}$ below (hence all throughout this section $\overline{\mathcal{K}}$ denotes residue field rather than algebraic closure).
A valued field is called *Henselian* when it satisfies Hensel's Lemma, hence when the following holds:

Let $f(x)$ be a monic polynomial in $R[x]$ and let $\overline{f}(x)$ denote its projection in $\overline{\mathcal{K}}[x]$. Then any decomposition of $\overline{f}(x)$ in $\overline{\mathcal{K}}[x]$ as a product of two relatively coprime monic polynomials

$$\overline{f}(x) = \gamma(x) \cdot \kappa(x)$$

lifts to a decomposition of $f(x)$ into the product of two monic polynomials in $R[x]$

$$f(x) = g(x) \cdot k(x)$$

where $\overline{g}(x) = \gamma(x)$ and $\overline{k}(x) = \kappa(x)$.

For every prime p, both $(\mathbf{Q}_p, \mathbf{Z}, v_p)$ and $(\mathbf{F}_p((t)), \mathbf{Z}, v_{\mathbf{F}_p})$ are Henselian valued fields (although they do not admit the same characteristic, and hence their valuation of p is not the same).
Now let us come back to our original framework and hence to p-adic numbers. Here Artin proposed a famous conjecture.

Conjecture 3.4.1 (Artin's Conjecture) *Let p be a prime. For all positive integers n and d with $n > d^2$, every homogeneous polynomial $f(x_1, \ldots, x_n) \in \mathbf{Q}_p[x_1, \ldots, x_n]$ of degree d has a nonzero root in $\mathbf{Q}_p$.*

The conjecture is inspired by the underlined resemblance between $\mathbf{Q}_p$ and $\mathbf{F}_p((t))$ for every prime p. Actually in the valued field $\mathbf{F}_p((t))$ the claim is true for every choice of p, n and d, as proved by Lang. But the behaviour of the p-adics in this setting is not the same, in fact Artin's Conjecture fails for some p, n and d. This was observed by Terjanian, who in [161] did exactly what one is expected to do in disproving a statement, and exhibited a counterexample for some suitable p, n and d.
However an asymptotic form of the conjecture is true: for any choice of n and d, Artin's statement is satisfied by all but finitely many values of p. This was the content of a celebrated theorem of Ax, Kochen and Ershov in 1965, which, combined with Terjanian's counterexample, provides a sufficiently complete answer to the question. The Ax-Kochen-Ershov approach is essentially model theoretic. Indeed, they developed a general analysis of the model theory of valued fields, and deduced the asymptotic form of Artin's Conjecture as a consequence, using compactness and transfer techniques. Let us briefly survey their work.
First of all, we have to clarify how to handle valued fields from a model theoretic point of view: which language to use, and so on. The more convincing approach views a valued field as a *two-sorted structure*, in other words as a structure with two sorts of variables, where

⋆ the former sort of variables concerns the elements of the field,

⋆ the latter sort of variables is devoted to the elements of the valuation group;

moreover there are the usual field symbols for the former sort, and (disjointly) the symbols of ordered groups for the latter; finally a 1-ary operation symbol v is reserved for the valuation map. Valued fields are easily axiomatized in a first order way in this language.
This approach emphasizes, within valued fields $(\mathcal{K}, \mathcal{G}, v)$, the role of three underlying structures:

the original field $\mathcal{K}$,

the ordered abelian group $\mathcal{G}$,

and finally, to capture the valuation map v,

the residue field $\overline{\mathcal{K}}$.

Ax, Kochen and Ershov show that these structures rule the behaviour of the whole valued field $(\mathcal{K}, \mathcal{G}, v)$ with respect to elementary equivalence. In fact, their main general result says

Theorem 3.4.2 *Let $(\mathcal{K}, \mathcal{G}, v)$ be an Henselian valued field, whose residue field has characteristic 0. Then the complete theory of $(\mathcal{K}, \mathcal{G}, v)$ is fully determined by the complete theories of the ordered valued group $\mathcal{G}$ and the residue field $\overline{\mathcal{K}}$.*

Warning: The theorem does not apply directly to $\mathbf{Q}_p$ or to $\mathbf{F}_p((t))$ for any prime p because these valued fields do not respect the hypothesis on the characteristic of the residue field. Nevertheless the Ax-Kochen-Ershov main theorem is enough to throw a bridge between the valued fields of Laurent series $\mathbf{F}_p((t))$ and the p-adic valued fields $\mathbf{Q}_p$ with respect to Artin's Conjecture, and to deduce its asymptotic solution.

Theorem 3.4.3 (Ax-Kochen-Ershov) *For all positive integers n, d with $n > d^2$, for all but finitely many primes p, every homogeneous polynomial $f(x_1, \ldots, x_n) \in \mathbf{Q}_p[x_1, \ldots, x_n]$ of degree d has a nontrivial root in $\mathbf{Q}_p$.*

Proof. (Sketch) We can limit ourselves to treat the case $n = d^2 + 1$. Otherwise put $x_i = 0$ for $n > i > d^2+1$ and work with $x_1, \ldots, x_{d^2+1}$: up to rearranging the indices of our unknowns, we can assume that what we get in this way is a homogeneous polynomial of degree d in $x_1, \ldots, x_{d^2+1}$, and every nontrivial zero of this polynomial clearly produces a nontrivial root of $f(x_1, \ldots, x_n)$. Needless to say, for any fixed d (and n), Artin's Conjecture becomes a first order sentence α_d in the language of valued fields: this is a routine exercise, easy to check. We know from Lang that this sentence α_d is true in the valued field $(\mathbf{F}_p((t)), \mathbf{Z}, v_{\mathbf{F}_p})$ *for all primes p*. This fact, in particular its uniform validity *for every p*, is noteworthy; actually, we are in a situation quite similar to the one described in Chapter 2, § 4, for algebraically closed fields. In particular, one can apply the same transfer machinery, combine the Ax-Kochen-Ershov main theorem and Lang's result, and in conclusion deduce that α_d is true in every valued field $(\mathcal{K}, \mathcal{G}, v)$ such that

⋆ $\mathcal{K}$ is Henselian,

⋆ $\mathcal{G}$ is elementarily equivalent (as an ordered group) to the integers ($\mathcal{G}$ is called a **Z**-group in this case),

⋆ the residue field $\overline{\mathcal{K}}$ is a pseudofinite field of characteristic 0.

Of course, this does not concern directly any p-adic field $\mathbf{Q}_p$, because its residue field has characteristic p. But, in proving α_d, only finitely many sentences about the characteristic 0 of the residue field are necessary. This implies that α_d is true in $\mathbf{Q}_p$ for almost all primes p, as claimed. ♣

Owing to the Terjanian counterexample quoted before, this Ax-Kochen-Ershov answer to the Artin Conjecture is the best possible.
None of these results refers directly to model completeness. Nevertheless the spirit and the techniques of the model theoretic approach of Ax, Kochen and Ershov clearly owe to A. Robinson's ideas, and are intimately related to his dream of linking Algebra and Mathematical Logic via Model Theory. One should also remember that the Ax-Kochen-Ershov main theorem does not apply to the p-adic fields $\mathbf{Q}_p$, because, as already observed, the characteristic of their residue fields is not 0. So one may wonder, for every prime p, how to characterize $\mathbf{Q}_p$, and even its twin $\mathbf{F}_p((t))$, up to elementary equivalence, in other words how to axiomatize in a first order way their complete theories (incidentally recall that $Th(\mathbf{Q}_p) \neq Th(\mathbf{F}_p((t))$, because the involved characteristics are not equal). Here model completeness sounds useful. Indeed, we already underlined that p-adics and reals share several relevant similarities: model completeness, and a precise first order axiomatization, are among them. In fact, the following theorem holds.

Theorem 3.4.4 (Ax-Kochen-Ershov) *For every prime p, let T_p be the theory of the valued fields $(\mathcal{K}, \mathcal{G}, v)$ such that*

⋆ *$\mathcal{K}$ is Henselian and has characteristic 0,*

⋆ *$\mathcal{G}$ is a **Z**-group,*

⋆ *the residue field $\overline{\mathcal{K}}$ is (elementarily equivalent to) the field with p elements,*

⋆ *$v(p)$ is 1 -the least positive element in G-.*

Then T_p is model complete and complete. In particular T_p is the theory of the valued field of p-adic numbers.

What can we say about definable sets in $\mathbf{Q}_p$, or also about quantifier elimination for p-adics? Macintyre showed quantifier elimination in a very natural language with additional relation symbols for the valuation ring and the set of n-th powers for every n. This provides, of course, some more information on definable sets; in particular, one can see that the p-adics satisfy

Theorem 3.4.5 (Macintyre) *Every infinite definable subset of $\mathbf{Q}_p$ has a nonempty interior.*

Notice that this is exactly what happens for the reals. Finally, let us remind that T_p is not the theory of $\mathbf{F}_p((t))$; actually, the question of determining the first order theory of this valued field seems still open.

3.5 Existentially closed structures

Example 3.5.1 The class of fields and the subclass of algebraically closed fields satisfy the following properties:

(i) every field embeds into an algebraically closed fields (for instance, into its algebraic closure);

(ii) every embedding between algebraically closed fields is elementary (in other words, the theory ACF is model complete).

Consequently (in our language L for fields)

(iii) for every embedding of an algebraically closed field $\mathcal{K}$ into a field $\mathcal{H}$, and for every quantifier free formula $\varphi(\vec{w})$ of $L(K)$, if $\mathcal{H} \models \exists\vec{w}\varphi(\vec{w})$, then $\mathcal{K} \models \exists\vec{w}\varphi(\vec{w})$ (in other words the embedding is existential).

In fact, owing to (i), $\mathcal{H}$ embeds in some algebraically closed field $\overline{\mathcal{H}}$; as $\mathcal{H} \models \exists\vec{w}\varphi(\vec{w})$, also $\overline{\mathcal{H}}$ satisfies $\exists\vec{w}\varphi(\vec{w})$. Clearly $\mathcal{K}$ embeds into $\overline{\mathcal{H}}$ through $\mathcal{H}$, and this embedding is elementary because of (ii). Hence $\mathcal{K} \models \exists\vec{w}\varphi(\vec{w})$. Notice that (iii) is not a secondary property, but does include the definition itself of an algebraically closed field $\mathcal{K}$. In fact, the latter requires that every non-constant polynomial $f(x) \in K[x]$ has a root $\mathcal{K}$, or, equivalently, that $\mathcal{K}$ satisfies all the $L(K)$-formulas

$$\exists w(f(w) = 0)$$

(true in some extension of $\mathcal{K}$). (iii) says that this still holds when we replace $f(w) = 0$ by any quantifier free $L(K)$-formula $\varphi(\vec{w})$ (with arbitrarily many variables).

From an algebraic point of view, (iii) implies that any finite system of equations

$$f_0(\vec{x}) = 0, \ldots, f_t(\vec{x}) = 0,$$

with $f_0(\vec{x}), \ldots, f_t(\vec{x}) \in K[\vec{x}]$, has a solution in our algebraically closed field $\mathcal{K}$ whenever it finds a zero in some extension of $\mathcal{K}$: to see this, just apply (iii) to

$$\varphi(\vec{v}) : \bigwedge_{i \leq t} f_i(\vec{v}) = 0.$$

Example 3.5.2 A similar analysis can be developed in the language L of ordered fields with respect to the subclass of real closed fields.
In fact

(i) any ordered field has some real closed extension (for instance, its real closure);

(ii) any embedding between real closed fields is elementary (as the theory RCF is model complete).

(i) and (ii) imply, just as in the previous case, that, in the language L of ordered fields,

(iii) for every embedding of a real closed field $\mathcal{K}$ into an ordered field $\mathcal{H}$ and for every quantifier free formula $\varphi(\vec{w})$ of $L(K)$, if $\mathcal{H} \models \exists \vec{w} \varphi(\vec{w})$, then $\mathcal{K} \models \exists \vec{w} \varphi(\vec{w})$.

Algebraically speaking, (iii) implies that, in a real closed field $\mathcal{K}$, any finite system of equations and disequations

$$f_0(\vec{x}) = 0, \ldots, f_t(\vec{x}) = 0, \quad g_0(\vec{x}) > 0, \ldots, g_s(\vec{x}) > 0,$$

with $f_0(\vec{x}), \ldots, f_t(\vec{x}), g_0(\vec{x}), \ldots, g_s(\vec{x}) \in K[\vec{x}]$, admitting a solution in some ordered field extending $\mathcal{K}$, does have some solution even in $\mathcal{K}$.
So (iii) is not a minor property of real closed fields but includes their definition itself, or, more precisely, their characterization saying that real closed fields are just the ordered fields $\mathcal{K}$ with no proper ordered algebraic extension: if $f(x) \in K[x]$ has a root in some ordered extension of $\mathcal{K}$, then it finds a solution also in $\mathcal{K}$.

Now consider any class $\mathbf{K}$ of structures in a language L. The previous examples suggest the following notion.

Definition 3.5.3 *A structure $\mathcal{A}$ in* **K** *is* **existentially closed (e. c.)** *if and only if every embedding of $\mathcal{A}$ in some $\mathcal{B} \in$* **K** *is existential (i. e., for every quantifier free $L(A)$-formula $\alpha(\vec{w})$, if $\mathcal{B} \models \exists \vec{w} \alpha(\vec{w})$, then $\mathcal{A} \models \exists \vec{w} \alpha(\vec{w})$).*

Let $\mathcal{E}(\mathbf{K})$ denote the class of e. c. structures in **K**. Clearly, among fields, e. c. structures are just the algebraically closed fields, and, among ordered fields, e. c. structures are just the real closed fields.
Given a class $\mathbf{K} \neq \emptyset$, first we can wonder whether there exist some e. c. structures in **K**, and hence whether $\mathcal{E}(\mathbf{K}) \neq \emptyset$. The answer is positive, at least under some simple conditions on **K**. For instance, one can see what follows.

Example 3.5.4 Let **K** be a class of structures. Assume that, whenever $(I, \leq)$ is a totally ordered set and each $i \in I$ indicates a structure $\mathcal{A}_i \in \mathbf{K}$ in such a way that, for $i \leq j$ in I, $\mathcal{A}_i$ is a substructure of $\mathcal{A}_j$, then the union $\mathcal{A} = \bigcup_{i \in I} \mathcal{A}_i$ -as defined in Chapter 1- is still in **K**. In this case, every $\mathcal{A} \in \mathbf{K}$ embeds in some structure of $\mathcal{E}(\mathbf{K})$; in particular, $\mathcal{E}(\mathbf{K})$ is not empty.

Notice also that, in the examples above, **K** is elementary, as well as $\mathcal{E}(\mathbf{K})$. Accordingly one can wonder whether, in general, if **K** is elementary, then $\mathcal{E}(\mathbf{K})$ is. The answer is negative.

Examples 3.5.5

(a) If **K** is the -elementary- class of groups, then $\mathcal{E}(\mathbf{K})$ is not elementary any more (this is a result of Eklof and Sabbagh).

(b) If **K** is the -elementary- class of commutative rings, then $\mathcal{E}(\mathbf{K})$ is not elementary any more (as shown by Cherlin).

The proofs mix compactness and some algebraic facts. We shall provide their details at the end of this section. It should be emphasized that, while e. c. fields are fully characterized in a first order way, e. c. rings are not. The same happens for groups. Conversely, among abelian groups, e. c. structures are an elementary class: they are exactly the divisible abelian groups admitting infinitely many elements of period p for every prime p; this is shown in [41], where e. c. closed modules over suitable rings are also discussed.
However, take an elementary class **K**; let T denote its theory. If $\mathcal{E}(\mathbf{K})$ is elementary, then $T^\star = Th(\mathcal{E}(\mathbf{K}))$ is said to be a *model companion* of T. Clearly $T \subseteq T^\star$. Moreover the following result holds (generalizing what was observed in our starting examples).

Theorem 3.5.6 (Eklof - Sabbagh) *Let* **K** *be an elementary class of L-structures,* $T = Th(\mathbf{K})$, $T^\star$ *be an L-theory containing T. Then* $T^\star$ *is a model companion of T (and* $\mathcal{E}(\mathbf{K}) = Mod(T^\star)$*) if and only if*

(i) *for every* $\mathcal{A} \in \mathbf{K}$, *there exists some* $\mathcal{B} \in Mod(T^\star)$ *where* $\mathcal{A}$ *embeds in;*

(ii) $T^\star$ *is model complete.*

Existentially closed structures, and model companions, were intensively studied in the sixties and in the seventies. Several -elementary- algebraic classes **K** were considered, to check whether existentially closed structures in **K** were or not an elementary class, to provide a satisfactory characterization in the positive case, and to analyse their complexity in the negative one. This was not (and is not) a barren and unproductive exercise: in fact, it brought to light some very interesting classes of (existentially closed) structures, such as differentially closed fields. We shall treat them (and other key examples) in the next sections; but we want to emphasize since now that the notion itself of differentially closed field -a quite algebraic concept- arises for the first time within the framework of e. c. structures: no algebraic treatment preceded the model theoretic approach. Moreover, it should be pointed out that new interesting elementary classes of existentially closed structures have been considered quite recently, for instance among fields with a distinguished automorphism. This framework, too, will be explained later in this chapter.
Also in the negative case, when existentially closed structures cannot form an elementary class, their analysis has some intriguing features. The main purpose here is to understand the reasons of nonelementarity and, hence, in some sense, to measure how complicated the class of e. c. objects is. To illustrate this point, we show, as promised, that existentially closed groups are not an elementary class, and the same is true for existentially closed commutative rings (with identity), and we discuss briefly these negative results. First let us deal with groups.

Theorem 3.5.7 (Eklof - Sabbagh) *Let* **K** *be the -elementary- class of groups. Then* $\mathcal{E}(\mathbf{K})$ *is not elementary.*

Proof. We work in the language $L = \{\cdot, {}^{-1}, 1\}$. We need two preliminary facts. $\mathcal{G}$ denotes an e. c. group.

Fact 3.5.8 *For every positive integer* n, $\mathcal{G}$ *admits some elements of period* n.

Proof. Take the direct product $\mathcal{G}'$ of $\mathcal{G}$ and a cyclic group of order n. Clearly $\mathcal{G}'$ has some elements of period n, in other words it satisfies the existential sentence $\exists w(w^n = 1 \wedge \bigwedge_{d|n,\, d\neq n} \neg(w^d = 1))$. As $\mathcal{G}$ is e. c., $\mathcal{G}$ satisfies the same condition, hence it has some elements as required.

Fact 3.5.9 *Two elements of infinite period in $\mathcal{G}$ are conjugate.*

Proof. This is a more delicate point, and refers to a theorem of Higman, B. H. Neumann and H. Neumann, building, for any two elements a and b of infinite period in $\mathcal{G}$, a group $\mathcal{G}'$ extending $\mathcal{G}$ and an element $t \in G'$ such that $tat^{-1} = b$. So $\mathcal{G}' \models \exists w(waw^{-1} = b)$. $\mathcal{G}$ inherits this property because it is existentially closed.
Now assume $\mathcal{E}(\mathbf{K})$ elementary, $\mathcal{E}(\mathbf{K}) = Mod(T)$ for a suitable L-theory T. Fact 2 says that, in a language L' extending L by two new constant symbols c and d,

$$T \cup \{\neg(c^n = 1), \neg(d^n = 1) \,:\, n \in \mathbf{N}, n > 0\} \models \exists w(wcw^{-1} = d).$$

By compactness there is some positive integer N for which

$$T \cup \{\neg(c^n = 1), \neg(d^n = 1) \,:\, 0 < n \leq N\} \models \exists w(wcw^{-1} = d).$$

Hence take an e. c. group $\mathcal{G}$ and two elements a, b in G of period $N+1$, $N+2$ respectively: owing to Fact 1, this can be done. Then a and b are conjugate in G: but this clearly contradicts the fact that their periods are different. ♣

Actually existentially closed groups are a very complicated class: [91] contains several results illustrating their wildness. In particular, we like to mention a noteworthy connection with the word problem for groups.

Theorem 3.5.10 (Macintyre, Neumann) *A finitely generated group can be presented with a solvable word problem if and only if it is embeddable in all e. c. groups.*

Now let us deal with rings.

Theorem 3.5.11 (Cherlin) *Let* $\mathbf{K}$ *be the -elementary- class of commutative rings with identity. Then $\mathcal{E}(\mathbf{K})$ is not elementary.*

Proof. Throughout the proof, ring abbreviates commutative ring with identity. Accordingly we work in the language $L = \{+, -, \cdot, 0, 1\}$. Again, we need two preliminary facts. $\mathcal{R}$ denotes an e. c. ring.

Fact 3.5.12 *The nilradical of $\mathcal{R}$ (i. e. the ideal of nilpotent elements) is $\emptyset$-definable in $\mathcal{R}$ in a uniform way (valid in any e. c. ring).*

Proof. We claim that, for every $a \in R$, a is nilpotent if and only if a does not divide any nonzero idempotent element. This is clearly enough for our purposes, because the latter condition can be easily written as a(n existential) first order formula in L

$$\exists w(v \cdot w = v^2 \cdot w^2 \wedge \neg(v \cdot w = 0)).$$

The implication from the left to the right is quite simple. For, let a divide some idempotent $e \in R$; as $a^n = 0$ for some integer $n \geq 2$, it follows $0 = e^n = e$. Conversely, assume that a is not nilpotent. Form the quotient ring $\mathcal{R}' = \mathcal{R}[x]/I$ where I is the ideal generated by the polynomial $ax - a^2x^2$. Then $\mathcal{R}'$ extends $\mathcal{R}$ in the obvious way, by the embedding of $\mathcal{R}$ in $\mathcal{R}[x]$ and the projection of $\mathcal{R}[x]$ onto $\mathcal{R}'$. In $\mathcal{R}'$ the image $I + a$ of a divides the idempotent $I + ax$; moreover $I + ax \neq 0$, otherwise in $\mathcal{R}[x]$

$$(\star) \qquad ax = (ax - a^2x^2)f(x)$$

for some $f(x) = \sum_{i \leq n} f_i x^i \in R[x]$. A comparison of the coefficients of the same degree in $(\star)$ yields

$$af_0 = a,$$

$$a^2 f_i = af_{i+1} \quad \forall i = 1, \ldots, n-1,$$

and, eventually,

$$a^2 f_n = 0.$$

In particular $a^{n+2} = 0$ -a contradiction-. So the image of a in R' divides a nonzero idempotent; as $\mathcal{R}$ is e. c., a satisfies the same condition in $\mathcal{R}$.

Fact 3.5.13 *For every positive integer n, $\mathcal{R}$ contains some nilpotent a satisfying $a^{n+1} = 0$ and $a^n \neq 0$.*

Proof. Again enlarge $\mathcal{R}$ to suitable quotients of the polynomial ring $\mathcal{R}[x]$. This time, for every positive integer n, form $\mathcal{R}' = \mathcal{R}[x]/I$ where I is the ideal generated by x^{n+1}. One easily checks that R embeds in $\mathcal{R}'$ by the map $a \mapsto I + a$ for every $a \in R$; moreover $(I + x)^{n+1} = I$, while $(I + x)^n \neq I$. Hence $\mathcal{R}'$ satisfies $\exists w(w^{n+1} = 0 \wedge \neg(w^n = 0))$. As $\mathcal{R}$ is e. c., the same is true in $\mathcal{R}$.

Now we can conclude our proof. Again we use compactness. Enlarge L by a new constant symbol c and, in the new language L', look at

$$T' = \mathrm{Th}(\mathcal{E}(\mathbf{K})) \cup \{\neg(c^n = 0) \, : \, n \in \mathbf{N}, n > 0\} \cup$$

$$\cup\{\exists w(c \cdot w = c^2 \cdot w^2 \wedge \neg(c \cdot w = 0))\}.$$

Every finite $T'_0 \subseteq T'$ has its own model: for, there is a positive integer N such that

$$T'_0 \subseteq \mathrm{Th}(\mathcal{E}(\mathbf{K})) \cup \{\neg(c^n = 0) : 0 < n \leq N\} \cup \{\exists w(c \cdot w = c^2 \cdot w^2 \wedge \neg(c \cdot w = 0))\},$$

and the latter set of sentences does admit a model: it suffices to take an e. c. ring and interpret c in a nilpotent element a satisfying $a^N \neq 0$ (Fact 2 ensures that a exists). By compactness T' has a model: this is a ring satisfying the same sentences as e. c. rings, but it is not e. c., because it contains a nonnilpotent element (the interpretation of c) dividing some nonzero idempotent. In conclusion, $\mathcal{E}(\mathbf{K})$ is not elementary, because its theory has non e. c. models. ♣

Notably, nilpotents are the key obstacle to the non elementarity of the class of e. c. rings. In fact, for the restricted class $\mathbf{K}$ of reduced rings (i. e. rings without nonzero nilpotents) $\mathcal{E}(\mathbf{K})$ is elementary. On the other side, if one allows nilpotent elements, even of bounded exponent (for instance one considers rings whose nilpotents a satisfy $a^2 = 0$), then the elementarity of e. c. objects gets lost [164].

3.6 DCF_0

A *differential ring* is a commutative ring $\mathcal{K}$ with identity, having an additional 1-ary operation D (called *derivation*) such that

$$D(a + b) = Da + Db,$$

$$D(a \cdot b) = a \cdot Db + b \cdot Da$$

for every a and b in K. A *differential field* is a differential ring which is also a field. So a suitable first order language L' for differential fields enlarges our language L for fields by a new 1-ary operation symbol (to be represented by D).

Differential fields include

- $(\mathcal{K}, D)$ where $\mathcal{K}$ is any field and $D = 0$,

but also more significant and interesting structures, like

- $(\mathcal{K}(x), \frac{d}{dx})$ where $\mathcal{K}$ is any field, $\mathcal{K}(x)$ is the field of rational functions with coefficients from K in the unknown x, and $\frac{d}{dx}$ is the derivative operation;

- for any nonempty connected open subset U of $\mathbf{C}$, the field of meromorphic functions from U to $\mathbf{C}$, with respect to the usual derivative.

Differential fields and rings were introduced by Ritt in the thirties, and differential algebra developed greatly since then. Now classical and excellent references, like [76] or the nimbler [67], expound its foundations. For our purposes in this section, we just need a few algebraic crumbs on differential fields $(\mathcal{K}, D)$.
First of all, notice that the usual derivation rules for powers and quotients, like

$$D(a^n) = na^{n-1}Da$$

for every element a in K and every integer n, still hold and can be easily deduced from the definition.
Moreover, the elements $a \in K$ such that $Da = 0$ form a subfield of $\mathcal{K}$, called the *constant subfield* and denoted $C(\mathcal{K})$.
Instead of the usual (algebraic) polynomials $f(x) \in K[x]$, now *differential polynomials* are considered: they are algebraic polynomials in the unknowns

$$x,\ Dx,\ \ldots,\ D^n x,\ \ldots$$

where n is a natural, and form a differential ring $\mathcal{K}\{x\}$ with respect to the obvious operations.
For every non-zero differential polynomial $f(x) \in K\{x\}$ we can define

- the *degree* of $f(x)$ (with respect to x, or Dx, and so on)

but also

- the *order* of $f(x)$: this is the maximal natural n such that $D^n x$ occurs in $f(x)$, if there is some n with this property, and -1 otherwise, so when $f(x) = a \in K$ (clearly one agrees $D^0 x = x$).

Differential polynomial rings in more variables $\mathcal{K}\{\vec{x}\}$ are introduced in the same way.
What is the role of model theory in this setting? As we shall see in a few lines, it is quite relevant, mainly with respect to existentially closed structures. Actually differential algebra did not provide, before model theory, any notion of differentially closed field -something resembling algebraically closed fields among fields, or real closed fields among ordered fields-. But the interest in existentially closed structures, and hence, particularly, in existentially closed differential fields, led A. Robinson to consider this question from the model theoretic point of view.

Hence take the -elementary- class of differential fields in our language L'. We wonder if the class of existentially closed objects is elementary, in other words if the theory of differential fields has a model companion. At this point it is advisable for us to distinguish in our treatment the characteristic 0 case from the prime characteristic case. We shall devote the next section to the latter; hence, now we limit ourselves to differential fields of characteristic 0. In this restricted framework, A. Robinson pointed out

Theorem 3.6.1 (A. Robinson, 1959) *The theory of differential fields of characteristic 0 has a model companion, and this eliminates the quantifiers in the language L'.*

This model companion was denoted DCF_0; its models were called differentially closed fields (of characteristic 0). Unfortunately, A. Robinson's approach was not able to determine any incisive first order axiomatization for DCF_0. Actually, what was lacking at that time was a suitable background. In fact, in the paradigmatical cases of ACF and RCF, the notions themselves of algebraically closed field and real closed field clearly preceded A. Robinson's treatment, and the existence and uniqueness of an algebraic closure and a real closure played a key role in the model completeness proofs. On the contrary, in the differential case, the concept of differentially closed fields was just rising within the model theoretic approach, and, consequently, nothing resembling a *differential closure* of a differential field (as a *minimal* differentially closed extension) was known, even in characteristic 0.
Accordingly one had to wait for new significant progress in model theory, mainly due to Michael Morley, before overcoming this algebraic gap. We shall refer in detail Morley's ideas, and their effects for differentially closed fields, in Chapter 6. But we wish to anticipate here a short report on the end of the affair (so far). In particular, we want to recall that in 1968 L. Blum, in her PhD thesis, found an elegant and nice axiomatization of DCF_0. What she showed was

Theorem 3.6.2 (L. Blum) *Among differential fields $(\mathcal{K}, D)$ in characteristic 0, the existentially closed objects are exactly those satisfying the following property:*

($\star$) *for every choice of $f(x)$ and $g(x)$ in $\mathcal{K}\{x\} - \{0\}$ such that the order of f is larger than the order of g, there is some $a \in K$ for which $f(a) = 0$ but $g(a) \neq 0$.*

It is easy to express ($\star$) in a first order way by infinitely many sentences in L'. L. Blum's work is described in Sacks's book [146]: she confirmed quantifier elimination, model completeness (and completeness) of DCF_0; but her analysis went farther and, combined with a quite abstract model theoretic result of Shelah, yielded

Theorem 3.6.3 *Any differential field* $(\mathcal{K}, D)$ *of characteristic 0 has a differential closure (a* **minimal** *differentially closed extension), and this is unique up to isomorphism fixing* K *pointwise.*

We shall refer in more detail on this point in Chapter 6. But we want to emphasize that, at last, *differential closures* do exist and are unique (in characteristic 0), and the model theoretic approach of A. Robinson and L. Blum was the very first not only in introducing differentially closed fields, but also in showing these existence and uniqueness results. Of course, some mystery still persists. Most notably, it is surprising, and perhaps regrettable, to learn that, presently, no explicit example of differentially closed field is known.
However, we conclude this section by discussing some minor but useful points.
First of all, notice that every differentially closed field $(\mathcal{K}, D)$ of characteristic 0 is algebraically closed: to see this, just apply Blum's theorem to the particular case when $f(x) \in K[x]$, $g(x) = 1$. Similarly the constant field $C(\mathcal{K})$ is algebraically closed (we will see why in Chapter 6).
Secondly, what can we say about definable sets in a differentially closed field $(\mathcal{K}, D)$ of characteristic 0? Here

($\star$) the basic definable sets are, of course, the sets of roots of finite systems of differential polynomials in $K\{\vec{x}\}$; they are just the closed sets in a Noetherian topology (the *Kolchin topology*) in K^n (where n is the length of $\vec{x}$);

($\star$) the Kolchin constructible sets -i. e. the finite Boolean combinations of Kolchin closed sets- are still definable, of course;

($\star$) owing to the elimination of quantifiers, no further definables occur; in other words, *definable* just means *Kolchin constructible.*

3.7 SCF_p and DCF_p

As promised, we want to deal here with differentially closed fields in prime characteristic p. But our treatment needs a preliminary remark. Indeed we underlined in the previous section that differentially closed fields of characteristic 0 are algebraically closed as well. This cannot hold any more in the prime characteristic case. In fact, take any differential field $(\mathcal{K}, D)$ in characteristic p: notice that $K^p \subseteq C(K)$ because

$$D(a^p) = p\, a^{p-1}\, Da = 0 \quad \forall a \in K.$$

Now any algebraically closed field of characteristic p is *perfect*; in other words, every element can be (uniquely) expressed as a p-th power. It follows that no differential field of characteristic p can be algebraically closed, except the trivial case when the derivation D is identically 0. In particular we cannot expect that the underlying field of a differentially closed field is still algebraically closed. This remark threatens some major complications with respect to the characteristic 0 case; and anyhow suggests a preliminary analysis on some possible weak closure notions for pure fields $\mathcal{K}$ in characteristic p, compatible with $K \neq K^p$.
Accordingly, take a field $\mathcal{K}$ (in any characteristic). A polynomial $f(x) \in K[x]$ of degree ≥ 1 is said to be *separable* if and only if $f(x)$ has no multiple roots (in the algebraic closure of $\mathcal{K}$).
When $\mathcal{K}$ has characteristic 0, every irreducible polynomial $f(x) \in K[x]$ is also separable. Otherwise a multiple root a of $f(x)$ annihilates also the formal derivative $f'(x)$ of $f(x)$ -still a polynomial in $K[x]$-, and hence the greatest common divisor $q(x)$ of $f(x)$ and $f'(x)$ (in $K[x]$). So $q(x)$ is not constant; however the degree of $q(x)$ is strictly smaller than the degree of $f(x)$, because it is less or equal to the degree of $f'(x)$. Hence $f(x)$ is reducible.
Now assume that $\mathcal{K}$ has a prime characteristic p. Recall the Frobenius morphism Fr in $\mathcal{K}$ (and in every extension of $\mathcal{K}$), the one sending any element a into its p-th power a^p. Fr is injective. This time, $K[x]$ may include some irreducible non-separable polynomials: for instance, given $a \in K - K^p$ and a positive integer h,

$$f(x) = x^{p^h} - a$$

is irreducible, but has a unique root in the algebraic closure of $\mathcal{K}$ because Fr is 1-1; notice that $f'(x) = 0$.
We say that $\mathcal{K}$ is *separably closed* if and only if every separable polynomial $f(x)$ in $K[x]$ has a root in $\mathcal{K}$.
Separably closed fields $\mathcal{K}$ form an elementary class in the language L of fields; in fact, they are axiomatized by the infinitely many sentences saying that every polynomial $f(x)$ with a non-zero derivative has a root in K.

However, in characteristic 0, separably closed just means algebraically closed. On the contrary, in a prime characteristic p, there exist some separably closed, non-algebraically closed fields; indeed one shows that a field $\mathcal{K}$ is

algebraically closed

if and only if

$\mathcal{K}$ is separably closed and Fr is onto (that is, $\mathcal{K} = \mathcal{K}^p$ is perfect).

If this is not the case, we are led to consider the field extension $\mathcal{K} \supseteq \mathcal{K}^p$ and hence $\mathcal{K}$ as a vectorspace over $\mathcal{K}^p$. A set $B \subseteq K$ is called a *p-basis* of $\mathcal{K}$ over $\mathcal{K}^p$ if the monomials

$$a_0^{e_0} \ldots a_n^{e_n},$$

(when n, e_0, ..., e_n range over natural numbers, a_0, ..., a_n are pairwise distinct elements in B and $e_i < p$ for every $i \leq n$) form a basis of $\mathcal{K}$ over $\mathcal{K}^p$ (as a vectorspace). One shows that a p-basis B always exists and its cardinality depends only on $\mathcal{K}$: it is called the *imperfection degree* of $\mathcal{K}$. The model theory of separably closed (possibly non algebraically closed) fields $\mathcal{K}$ of prime characteristic p was investigated by Ershov in 1967 [46]. Let SCF_p denote their theory. Of course, SCF_p is not complete, because the imperfection degree of these fields may change, and suitable first order formulas in the language of fields can express its value, when finite, or witness its infinity otherwise. But Ershov observed that this imperfection degree is the only obstruction to completeness; in fact, he showed

Theorem 3.7.1 (Ershov) *Let $\mathcal{K}$ be a separably closed field of prime characteristic p. Then the elementary equivalence class of $\mathcal{K}$ is fully determined by its imperfection degree, if finite, or by the fact that this degree is infinite, otherwise.*

Ershov's proof shows (and uses) model completeness in a suitable enriched language capturing the notion of p-basis. Notice that separably closed fields do have some noteworthy algebraic connections with differential fields. For instance, given any separably closed field $\mathcal{K}$ and a p-basis B of $\mathcal{K}$, one can see that any function δ from B to K enlarges uniquely to a derivation D of $\mathcal{K}$, and so equips $\mathcal{K}$ with a structure of differential field.

This relationship between separably closed fields and differential fields in prime characteristic p gets stronger if we enter the model theoretic framework and we consider differentially closed fields. But, before providing more details, we have to explain what a differentially closed field in characteristic

p is. Indeed, we still have to clarify if, even in this setting, existentially closed objects are an elementary class: so far, we have simply realized that the analysis should be more complicated than in characteristic 0, because no existentially closed differential field can be algebraically closed.
Well, the answer is again positive: the theory of differential fields of characteristic p does admit a model companion. This is the content of a theorem of Carol Wood, who also found a nice axiomatization of existentially closed objects.

Theorem 3.7.2 (C. Wood) *A differential field $(\mathcal{K}, D)$ in characteristic p is existentially closed if and only if*

(i) *$C(K)$ equals K^p,*

(ii) *for every choice of two differential polynomials $f(x)$ and $g(x)$ in $K\{x\}-\{0\}$ such that the order of $f(x)$ is greater than the order of $g(x)$ and the formal derivative of $f(x)$ with respect to $D^n x$ is not 0, there is some $a \in K$ such that $f(a) = 0$ and $g(a) \neq 0$.*

The model companion of the theory of differential fields in characteristic p is usually denoted DCF_p; its models -hence the existentially closed differential fields- are again called *differentially closed fields.* Notably, they are separably closed: this is implicitly said in (ii), provided that we restrict to polynomials $f(x)$ of order 0 and we take $g(x) = 1$. Moreover their imperfection degree is infinite, because the dimension itself of $\mathcal{K}$ over $C(K) = K^p$ is infinite: to see this, just use (ii) again and take, for every positive integer m, an element x_m in K satisfying $D^m x_m = 0$ and $D^{m-1} x_m \neq 0$; notice that the x_m's are linearly independent over $C(K)$ (this is an easy exercise).
DCF_p is model complete and complete, but does not eliminate the quantifiers in the language L': quantifier elimination needs a larger setting, with a further operation extracting p-th roots when possible, and valuing 0 otherwise.
Again, general results of pure model theory ensure the existence and uniqueness of a differential closure for differential fields $(\mathcal{K}, D)$ satisfying $C(K) = K^p$.

3.8 ACFA

In this section we deal with difference fields. These are structures $(\mathcal{K}, \sigma)$ where $\mathcal{K}$ is a field and σ is a distinguished (surjective) automorphism. More generally, difference rings can be introduced in the same way. Difference fields include some trivial examples like

$(\mathcal{K}, id)$ for every field $\mathcal{K}$

but also

$(\mathcal{K}, Fr)$

where $\mathcal{K}$ is a perfect field of prime characteristic p and Fr is the Frobenius morphism in $\mathcal{K}$: for every $a \in K$, $Fr(a) = a^p$. Notice that these examples include any finite field $\mathcal{K}$.
At a superficial sight, this setting resembles differential fields. In both cases the underlying field is enriched by a 1-ary operation, and the only dissonance concerns the rules that this new function has to satisfy: the derivation laws in the differential case, and the morphism laws presently. Of course, this connection is shallow, and sharp distinctions arise in examining these structures in a deeper way. Nevertheless some (minor) similarities persist. For instance, given a difference field $(\mathcal{K}, \sigma)$, one can look at the fixed subfield

$$Fix(\sigma) = \{a \in K : \sigma(a) = a\}.$$

This resembles in some way the constant subfield of a differential field.
Similarly, instead of algebraic polynomials $f(x) \in K[x]$, one can form difference polynomials in x (or, possibly, in more unknowns): they are algebraic polynomials in

$$x, \sigma(x), \sigma^2(x), \ldots, \sigma^n(x), \ldots$$

(with n natural). So formally this is the same set as in the differential case; but, of course, the new rules relating σ and the operations of addition and multiplication dictate a different ring structure. This ring is usually denoted $\mathcal{K}\langle x\rangle$ and gets a difference ring structure extending $(\mathcal{K}, \sigma)$ in the obvious way.
Difference algebra was began by Ritt in the thirties; now it is largely developed and includes some fundamental references, such as [28] (warning: the terminology used in [28] is not the same as here; in fact in that book a difference field is a field with a distinguished monomorphism σ; when σ is also surjective, the difference field is called inversive). What can we say within the model theoretic approach? First of all, notice that the language for difference fields enlarges $\{+, \cdot, -, 0, 1\}$ by a new 1-ary operation symbol. So the resulting setting again reminds differential fields, although now we prefer to denote the new symbol by σ. Difference fields are easily axiomatized in a first order way in this language.

But our main question is how to characterize existentially closed difference fields. Is their class axiomatizable? In other words, does the theory of difference fields admit a model companion?
We should emphasize that this interest in existentially closed difference fields is comparatively recent. Indeed, at the beginning of the nineties, Macintyre, Van den Dries and Wook showed their elementarity and found a nice (although non trivial) axiomatization: this is reported in [95]. In detail

Theorem 3.8.1 *A difference field* $(\mathcal{K}, \sigma)$ *is existentially closed if and only if*

(i) $\mathcal{K}$ *is algebraically closed;*

(ii) *for every irreducible affine variety* U *over* K *and for every subvariety* V *of* $U \times \sigma(U)$ *such that both the projections of* V *into* U *and* $\sigma(U)$ *are dense with respect to the Zariski topology, there is a point* a *of* U *(over* K*) for which* $(a, \sigma(a))$ *is in* V.

Expressing (ii) in a first order way is not immediate, because it requires, after all, to quantify with respect to irreducible varieties. However this can be done, owing to general boundedness results for (algebraic) polynomials over fields [172].
The theory of existentially closed difference fields is denoted $ACFA$; their models are called *algebraically closed fields with an automorphism*, although this name is a little misleading, and we have to be careful about it. So recall that they are not simply algebraically closed fields enriched by any arbitrary automorphism, but just existentially closed structures among fields with an automorphism: (ii) has to be respected. Of course $ACFA$ is model complete, as a model companion. But $ACFA$ is not complete. However one shows that the key features fully determining the elementary equivalence class of a model $(\mathcal{K}, \sigma)$ of $ACFA$ are

($\star$) the characteristic of the underlying field $\mathcal{K}$,

and

($\star$) the action of σ on the algebraic closure of the prime subfield (up to isomorphism).

The interest in $ACFA$ is also related to its role towards a model theoretic proof of classical questions in Algebraic Geometry, like the Manin-Mumford Conjecture. We will discuss this point in Chapter 8. However these intriguing connections led to a systematic study of $ACFA$, mainly pursued by

Chatzidakis and Hrushovski in [20], and later by Chatzidakis, Hrushovski and Peterzil in [21]. Here we limit ourselves to some very basic information. First of all, it turns out that, for any algebraically closed field with an automorphism $(\mathcal{K}, \sigma)$, the fixed subfield $Fix(\sigma)$ is pseudofinite, so it satisfies the axioms of finite fields, but (consequently) it is not algebraically closed. What about definable sets?

- First one meets the zero sets of finite systems of difference polynomials in $\mathcal{K}\langle\vec{x}\rangle$; they are, again, the closed sets in a Noetherian topology for K^n (where n is the length of $\vec{x}$). $Fix(\sigma)$ is an example, because it is defined by $\sigma(v) = v$.

- Secondly, one has to include the constructible sets in this topology, i. e. the finite Boolean combinations of closed sets.

- However, this is not enough, because quantifier elimination fails for $ACFA$ in its original language. So *definable* does not imply *constructible*. Here is a possible counterexample. We work inside an "algebraically closed field with an automorphism" $(\mathcal{K}, \sigma)$ of characteristic 0 and suitably large cardinality. Due to this assumption and existential closedness, we can find inside K two elements a and b transcendental over the prime subfield $\mathbf{Q}$, a and b in $Fix(\sigma)$, and two square roots a' and b' of a, b, respectively, such that

  $$\sigma(a') = a', \qquad \sigma(b') = -b'.$$

  a and b satisfy the same quantifier free formulas, because they are fixed by σ and $f(a), f(b) \neq 0$ for every nonzero polynomial $f(x)$ over the rationals. Nevertheless

  $$\sigma(v) = v \wedge \exists w(w^2 = v \wedge \neg(\sigma(w) = w))$$

  holds for b and not for a, and hence defines a non-constructible set. Of course, owing to model completeness, the projections of constructible sets exhaust the definable ones.

  Of course, owing to model completeness, the projections of constructible sets exhaust the definable ones.

A final important remark. We saw that no example of differentially closed field (even in characteristic 0) is known so far. This is not the case for $ACFA$. In fact, Hrushovski and, independently, Macintyre built some models $(\mathcal{K}, \sigma)$ of $ACFA$ explicitly; they are obtained by considering suitable pseudofinite fields with a rather natural extension of the Frobenius morphism.

3.9 References

An excellent introduction to model completeness is in Macintyre expository paper [94]. But also the A. Robinson books [141] and [143] are still an enjoyable reading. Hilbert Nullstellensatz and Hilbert Seventeenth Problem are described in [94]; p-adic numbers are also the matter of a (particular) chapter in [36]; the Ax-Kochen-Ersov theorems are in [4, 5, 6] and [44, 45], while Macintyre's quantifier elimination results on p-adics are in [93].
A very good reference for existentially closed structures (and model completeness, too) is Cherlin's book [23]. See also the Eklof-Sabbagh paper [41]. Macintyre's analysis of existentially closed groups is in [91], and Cherlin's treatment of e.c. commutative rings in [22]: in the latter setting, the role of nilpotent elements is also discussed by Carson [17], Lipshitz-Saracino [86], and Macintyre again [92], who observed that the theory of commutative rings without nonzero nilpotent elements do have a model companion. See also [164]. Differentially closed fields are introduced in [142], but their model theory (in characteristic 0) is principally developed in L. Blum's Ph.D. Thesisi [12], also related in the Sacks book [146]. By the way, both [76] and [67] provide complete treatments of the basic differential algebra.
The Ershov model theoretic analysis of separably closed fields is in [46], while C. Wood's theorem on differentially closed fields of prime characteristic is in [179] and, with respect to prime models, in [180]; the Marker and Messmer contributions to [110] provide a complete and excellent exposition of these matters.
The Macintyre - Van den Dries - Wood axiomatization of existentially closed fields with an automorphism is related in [95]; their large model theoretic analysis due to Chatzidakis, Hrushovski and (later) Peterzil is in [20] and in [21]. The Hrushovski and Macintyre explicit example of an e.c. field with an automorphism can be found in [88] and [58].

Chapter 4

Elimination of imaginaries

4.1 Interpretability

Let $\mathcal{A}$ be a structure for a language L. We already dealt with definability in $\mathcal{A}$ when we introduced definable sets and, more generally, definable structures in $\mathcal{A}$ (see example 1.7.3, 4). The latter are the structures $\mathcal{A}'$ for a language L' (possibly different from L) such that both the universe A' of $\mathcal{A}'$ and the interpretations of symbols of L' in $\mathcal{A}'$ are definable in $\mathcal{A}$. As in the case of sets, we can introduce also the concept of *X-definable structure* for $X \subseteq A$. In the quoted example, we observed that $(\mathbf{N}, +, \cdot)$ is a structure definable in $(\mathbf{Z}, +, \cdot)$. Here we provide some further examples.

Examples 4.1.1 1. Let $L = \{\times, e, ^{-1}\}$ be the language for groups, $\mathcal{G}$ be a group. The centre $Z(\mathcal{G})$ of $\mathcal{G}$ is the set of elements of G commuting with any element of G, and so it is $\emptyset$-definable as a set, by the formula

$$\forall w(v \times w = w \times v).$$

But $Z(\mathcal{G})$ is also a subgroup of $\mathcal{G}$, and hence a group with respect to the restrictions to $Z(\mathcal{G})$ of the operations of $\mathcal{G}$. Clearly these operations are $\emptyset$-definable in $\mathcal{G}$. It follows that $Z(\mathcal{G})$ is $\emptyset$-definable in $\mathcal{G}$ also as a group (and hence as an L-structure).

2. Let $L = \{0, 1, +, \cdot, -\}$ be the language for fields, $\mathcal{K}$ be a field. Let us consider, in the language $L' = \{\times, e, ^{-1}\}$ for groups, the linear group of degree n (n a positive integer) over $\mathcal{K}$

$$GL(n, \mathcal{K}),$$

in other words the group of $n \times n$ matrices with entries in K and determinant $\neq 0$, with the usual row-by-column product. Then $GL(n, \mathcal{K})$ is a structure $\emptyset$-definable in $\mathcal{K}$.

In fact, the $n \times n$ matrices with entries in K can be viewed as tuples $\vec{a} = (a_{ij})_{i,j}$ in K^{n^2}. So the elements of $GL(n, \mathcal{K})$ are exactly those tuples in K^{n^2} satisfying

$$\neg(det\, \vec{v} = 0)$$

(here $\vec{v}$ abridges $(v_{ij})_{i,j}$): recall that det is a homogeneus polynomial of degree n with coefficients in the substructure of $\mathcal{K}$ generated by $0, 1$. The multiplication is defined by

$$\vec{v} \times \vec{w} = \vec{z} \; : \; \bigwedge_{1 \leq i,j \leq n} (z_{ij} = \sum_{h=1}^{n} v_{ih}\, w_{hj}),$$

the identical matrix I_n by

$$\vec{v} = I_n \; : \; \bigwedge_{1 \leq i \leq n} v_{ii} = 1 \wedge \bigwedge_{1 \leq i,j \leq n, i \neq j} v_{ij} = 0,$$

and, finally, the inverse operation by

$$\vec{v} = \vec{w}^{-1} \; : \; \text{“}\vec{v} \times \vec{w} = I_n\text{”} \,.$$

But Algebra deals not only with substructures, but also with quotient structures. For instance, in the examples quoted before, one can observe what follows.

1. Look at the quotient group $\mathcal{G}/Z(\mathcal{G})$. As $Z(\mathcal{G})$ is $\emptyset$-definable in $\mathcal{G}$, the equivalence relation in G

$$a \sim b \Leftrightarrow a \times b^{-1} \in Z(\mathcal{G})$$

whose equivalence classes are just the elements of $\mathcal{G}/Z(\mathcal{G})$ is also $\emptyset$-definable. It follows that $\mathcal{G}/Z(\mathcal{G})$, as a quotient set, but also as a quotient group, "lives" in $\mathcal{G}$.

2. The special linear group of degree n over $\mathcal{K}$

$$SL(n, \mathcal{K})$$

(a subgroup of $GL(n, \mathcal{K})$) is $\emptyset$-definable in $\mathcal{K}$: just consider the formula

$$det(\vec{v}) = 1.$$

Accordingly, just as in the previous example, the quotient group $GL(n, \mathcal{K})/SL(n, \mathcal{K})$ can be recovered in $\mathcal{K}$ by a suitable $\emptyset$-definable equivalence relation. On the other side $GL(n, \mathcal{K})/SL(n, \mathcal{K})$ is isomorphic to the multiplicative group $\mathcal{K}^*$ of nonzero elements of K, and $\mathcal{K}^*$ itself is directly $\emptyset$-definable in $\mathcal{K}$ (as a group, without involving any quotient construction).

These remarks introduce the following

Definition 4.1.2 *Let $\mathcal{A}$ be a structure for L. A structure $\mathcal{A}'$ for L' is said to be* **interpretable** *in $\mathcal{A}$ if and only if there exist a positive integer n, a subset $S \subseteq A^n$ definable in $\mathcal{A}$, an equivalence relation E over A^n definable in $\mathcal{A}$ such that $A' = S/E$ and*

(i) *for every costant symbol c of L', $\{\vec{a} \in A^n : \vec{a}|E = c^{\mathcal{A}'}\}$ is definable in $\mathcal{A}$;*

(ii) *for every k-ary function symbol f of L', $\{(\vec{a_1}, \ldots, \vec{a_n}, \vec{a}) \in A^{n(k+1)} : f^{\mathcal{A}'}(\vec{a_1}|E, \ldots, \vec{a_k}|E) = \vec{a}|E\}$ is definable in $\mathcal{A}$;*

(iii) *for every k-ary relation symbol R of L', $\{(\vec{a_1}, \ldots, \vec{a_k}) \in A^{nk} : (\vec{a_1}|E, \ldots, \vec{a_k}|E) \in R^{\mathcal{A}'}\}$ is definable in $\mathcal{A}$.*

The concept of X-interpretable structure (for $X \subseteq A$) is defined in the usual way. Every structure definable in $\mathcal{A}$ is interpretable in $\mathcal{A}$ (through the equality relation).

Example 4.1.3 $GL(n, \mathcal{K})/SL(n, \mathcal{K})$ is $\emptyset$-interpretable in $\mathcal{K}$.

4.2 Imaginary elements

The examples of the previous section show that, in a given L-structure $\mathcal{A}$, one meets not only

real elements

(those of the domain A), but even

imaginary elements,

i.e. equivalence classes of $\emptyset$-definable equivalence relations E.

The following technique, essentially due to Shelah, shows how to expand in a natural way the structure $\mathcal{A}$ (and its language L) in order to make the imaginary elements real.

Definition 4.2.1 *L^{eq} is the language obtained by L taking, for every equivalence relation E over A^n (for some positive integer n) $\emptyset$-definable in $\mathcal{A}$*

- *a 1-ary relation symbol A_E (and among these also $A_=$);*
- *an n-ary function symbol π_E.*

Definition 4.2.2 *$\mathcal{A}^{eq}$ is the structure for L^{eq} such that*

- *the universe of $\mathcal{A}^{eq}$ is the disjoint union of the interpretations in $\mathcal{A}^{eq}$ of the relation symbols A_E;*
- *for every equivalence relation E, A_E is interpreted as the quotient set A^n/E (in particular $A_=$ as $\{\{a\} : a \in A\}$, which can be canonically identified with A) and π_E as the natural projection from A^n onto A^n/E;*
- *the symbols from L have the same interpretation as in $\mathcal{A}$ (after identifying A and the interpretation of $A_=$).*

One can check that, if $\mathcal{A}$ and $\mathcal{B}$ are structures for L and $\mathcal{A} \equiv \mathcal{B}$, then $\mathcal{A}^{eq} \equiv \mathcal{B}^{eq}$. This allows to define, $T^{eq} = \mathrm{Th}(\mathcal{A}^{eq})$ when $T = \mathrm{Th}(\mathcal{A})$. Many significant properties of T are preserved under passing from T to T^{eq}. On the other hand, in $T^{eq} = \mathrm{Th}(\mathcal{A}^{eq})$ the imaginary elements of $\mathcal{A}$ get real. For example, take a positive integer n and the relation $=$ in A^n; $=$ is $\emptyset$-definable, and hence, for every $\vec{a} = (a_1, \ldots, a_n) \in A^n$, the class $\{\vec{a}\}$ of $\vec{a}$ modulo $=$, is a real element of $\mathcal{A}^{eq}$; so we can view any n-tuple in A^n as a real element. Indeed it is a common agreement in Model Theory to consider the tuples from a structure $\mathcal{A}$ as elements of A: they are imaginary elements in $\mathcal{A}$, and hence real elements in $\mathcal{A}^{eq}$.

Sometimes imaginary elements can be naturally identified with real elements of $\mathcal{A}$ (or with finite sequences of real elements of $\mathcal{A}$), and hence referring to $\mathcal{A}^{eq}$ is no more necessary. Let us propose a simple example.

Example 4.2.3 Let $\mathcal{K}$ be a field. We already pointed out that $GL(n, \mathcal{K})/SL(n, \mathcal{K})$ is isomorphic to the multiplicative group $\mathcal{K}^*$ ($\emptyset$-definable in $\mathcal{K}$). It follows that the imaginary elements $SL(n, \mathcal{K})\vec{a}$ with $\vec{a} \in GL(n, \mathcal{K})$ can be identified with the real elements $det(\vec{a})$ of K^*, by the determinant function (more precisely by the isomorphism from $GL(n, \mathcal{K})/SL(n, \mathcal{K})$ to K^* induced by det).

In general:

Definition 4.2.4 *A structure $\mathcal{A}$ for L* **has the elimination of imaginaries** *if and only if, for every equivalence relation E over A^n (n a positive integer) $\emptyset$-definable in $\mathcal{A}$ and $\vec{a} \in A^n$, there are a formula $\varphi(\vec{v}, \vec{w})$ of L and a unique sequence $\vec{b}$ in A such that*

$$\vec{a}|E = \varphi(\mathcal{A}^n, \vec{b}).$$

A structure $\mathcal{A}$ for L **has a uniform elimination of imaginaries** *if and only if, for every E as above, there exist a formula $\varphi(\vec{v}, \vec{w})$ of L and, for each $\vec{a} \in A^n$, a unique sequence $\vec{b}$ in A such that*

$$\vec{a}|E = \varphi(\mathcal{A}^n, \vec{b}).$$

Theorem 4.2.5 *Let $\mathcal{A}$ be a structure for L. Then the following propositions are equivalent.*

(i) *$\mathcal{A}$ has a uniform elimination of imaginaries;*

(ii) *for every positive integer n and equivalence relation E over A^n $\emptyset$-definable in $\mathcal{A}$, there is a map F_E $\emptyset$-definable in $\mathcal{A}$, with domain A^n and range $\subseteq A^m$ (for some positive integer m), such that, $\forall \vec{a}, \vec{a}' \in A^n$,*

$$\vec{a}, \vec{a}' \text{ are equivalent in } E \quad \Leftrightarrow \quad F_E(\vec{a}) = F_E(\vec{a}').$$

Hence, if $\mathcal{A}$ has a uniform elimination of imaginaries, then, for every equivalence relation E as above, the equivalence classes of E can be thought as tuples of real elements of A, provided we identify, $\forall \vec{a}$ in A^n,

$$\vec{a}|E \qquad \text{with} \qquad F_E(\vec{a}).$$

It follows that referring to $\mathcal{A}^{eq}$ is not necessary, because what is $\emptyset$-interpretable in $\mathcal{A}$ is even $\emptyset$-definable in $\mathcal{A}$.

Proof. (i)$\Rightarrow$(ii) For each $\vec{a} \in A$, define $F_E(\vec{a})$ as the unique element $\vec{b}$ such that $\vec{a}|E = \varphi(\mathcal{A}^n, \vec{b})$.
(ii)$\Rightarrow$(i) Take $\varphi(\vec{v}, \vec{z})$: "$F_E(\vec{v}) = \vec{z}$" e $\vec{b} = F_E(\vec{a})$. ♣

The theory of elimination of imaginaries was essentially developed by Poizat. Actually Poizat's treatment is slightly different from ours. They do coincide for structures admitting at least two constants, or even two distinct definable elements (see [131], Theorem 16.16). As we are mainly interested in fields, and fields clearly satisfy this condition, we can proceed with no anguishes.

4.3 Algebraically closed fields

Notably, many familiar structures admit a uniform elimination of imaginaries. Let us propose some examples. The first case we deal with concerns algebraically closed fields: the aim of this section is just to prove

Theorem 4.3.1 *Every algebraically closed field $\mathcal{K}$ has a uniform elimination of imaginaries.*

The proof is a direct consequence of the following two lemmas.

Lemma 4.3.2 *If $\mathcal{K}$ is an algebraically closed field, then $\mathcal{K}$ has the elimination of imaginaries.*

Lemma 4.3.3 *Let L be a language with (at least) two constant symbols, $\mathcal{K}$ be an L-structure interpreting these two constant symbols in different elements. If $\mathcal{K}$ has the elimination of imaginaries, then $\mathcal{K}$ has a uniform elimination of imaginaries.*

Obviously an algebraically closed field $\mathcal{K}$ satisfies also the hypotheses of lemma 4.3.3: the language for fields has two constant symbols $0, 1$ interpreted in $\mathcal{K}$ as two different elements. Hence $\mathcal{K}$ uniformly eliminates imaginaries (provided it eliminates them).

Now we show Lemma 4.3.2. The proof we provide here uses the minimality of any algebraically closed field $\mathcal{K}$, hence the fact that every definable subset of K is finite or cofinite. An alternative approach (working even for differentially closed fields) will be produced in Chapter 6.

Proof. Let n be a positive integer, E be an equivalence relation $\emptyset$-definable in K^n. Let $E(\vec{v}, \vec{w})$ indicate the formula defining E. Fix $\vec{a} \in K^n$. Consider the formula

$$E_1(v_1, \vec{w}) : \exists v_2 \ldots \exists v_n E(\vec{v}, \vec{w}).$$

$E_1(\mathcal{K}, \vec{a})$ is a definable set, and hence by Corollary 2.4.8 is either finite or cofinite. If $E_1(\mathcal{K}, \vec{a})$ is finite, then it contains only elements algebraic over (the subfield generated by) $\vec{a}$; in fact two elements transcendental over (the subfield generated by) $\vec{a}$ are linked by some automorphism of $\mathcal{K}$ fixing every element of $\vec{a}$; hence if one of them occurs in $E_1(\mathcal{K}, \vec{a})$, then the latter is in $E_1(\mathcal{K}, \vec{a})$ as well, and hence all the elements of $\mathcal{K}$ transcendental over $\vec{a}$ are in $E_1(\mathcal{K}, \vec{a})$, and $E_1(\mathcal{K}, \vec{a})$ is infinite. On the other side, if $E_1(\mathcal{K}, \vec{a})$ is cofinite, then $E_1(\mathcal{K}, \vec{a})$ contains at least an algebraic element, because $\mathcal{K}$ has finitely many elements algebraic over the subfield generated by $\vec{a}$.

In any case we can choose $c_1 = c_1(\vec{a}) \in E_1(\mathcal{K}, \vec{a})$ algebraic over $\vec{a}$. Now let

$$E_2(v_1, v_2, \vec{w}) : \exists v_3 \ldots \exists v_n E(\vec{v}, \vec{w}).$$

$E_2(c_1, \mathcal{K}, \vec{a})$ is either finite or cofinite, and as above we can pick out in $E_2(c_1, \mathcal{K}, \vec{a})$ an element algebraic over $(\vec{a}, c_1)$, and hence over $\vec{a}$. Repeating the procedure we build $\vec{c} = (c_1, \ldots, c_n) \in E(K^n, \vec{a})$ such that $c_1, \ldots, c_n$ are algebraic over $\vec{a}$. We can even form a finite set $X(\vec{a}) \subseteq K^n$ $\vec{a}$-definable such that $\vec{c} \in X(\vec{a})$; it suffices to consider the set $X(\vec{a})$ defined by

$$\bigwedge_{1 \leq i \leq n} \text{“} f_i(v) = 0 \text{”}$$

where, for each $i = 1, \ldots, n$, $f_i(x)$ is the minimum polynomial of c_i over the subfield generated by $\vec{a}$. Without loss of generality, $X(\vec{a}) \subseteq E(K^n, \vec{a})$ (otherwise substitute $X(\vec{a})$ for $X(\vec{a}) \cap E(K^n, \vec{a})$).
Suppose $X(\vec{a}) = \{\vec{c}^{(0)}, \ldots, \vec{c}^{(m)}\}$ where $\vec{c} = \vec{c}^{(0)}$ and $\vec{c}^{(j)} = ({c_1}^{(j)}, \ldots, {c_n}^{(j)})$ for every $j \leq m$. Then we have

$$E(K^n, \vec{a}) = E(K^n, \vec{c}^{(j)})$$

for every $j \leq m$. Recall that we are looking for a formula $\varphi(\vec{v}, \vec{z})$ and a unique sequence $\vec{b}$ such that

$$E(K^n, \vec{a}) = \varphi(K^n, \vec{b}).$$

Then consider the following polynomial $f(y, \vec{x}) \in \mathcal{K}[y, \vec{x}]$ (with $\vec{x} = (x_1, \ldots, x_n)$):

$$f(y, \vec{x}) = \prod_{j \leq m} (y - \sum_{i=1}^{n} {c_i}^{(j)} x_i).$$

Let $\vec{b}$ be the sequence of coefficients of $f(y, \vec{x})$ (with respect to some pre-established ordering). It is clear that $\vec{b}$ is uniquely defined by $X(\vec{a}) = \{\vec{c}^{(0)}, \ldots, \vec{c}^{(m)}\}$. Conversely, as $\mathcal{K}[y, \vec{x}]$ is a unique factorization domain, given the sequence $\vec{b}$ of the coefficients of $f(y, \vec{x})$, in other words given $f(y, \vec{x})$, we know that $f(y, \vec{x})$ decomposes in at most one way as

$$\prod_{j \leq m} (y - \sum_{i=1}^{n} {c_i}^{(j)} x_i)$$

(up to permuting factors). So $\vec{b}$ lets recover $X(\vec{a}) = \{\vec{c}^{(0)}, \ldots, \vec{c}^{(m)}\}$ (although it may not provide the single sequences $\vec{c}^{(0)}, \ldots, \vec{c}^{(m)}$). Then we can pick out $\vec{b}$ as the sequence of coefficients of $f(y,\vec{x})$, and define $\varphi(\vec{v},\vec{z})$ in such a way that

$$\varphi(\vec{v},\vec{b})$$

is the formula

$$\forall \vec{w}\,(\text{``}(y-\sum_{i=1}^{n} w_i\,x_i)\text{ divides } f(y,\vec{x})\text{''} \;\rightarrow\; E(\vec{v},\vec{w})).$$

Obviously $\varphi(\vec{v},\vec{z})$ depends on $\vec{a}$. ♣

Let us prove now Lemma 4.3.3.

Proof. Let 0, 1 be two different constant symbols in L such that $\mathcal{K}\models\neg(0=1)$. We know that, if E is an equivalence relation over K^n $\emptyset$-definable in $\mathcal{K}$, then for every $\vec{a}\in K^n$ there exist a formula $\varphi_{\vec{a}}(\vec{v},\vec{z})$ and a unique sequence $\vec{b}=\vec{b}(\vec{a})$ in $\mathcal{K}$ such that

$$E(K^n,\vec{a})=\varphi_{\vec{a}}(K^n,\vec{b}).$$

By compactness there exist a natural number h and $\vec{a}_0,\ldots,\vec{a}_h\in K^n$ such that, for each $\vec{a}\in K^n$,

$$\mathcal{K}\models\bigvee_{i\le h}\exists!\,\vec{z}_i\,\forall\vec{v}(E(\vec{v},\vec{a})\;\leftrightarrow\;\varphi_{\vec{a}_i}(\vec{v},\vec{z}_i)).$$

Without loss of generality, we can assume that $\vec{z}_0,\ldots,\vec{z}_h$ have the same length m (otherwise add some 0's to the shorter sequences). Consider the formula

$$\varphi'(\vec{v},\vec{u},\vec{z})$$

(where $\vec{u}=(u_0,\ldots,u_h)$ and $\vec{z}$ has length m) conjuncting:

- the formula saying that, $\forall i\le h$, $u_i\in\{0,1\}$ and there exists a unique $i\le h$ such that $u_i=1$;
- the formula saying that, $\forall i\le h$, $u_i=1$ if and only if $\varphi_{\vec{a}_i}(\vec{v},\vec{z}))$, $\vec{z}$ is the unique sequence $\vec{t}$ such that $\varphi_{\vec{a}_i}(K^n,\vec{z})=\varphi_{\vec{a}_i}(K^n,\vec{t})$ and, $\forall j\le h$ with $j<i$, it is not true that there is a unique sequence $\vec{t}$ such that $\varphi_{\vec{a}_i}(K^n,\vec{z})=\varphi_{\vec{a}_j}(K^n,\vec{t})$.

It follows that for every $\vec{a}\in K^n$, there exists a unique $\vec{b}$ in K^{m+h+1} such that

$$\mathcal{K}\models\forall\vec{v}\,(E(\vec{v},\vec{a})\;\leftrightarrow\;\varphi'(\vec{v},\vec{b})).$$

The formula $\varphi'(\vec{v},\vec{u},\vec{z})$ depends only on E. Hence $\mathcal{K}$ has a uniform elimination of imaginaries. ♣

4.4 Real closed fields

The aim of this section is to show that real closed fields uniformly eliminate imaginaries.

Theorem 4.4.1 *In every real closed field $\mathcal{K}$ imaginary elements can be uniformly eliminated.*

Proof. First of all, let us emphasize a relevant property of definable sets in real closed fields, essentially related to their o-minimality. Take any formula $\theta(v, \vec{w})$ in the language $L = \{+, \cdot, -, 0, 1, \leq\}$ of ordered fields. We know that $\vartheta(v, \vec{w})$ is (equivalent in RCF to) a finite boolean combination of formulas

$$(*) \qquad f_n(\vec{w})v^n + \ldots + f_1(\vec{w})v + f_0(\vec{w}) \geq 0$$

where n is a natural number and $f_0(\vec{x}), \ldots, f_n(\vec{x})$ are polynomials with integral coefficients. Let $q(y, \vec{x}) = \sum_{i \leq n} f_i(\vec{x})y^i$, $q(y, \vec{x})$ - as a polynomial in y - has at most n roots denoted by $s_0(\vec{x}), \ldots, s_{r-1}(\vec{x})$ (where clearly r depends on $\vec{x}$, but $r \leq n$). Then $(*)$ is equivalent to the formula saying that either w is equal to some $s_0(\vec{v}), \ldots, s_{r-1}(\vec{v})$ or it is in a suitable union of intervals among

$$]-\infty, s_0(\vec{v})[,\ ]s_0(\vec{v}), s_1(\vec{v})[, \ldots,\ ]s_{r-1}(\vec{v}), +\infty[.$$

It follows that in a real closed field $\mathcal{K}$, for every $\vec{a}$ in K, $\vartheta(\mathcal{K}, \vec{a})$ is a finite union of intervals whose endpoints annihilate some polynomials $q(y, \vec{a})$; hence these intervals are $\vec{a}$-definable by a formula only depending on $\vartheta(v, \vec{w})$; even the number of these intervals is uniformly bounded with respect to $\vartheta(v, \vec{w})$. Let $\vec{b}$ denote the sequence of these endpoints: $\vec{b}$ depends on $\vec{a}$, as said before, indeed every element in $\vec{b}$ is $\vec{a}$-definable. At this point we may wonder how to determine, for a given $\vec{a}$, the zeros $s_0(\vec{a}), \ldots, s_{r-1}(\vec{a})$ (with $r = r(\vec{a}) \leq n$) belonging to $\theta(\mathcal{K}, \vec{a})$ and, above all, the intervals among

$$]-\infty, s_0(\vec{a})[,\ ]s_0(\vec{a}), s_1(\vec{a})[, \ldots,\ ]s_{r-1}(\vec{a}), +\infty[$$

contained in $\theta(\mathcal{K}, \vec{a})$. Of course, this depends on the sign ($+$, $-$ or 0) of the polynomials $q(y, \vec{a})$ in each point of $\vec{b}$, and inside the intervals. This can be checked directly for the roots, and choosing in each interval a $\vec{b}$-definable (hence $\vec{a}$-definable) witness c to test. The latter operation can be easily done for every possible choice of the endpoints $-\infty \leq d < e \leq +\infty$. For, take

$$\begin{array}{lll} c = \frac{d+e}{2} & \text{if} & -\infty < d < e < +\infty, \\ c = d+1 & \text{if} & -\infty < d < e = +\infty, \\ c = e-1 & \text{if} & -\infty = d < e < +\infty, \\ c = 0 & \text{if} & -\infty = d < e = +\infty. \end{array}$$

and notice that c is $\vec{a}$-definable when d and e are $\vec{a}$-definable (or infinite). In particular, observe that we can pick a unique $\vec{a}$-definable c in $\theta(\mathcal{K}, \vec{a})$: just choose the lowest root, or the witness in the leftmost interval. Hence, what we have seen so far is that, given a formula $\theta(v, \vec{w})$, one can build a formula $\sigma(v, \vec{z})$ and, for every $\mathcal{K} \models RCF$ and every sequence $\vec{a}$ in K, a unique sequence $\vec{b}$ in K for which $\sigma(\mathcal{K}, \vec{b}) = \theta(\mathcal{K}, \vec{a})$: $\vec{b}$ is the ordered tuple of the roots of the polynomials $q(y, \vec{a})$ (and so is $\vec{a}$-definable), $\sigma(v, \vec{b})$ specifies which roots and which intervals with endpoints among $\vec{b}$ and $\pm\infty$ form $\theta(\mathcal{K}, \vec{a})$. A patient reader may write down $\sigma(v, \vec{z})$ in detail as an exercise.

Now let us come back to elimination of imaginaries. Let E be an equivalence relation on K^n, and assume E $\emptyset$-definable in $\mathcal{K}$. Let $E(\vec{v}, \vec{w})$ the formula defining it. We are looking for a formula $\varphi(\vec{v}, \vec{z})$ and, for every $\vec{a} \in K^n$, a sequence $\vec{b}$ such that $E(K^n, \vec{a}) = \varphi(K^n, \vec{b})$, and $\vec{b}$ is the unique sequence with this property.
When $n = 1$, our claim is a direct consequence of our preliminary work: just let $\theta(v, w)$ be the defining formula $E(v, w)$. Now let $n > 1$. Put

$$\theta_1(v, \vec{w}) \quad : \quad \exists v_1 \ldots \exists v_n \, E(\vec{v}, \vec{w}).$$

Again using our preliminary work, and applying it to $\theta_1(v_1, \vec{w})$, we deduce that there are a formula $\sigma_1(v_1, \vec{z_1})$ and, for each $\vec{a} \in K^n$, a unique sequence $\vec{b_1}$ in K such that

$$\sigma_1(\mathcal{K}, \vec{b_1}) = \theta_1(\mathcal{K}, \vec{a}).$$

Moreover there is a formula $\chi_1(v_1, \vec{z_1})$ such that $\chi_1(v_1, \vec{b_1})$ picks a unique element c_1 in $\sigma_1(\mathcal{K}, \vec{b_1}) = \theta_1(\mathcal{K}, \vec{a})$. Now consider

$$\theta_2(v_1, v_2, \vec{w}) : \exists v_3 \ldots \exists v_n \; E(\vec{v}, \vec{w}).$$

As before we obtain the existence of a formula $\sigma_2(v_2, \vec{z_2})$ and, for every $\vec{a} \in K^n$, of a unique sequence $\vec{b_2}$ in K such that

$$\sigma_2(\mathcal{K}, \vec{b_2}) = \theta_2(c_1, \mathcal{K}, \vec{a});$$

as above, there is a formula $\chi_2(v_2, \vec{b_2})$ selecting a unique element c_2 in $\sigma_2(\mathcal{K}, \vec{b_2}) = \theta_2(c_1, \mathcal{K}, \vec{a})$.
Continuing the procedure one defines a formula $\chi(\vec{v}, \vec{z})$ and, for each $\vec{a}$ in K, a unique sequence $\vec{b}$ such that

$$\chi(\mathcal{K}^n, \vec{b}) \text{ consists of a unique element } \vec{c} = (c_1, \ldots, c_n)$$

and

$$\mathcal{K} \models E(\vec{c}, \vec{a}), \text{ hence } E(\mathcal{K}^n, \vec{a}) = E(\mathcal{K}^n, \vec{c}).$$

Hence

$$\varphi(\vec{v}, \vec{z}) : \exists \vec{w} \, (\chi(\vec{w}, \vec{z}) \wedge E(\vec{v}, \vec{w}))$$

is the required formula. ♣

4.5 The elimination of imaginaries sometimes fails

Differentially closed fields eliminate imaginaries. This will be shown in Chapter 6, in the section devoted to developing in detail their model theory. Similarly, algebraically closed fields with an automorphism (i. e. models of $ACFA$) eliminate imaginaries [20], as well as separably closed fields of finite imperfection degree in an enriched language capturing p-bases (see [110]). But there do exist structures without this property. This is the case, for instance, of the separably closed fields themselves in the pure language of fields (again, see [110]). Let us propose here a simpler example.

Proposition 4.5.1 *Let $\mathcal{K}$ be a finite field with at least 3 elements and $\mathcal{V}$ be a vectorspace of dimension ≥ 2 over $\mathcal{K}$. Then $\mathcal{V}$ does not eliminate the imaginaries.*

Proof. Suppose yes. As $\mathcal{K}$ is finite, one can consider in L_K the formula

$$E(v, w) : \bigvee_{k \in K - \{0\}} (v = kw)$$

This formula defines in $\mathcal{V}$ an equivalence relation E, identifying two elements $a, a' \in V$ if and only if a, a' generate the same subspace. Take $a \in V$, $a \neq 0$. Then there exist a formula $\varphi(v, \vec{z})$ of L_K and a unique sequence $\vec{b}$ in V such that $E(\mathcal{V}, a) = \varphi(\mathcal{V}, \vec{b})$. As K has at least three elements, fix $k \in K$, $k \neq 0, 1$; for every $x \in V$,

$$k^{-1}x \in E(\mathcal{V}, a) \Leftrightarrow k^{-1}x \in \varphi(\mathcal{V}, \vec{b}).$$

Since the multiplication by k is an automorphism of $\mathcal{V}$, we have, for every $x \in V$,

$$x \in E(\mathcal{V}, ka) \Leftrightarrow x \in \varphi(\mathcal{V}, k\vec{b}).$$

Then $E(\mathcal{V}, a) = E(\mathcal{V}, ka) = \varphi(\mathcal{V}, k\vec{b})$. It follows $\vec{b} = k\vec{b}$; but $k \neq 1$, hence necessarily $\vec{b} = \vec{0}$. In other words, we can assume that $\vec{z} = \emptyset$ and $\varphi(v, \vec{z})$ reduces to a L_K-formula $\varphi(v)$ with at most one free variable. The theory

${}_K T$ has the quantifier elimination in L_K, and hence $\varphi(v)$ is equivalent to a boolean combination of formulas

$$v = v, \quad v = 0.$$

But $\mathcal{V}$ has dimension at least 2, therefore no such boolean combination can equal $E(\mathcal{V}, a)$. ♣

4.6 References

An exhaustive introduction to $\mathcal{A}^{eq}$ can be found in Hodges' book [56]. The elimination of imaginaries was introduced by Poizat [130]; a complete treatment can be found in Poizat's book [131], recently translated in english [134].
The results on the elimination of quantifiers for separably closed fields can be found in [110], while the corresponding analysis for e.c. fields with an automorphism is provided in [20].

Chapter 5

Morley rank

5.1 A tale of two chapters

In 1965, M. Morley's work [116] proposed new ideas, new tools and, altogether, new fertile perspectives in Model Theory. Actually Morley's main theorem is very simple to state: for, it says that a theory T categorical in some uncountable power is, consequently, categorical in every uncountable power. This is noteworthy, but perhaps not so dramatic and relevant. However the germs of Morley's ideas went much further, and their richness permeated the development of Model Theory for several years.
Accordingly, this chapter, and the following one, will be devoted to preparing a proof of Morley's theorem (although this proof will be done only in chapter 7); but our main intent throughout these pages will be to introduce, to discuss and to apply several Morley tools: types, saturated models, algebraic and definable closure, totally transcendental theories.
We will deal also with some related questions, both model theoretical (like prime models) and algebraic (such as differentially closed fields, or ω-stable groups and fields).

5.2 Definable sets

Definable sets were introduced in Chapter 1, and were a constant leitmotiv throughout the subsequent pages. They arise quite naturally within a given structure $\mathcal{A}$ from the basic operations and relations in $\mathcal{A}$, and fully characterize $\mathcal{A}$. Indeed the structure $\mathcal{A}$ could be thought as a non empty set A plus the collection of its definable sets. This new outlook is less formal than the traditional definition given in Chapter 1, and may sound a little puzzling. However it is quite free and easy. Just to refer to our basic examples, it is instinctive to view an algebraically closed field as naturally endowed with the collection of its constructible sets (including varieties), or, in the same way, a real closed field with the collection of its semialgebraic sets, or a differentially closed field with the family of its Kolchin constructible sets. Besides, should you like to fix exactly this new perspective in introducing structures, you could take note that the definable sets in a given $\mathcal{A}$ can be characterized in a formally precise way, as follows.

Theorem 5.2.1 *The sets definable in a structure $\mathcal{A}$ are the smallest family D of subsets of $\bigcup_{n>0} A^n$ such that*

(i) *A^n is in D for every positive integer n; more generally for $i \leq j \leq n$ positive integers, $\{\vec{a} \in A^n : a_i = a_j\}$ is in D;*

(ii) *every relation of $\mathcal{A}$, as well as the graph of every operation in $\mathcal{A}$, is in D;*

(iii) *D is closed under union, intersection and complement;*

(iv) *D is closed under projections: if $X \subseteq A^n$ is in D, then the image of X by the projection of A^n onto any $i \leq n$ coordinates is also in D;*

(v) *D is closed under fibres: if $X \subseteq A^n$ is in D, i is a positive integer smaller than n and $\vec{b} \in A^{n-i}$, then the set of the sequences $\vec{a} \in A^i$ for which $(\vec{a}, \vec{b}) \in X$ is in D.*

Just a few words to comment. A careful and straightforward check easily confirms that these conditions are satisfied by the definable sets in $\mathcal{A}$ and even characterize them. Indeed, (i)-(v) allow sets defined by atomic formulas (i)-(ii), Boolean combinations (iii), quantifiers (iv), parameters (v) and nothing else.

In order to realize how complicated a structure or, more generally, a class of structures -for instance, the models of a given theory T- is, one can try to measure the complexity of its, or their, definable sets. A possible way to accomplish this program is to assign, to every definable, a value (such as an ordinal, or something similar) satisfying some reasonable assumptions, like monotonicity (for $C \subseteq D$ definable sets, the measure of D should not be smaller than the measure of C), and so on.
To prepare this assignment, let us consider again definables. Fix a set X of parameters. Correspondingly one can form the Boolean algebra of X-definable sets, as we saw in Chapter 1. In introducing our measure, we

may refer to this algebraic framework. But we have to be careful; for, the Boolean algebra of X-definable sets, as sketched in Chapter 1, seems to depend on the structure $\mathcal{A}$ where X lies, and clearly, even among the models of a given complete theory T, the choice of $\mathcal{A}$ may vary in an essential way. Also, if we enlarge X to a bigger set of parameters X', the X-definable sets become automatically X'-definable, but in this extended setting they gain more complexity because they have to be compared with more objects; so a finer analysis must be expected.

To clarify these doubts (at least the former one), consider our set X, a structure $\mathcal{A}$ containing X and a positive integer n. Take a model $\mathcal{B}_{f(X)}$ of $Th(\mathcal{A}_X)$; accordingly $\mathcal{B}$ is a structure for L and f is an elementary function from X into B, in particular $f(X)$ is a subset of B and $\mathcal{A}_X \equiv \mathcal{B}_{f(X)}$. As already said, for every positive integer n we wish to compare the algebras $\mathcal{B}_n(X, \mathcal{A})$ and $\mathcal{B}_n(f(X), \mathcal{B})$ introduced before. So consider any $L(X)$-formula $\varphi(\vec{v}, \vec{x})$ ($\vec{v}$ denots here the sequence $(v_1, \ldots, v_n)$). Then $\varphi(\vec{v}, \vec{x})$ defines two sets

$$\varphi(\mathcal{A}^n, \vec{x}) \text{ in } \mathcal{A}, \text{ and } \varphi(\mathcal{B}^n, f(\vec{x})) \text{ in } \mathcal{B}.$$

It is easy to realize that they satisfy the following properties.

1. If $\mathcal{A}$ is an elementary substructure of $\mathcal{B}$ (and f is the inclusion of X in B), then $\varphi(\mathcal{A}^n, \vec{x}) \subseteq \varphi(\mathcal{B}^n, \vec{x})$.

   In fact, for every $\vec{a} \in \mathcal{A}^n$, if $\mathcal{A} \models \varphi(\vec{a}, \vec{x})$, then $\mathcal{B} \models \varphi(\vec{a}, \vec{x})$, just because $\mathcal{B}$ elementarily extends $\mathcal{A}$.

2. If $\varphi(\mathcal{A}^n, \vec{x})$ is finite (of power k), then $|\varphi(\mathcal{B}^n, f(\vec{x}))| = k$ as well.

   In fact, as $\mathcal{A}_X \equiv \mathcal{B}_{f(X)}$, $\mathcal{A}_X$ and $\mathcal{B}_{f(X)}$ satisfy the sentence $\exists! k\, \vec{v} \varphi(\vec{v}, \vec{x})$. In particular, under the assumptions in 1, $\varphi(\mathcal{A}^n, \vec{x}) = \varphi(\mathcal{B}^n, \vec{x})$.

On the contrary, for an infinite $\varphi(\mathcal{A}^n, \vec{x})$, $\varphi(\mathcal{B}^n, f(\vec{x}))$ is infinite as well, owing to elementary equivalence, but is possibly larger when $\mathcal{B}$ is an elementary extension of $\mathcal{A}$. What we want to emphasize here is that, anyway, the Boolean algebras $B_n(X, \mathcal{A})$ and $B_n(f(X), \mathcal{B})$ are isomorphic; in this sense our analysis of X-definable sets and the corresponding assignment of a measure does not depend on the choice of $\mathcal{A}$ or $\mathcal{B}$.

For this purpose, we need a preliminary analysis of $\mathcal{B}_n(X, \mathcal{A})$ (and, in parallel, of $\mathcal{B}_n(f(X), \mathcal{B})$). We know that every D in $\mathcal{B}_n(X, \mathcal{A})$, namely every subset $D \subseteq A^n$ X-definable in $\mathcal{A}$, can be viewed as $D = \varphi(A^n, \vec{x})$ for some $L(X)$-formula $\varphi(\vec{v}, \vec{x})$. Of course, it may happen that two different $L(X)$-formulas $\varphi(\vec{v})$ and $\psi(\vec{v})$ define the same subset of A^n. This means that $\varphi(\vec{v})$

and $\psi(\vec{v})$ are logically equivalent within $Th(\mathcal{A}_X)$ in the usual sense

$$\forall \vec{v}\,(\varphi(\vec{v}) \leftrightarrow \psi(\vec{v})) \in Th(\mathcal{A}_X).$$

Let $\sim$ denote this (equivalence) relation. Elementary mathematical logic tells us that the quotient set of $L(X)$-formulas $\varphi(\vec{v})$ with respect to $\sim$ gets a natural Boolean algebra structure, provided we put, for $\varphi(\vec{v})$ and $\psi(\vec{v})$ in $L(X)$:

* the meet of the $\sim$-classes of $\varphi(\vec{v})$ and $\psi(\vec{v})$ is the $\sim$-class of their conjunction,
* the join of the $\sim$-classes of $\varphi(\vec{v})$ and $\psi(\vec{v})$ is the $\sim$-class of their disjunction,
* the complement of the $\sim$-class of $\varphi(\vec{v})$ is the $\sim$-class of its negation,

so the bottom element is the $\sim$-class of $\neg(v_1 = v_1)$ or any contradiction, and the top element is the $\sim$-class of $v_1 = v_1$ or any tautology; in particular the $\sim$-class of $\varphi(\vec{v})$ is smaller or equal to the $\sim$-class of $\psi(\vec{v})$ if and only if $\varphi(A^n) \subseteq \psi(A^n)$, and hence if and only if

$$\forall \vec{v}\,(\varphi(\vec{v}) \rightarrow \psi(\vec{v})) \in Th(\mathcal{A}_X).$$

Therefore it is easy to check that this quotient algebra is isomorphic to $B_n(X, \mathcal{A})$ via the function mapping the $\sim$-class of $\varphi(\vec{v})$ into $\varphi(A^n)$. The same applies to $B_n(f(X), \mathcal{B})$, of course, and hence, in conclusion, $B_n(X, \mathcal{A})$ and $B_n(f(X), \mathcal{B})$ are isomorphic, as claimed, by the function mapping $\varphi(A^n)$ into $\varphi(B^n)$ for every $\varphi(\vec{v})$. Hence the isomorphism type of $B_n(X, \mathcal{A})$ depends on the theory of $\mathcal{A}_X$ rather than the mere structure $\mathcal{A}_X$. Moreover we can view the elements of $B_n(X, \mathcal{A})$ not only as X-definable subsets of A^n, but also as $\sim$-classes of formulas in $L(X)$. We will often confuse these different points of view below, and even we will directly think of the elements in $B_n(X, \mathcal{A})$ as $L(X)$-formulas (identifying $\sim$-equivalent formulas).

5.3 Types

People acquainted with Boolean algebras $\mathcal{B}$ know what an ultrafilter is and remember that the ultrafilters in a given $\mathcal{B}$ can be naturally endowed with a topology which makes their set a Boolean (i.e. Hausdorff, compact and totally disconnected) space: the dual space of $\mathcal{B}$. Of course this procedure applies to the Boolean algebras of definable sets $B_n(X, \mathcal{A})$. Even in this particular setting one can look at the ultrafilters: they are called *complete*

n-types over X and their space is usually denoted $S_n(X, \mathcal{A})$. That's all about types.
But who is not familiar with Boolean algebras may appreciate some more details. It is worthy satisfying her (him), also in order to realize explicitly what a type is. This is the aim of this section.
Let us begin with a simple example. Everybody knows how real numbers are introduced starting from the rationals and their usual order. But let us summarize very briefly their construction (in our style using formulas and structures). Accordingly consider the language $L = \{\leq\}$ and the L-structure $(\mathbf{Q}, \leq)$. Partition $\mathbf{Q}$ in two non-empty subsets A, B such that, for every $a \in A$ and $b \in B$, $a < b$, A has no maximum and B has no minimum. Then consider the following set of $L(\mathbf{Q})$-formulas

$$p = \{a < v : a \in A\} \cup \{v < b : b \in B\}.$$

No element $r \in \mathbf{Q}$ can satisfy all the formulas in p. However, for every finite conjunction $\varphi(v)$ of formulas of p, there does exist $r \in \mathbf{Q}$ for which $\mathbf{Q} \models \varphi(r)$; in fact, let $a_0, \ldots, a_h \in A$, $b_0, \ldots, b_m \in B$, a be the maximal element among $a_0, \ldots, a_h$ and b be the minimal element among $b_0, \ldots, b_m$; then $a < b$ and, as the order of $\mathbf{Q}$ is dense, there is some rational r larger than a, and consequently than $a_0, \ldots, a_h$, and smaller than b, hence than $b_0, \ldots, b_m$. Now a simple application of Compactess Theorem (to be explained in detail in the next Theorem 5.3.4) proves that in some elementary extension of $(\mathbf{Q}, \leq)$ there is some element realizing p. On the other hand, even forgetting Compactness Theorem and Model Theory, we do know that p is realized in $(\mathbf{R}, \leq)$ just by the real irrational number corresponding to the section (A, B) in $\mathbf{Q}$.
The notion of type provides an abstract framework where to study the situation just sketched in the case of $(\mathbf{Q}, \leq)$ and $(\mathbf{R}, \leq)$. Accordingly take any structure $\mathcal{A}$ for a language L, a subset X of A and a positive integer n.

Definition 5.3.1 *A* **consistent** n**-type** *over X in $Th(\mathcal{A}_X)$ is a set p of $L(X)$-formulas $\varphi(\vec{v})$ (where $\vec{v}$ abbreviates $(v_1, \ldots, v_n)$) such that every finite conjunction of formulas of p is satisfied in $\mathcal{A}$ - more precisely in $\mathcal{A}_X$ - by some suitable tuple $\vec{a} \in A^n$.*

Definition 5.3.2 *A* **complete** n**-type** *over X in $Th(\mathcal{A}_X)$ is a consistent n-type maximal with respect to inclusion. $S_n(X, \mathcal{A})$ denotes the set of complete n-types over X in $Th(\mathcal{A}_X)$.*

In the sequel "n-type" will abbreviate "complete n-type", unless otherwise stated.

Example 5.3.3 Consider a model $\mathcal{B}_{f(X)}$ of $\mathrm{Th}(\mathcal{A}_X)$. Accordingly $\mathcal{B} \equiv \mathcal{A}$ and f is an elementary function of X into $\mathcal{B}$. For simplicity, assume that $X \subseteq B$ and f is the inclusion of X into B. Let $\vec{b} \in B^n$, p be the set of the $L(X)$-formulas $\varphi(\vec{v})$ such that $\mathcal{B} \models \varphi(\vec{b})$. It is easy to check that p is a complete n-type over X. Let us see why.

- For every finite conjunction $\varphi(\vec{v})$ of formulas of p, $\mathcal{B} \models \varphi(\vec{b})$, whence $\mathcal{B} \models \exists \vec{v} \varphi(\vec{v})$ and $\mathcal{A} \models \exists \vec{v} \varphi(\vec{v})$.
- Enlarge p by a new formula $\psi(\vec{v})$; $\psi(\vec{v}) \notin p$ implies $\mathcal{B} \not\models \psi(\vec{b})$ and so $\mathcal{B} \models \neg\psi(\vec{b})$ and $\neg\psi(\vec{v}) \in p$; hence no set of formulas extending $p \cup \{\psi(\vec{v})\}$ is a consistent n-type any more, because it contains two formulas - $\psi(\vec{v})$ and its negation - whose conjunction cannot be satisfied by any tuple in A^n.

p is called the *type of* $\vec{b}$ *over* X and is denoted $tp(\vec{b}/X)$.
The next theorem shows that any n-type over X can be obtained in this way.

Theorem 5.3.4 *Let p be a set of $L(X)$-formulas $\varphi(\vec{v})$ (where $\vec{v}$ abbreviates $(v_1, \ldots, v_n)$ as before). The following propositions are equivalent:*

(i) $p \in S_n(X, \mathcal{A})$;

(ii) *there are a model $\mathcal{B}$ of $Th(\mathcal{A})$ such that $X \subseteq B$ and the inclusion of X into B is an elementary function, and a tuple $\vec{b} \in B^n$ such that $tp(\vec{b}/X) = p$.*

Proof. (ii)$\Rightarrow$(i) was shown before.
(i)$\Rightarrow$(ii) Enlarge $L(X)$ by a tuple $\vec{c}$ of n new constant symbols. Let L' be the language obtained in this way. Consider the following set of sentences in L'

$$T' = \mathrm{Th}(\mathcal{A}_X) \cup \{\varphi(\vec{c}) : \varphi(\vec{v}) \in p\}.$$

Every finite subset T'_0 of T' has a model. In fact

$$T'_0 \subseteq Th(\mathcal{A}_X) \cup \{\varphi(\vec{c}) : \varphi(\vec{v}) \in p_0\}$$

for some finite subset p_0 of p. As p is consistent, there is some $\vec{a} \in A^n$ satisfying in $\mathcal{A}$ the conjunction of the formulas of p_0. At this point look at the L'-structure expanding $\mathcal{A}$ and interpreting the new constants $\vec{c}$ in $\vec{a}$ and notice that this provides a model of T'_0.
Now apply Compactness Theorem and deduce that there is a model $\mathcal{B}'$ of T'. Restrict $\mathcal{B}'$ to $L(X)$ and get a model $\mathcal{B}_{f(X)}$ of $Th(\mathcal{A}_X)$ (here $\mathcal{B}$ denotes

a model of $Th(\mathcal{A})$ and f is an elementary function of X into $\mathcal{B}$, but there is no loss of generality in supposing $X \subseteq B$, $f =$ inclusion of X into B). Let $\vec{b} \in B^n$ interpret $\vec{c}$ in $\mathcal{B}'$. In $\mathcal{B}_X$, $p \subseteq tp(\vec{b}/X)$. Owing to the maximality of p, $p = tp(\vec{b}/X)$. ♣

When $p = tp(\vec{b}/X)$ for $\vec{b} \in B^n$, one says that $\vec{b}$ is a *realization* of p (or also that $\vec{b}$ *realizes* p), and one writes $\vec{b} \models p$. Notice that the argument in (i)⇒(ii) actually works also when p is a consistent n-type over X, with the only exception of the last point (deducing $p = tp(\vec{b}/X)$ from $p \subseteq tp(\vec{b}/X)$); accordingly, if p is a consistent n-type over X, then there are $\mathcal{B}$ and $\vec{b}$ such that $p \subseteq tp(\vec{b}/X)$, and we can state:

Theorem 5.3.5 *Let p be a consistent n-type over X. Then p enlarges to a complete n-type over X (possibly in several ways).*

In the particular case when $X = A$, Theorem 5.3.4 says that every complete n-type over A is realized in some elementary extension of $\mathcal{A}$. Notice also that, when $X \subseteq A$, any complete n-type over X can be viewed as a consistent n-type over A and can be enlarged to a complete n-type over A, whence it is realized in some elementary extension of $\mathcal{A}$.
Theorem 5.3.4 (together with the definition of complete type) also implies:

Corollary 5.3.6 *Let $p \in S_n(X, \mathcal{A})$, $\varphi(\vec{v})$, $\psi(\vec{v})$ be $L(X)$-formulas.*

(i) *If $\varphi(\vec{v}) \in p$ and $\psi(\vec{v}) \in p$, then $\varphi(\vec{v}) \wedge \psi(\vec{v}) \in p$;*

(ii) *if $\varphi(\vec{v}) \in p$ and $\forall \vec{v}(\varphi(\vec{v}) \to \psi(\vec{v})) \in Th(\mathcal{A}_X)$, then $\psi(\vec{v}) \in p$.*

(iii) *either $\varphi(\vec{v}) \in p$ or $\neg\varphi(\vec{v}) \in p$ (and each case excludes the other).*

Just to summarize, we might say that, as the rational order $(\mathbf{Q}, \leq)$ implicitly contains through its sections all the real numbers -even the irrational ones- as *ideal elements*, similarly, for any $\mathcal{A}$, X and n, the n-types over X in $Th(\mathcal{A}_X)$ tell us which new n-tuples of elements can arise in the structures $\mathcal{B}_{f(X)} \equiv \mathcal{A}_X$, in particular in the elementary extensions of $\mathcal{A}$. Under this point of view, the notion of n-type seems to deserve a good deal of attention. So let us explore it. First we consider the set $S_n(X, \mathcal{A})$. We already linked types and topology at the beginning of this section. Let us examine this connection in more detail.
For every $L(X)$-formula $\varphi(\vec{v})$ di $L(X)$, put

$$\mathcal{U}_{\varphi(\vec{v})} = \{p \in S_n(X, \mathcal{A}) : \varphi(\vec{v}) \in p\}.$$

Notice that, for $\varphi(\vec{v})$, $\psi(\vec{v})$ $L(X)$-formulas,

$$\mathcal{U}_{\varphi(\vec{v})} \cap \mathcal{U}_{\psi(\vec{v})} = \mathcal{U}_{\varphi(\vec{v}) \wedge \psi(\vec{v})}.$$

In other words, for every type $p \in S_n(X, \mathcal{A})$,

$$\varphi(\vec{v}) \in p, \ \psi(\vec{v}) \in p \quad \Leftrightarrow \quad \varphi(\vec{v}) \wedge \psi(\vec{v}) \in p.$$

In fact ($\Rightarrow$) is just (i) in Corollary 5.3.6, while ($\Leftarrow$) is a simple consequence of (ii).
Consequently the sets $\mathcal{U}_{\varphi(\vec{v})}$ form a basis of open neighbourhoods for a topology on $S_n(X, \mathcal{A})$ when $\varphi(\vec{v})$ ranges over $L(X)$-formulas. Notice also that

$$\mathcal{U}_{\neg(v_1 = v_1)} = \emptyset, \quad \mathcal{U}_{v_1 = v_1} = S_n(X, \mathcal{A}).$$

Furthermore, for every $L(X)$-formula $\varphi(\vec{v})$,

$$S_n(X, \mathcal{A}) - \mathcal{U}_{\varphi(\vec{v})} = \mathcal{U}_{\neg\varphi(\vec{v})}$$

(as implicitly stated in Corollary 5.3.6, (iii)).
So the topological space $S_n(X, \mathcal{A})$ is:

- *Hausdorff* (in fact, for $p, q \in S_n(X, \mathcal{A})$ and $p \neq q$, choose $\varphi(\vec{v}) \in q - p$ and observe $\neg\varphi(\vec{v}) \in p$, whence $p \in \mathcal{U}_{\neg\varphi(\vec{v})}$, while $q \in \mathcal{U}_{\varphi(\vec{v})}$));
- *totally disconnected* (because every open set of the given basis is also closed).

$S_n(X, \mathcal{A})$ is also *compact*. In fact, take an open covering of $S_n(X, \mathcal{A})$. With no loss of generality we can assume that every open set of this covering is basic. So for some suitable set I of indexes our covering is just $\{U_{\varphi_i(\vec{v})} : i \in I\}$ where, for any $i \in I$, $\varphi_i(\vec{v})$ is an $L(X)$-formula. Notice that $\{U_{\varphi_i(\vec{v})} : i \in I\}$ covers $S_n(X, \mathcal{A})$ if and only if, for every model $\mathcal{B}_X$ of $\mathrm{Th}(\mathcal{A}_X)$ and $\vec{b} \in B^n$, there is some $i \in I$ such that $tp(\vec{b}/X) \in U_{\varphi_i(\vec{v})}$, namely $\mathcal{B} \models \varphi_i(\vec{b})$, and hence if and only if, in a language with an n-tuple $\vec{c}$ of new constants,

$$\mathrm{Th}(\mathcal{A}_X) \cup \{\neg\varphi_i(\vec{c}) : i \in I\}$$

has no model. So, by Compactness Theorem, there exists a finite subset I_0 of I for which

$$\mathrm{Th}(\mathcal{A}_X) \cup \{\neg\varphi_i(\vec{c}) : i \in I_0\}$$

has no model, equivalently, $\{\mathcal{U}_{\varphi_i} : i \in I_0\}$ is a (finite) subcovering of $S_n(X, \mathcal{A})$.
By the way, this topological application of Compactness Theorem is just the reason of its name. But the topological framework also suggests the following definition.

Definition 5.3.7 *A type $p \in S_n(X, \mathcal{A})$ is said to be* **isolated** *if and only if p is isolated as a point of the topological space $S_n(X, \mathcal{A})$, and so if and only if there is some $L(X)$-formula $\varphi(\vec{v})$ such that p is the only element in $\mathcal{U}_{\varphi(\vec{v})}$, namely the only n-type over X containing $\varphi(\vec{v})$.*

The next notion has a prevalent model theoretic flavour.

Definition 5.3.8 *An n-type $p \in S_n(X, \mathcal{A})$ is said to be* **algebraic** *if and only if there is a formula $\varphi(\vec{v}) \in p$ such that $\varphi(\mathcal{A}^n)$ is finite (and hence $\varphi(\mathcal{B}^n)$ is also finite, and even of the same cardinality as $\varphi(\mathcal{A}^n)$, for every model $\mathcal{B}_{f(X)}$ of $Th(\mathcal{A}_X)$).*

Let us compare these notions. p is any type in $S_n(X, \mathcal{A})$.

(i) If p is algebraic, then p is isolated.

In fact take a formula $\varphi(\vec{v})$ in p such that $\varphi(\mathcal{A}^n)$ is finite of minimal size k. We claim that p is the only type in $S_n(X, \mathcal{A})$ containing $\varphi(\vec{v})$. Let q be another type over X including $\varphi(\vec{v})$. For every formula $\vartheta(\vec{v}) \in p$, $\varphi(\vec{v}) \wedge \vartheta(\vec{v})$ is also in p, and so, owing to the choice of $\varphi(\vec{v})$, $\varphi(\mathcal{A}^n) \cap \vartheta(\mathcal{A}^n) = \varphi(\mathcal{A}^n)$, namely $\varphi(\mathcal{A}^n) \subseteq \vartheta(\mathcal{A}^n)$. Hence $\vartheta(\vec{v}) \in q$. It follows $p \subseteq q$, and so $p = q$ by the maximality of p.

(ii) If $X = A$ and $p \in S_n(A, \mathcal{A})$ is isolated, then p is algebraic.

Let $\varphi(\vec{v})$ be an $L(A)$-formula isolating p. p is realized in some elementary extension $\mathcal{B}$ of $\mathcal{A}$; consequently $\mathcal{B} \models \exists \vec{v} \varphi(\vec{v})$. As an elementary substructure, $\mathcal{A}$ satisfies $\exists \vec{v} \varphi(\vec{v})$ as well, and so includes a realization $\vec{a}$ of p, and $p = tp(\vec{a}/A)$. But $tp(\vec{a}/A)$ contains the formula $\vec{v} = \vec{a}$ and so $\vec{a}$ is its only realization.

However we will see within a few lines that an isolated type over an arbitrary subset X may be non-algebraic. But, more generally, let us propose now some specific examples of types.

Examples 5.3.9 1. Let $L = \emptyset$, A be an infinite set (viewed as a structure for L), $X \subseteq A$. For every $a \in X$, $tp(a/X)$ is the only 1-type containing $v = a$, and so is both isolated and algebraic. Now take two elements a and a' in $A - X$; there does exist a bijection from A onto A, hence an automorphism of A, fixing X pointwise and mapping a in a'. So, for every $L(X)$-formula $\varphi(v)$, $A \models \varphi(a)$ if and only if $A \models \varphi(a')$; in other words, $tp(a/X) = tp(a'/X)$. This shows that all the elements in $A - X$ realize the same type over X. Notice that, for a finite $X = \{x_0, \ldots, x_n\}$, this type is isolated, for instance by $\bigwedge_{i \leq k} \neg(v = x_i)$;

however, it has infinitely many realizations (the elements in $A - X$), whence it cannot be algebraic.

2. Now take the language $L = \{+, \cdot, -, 0, 1\}$ for fields and an algebraically closed field $\mathcal{K}$. Let X be a subset of K. For simplicity we assume that X is the domain of some subfield $\mathcal{H}$ of $\mathcal{K}$. This is not so restrictive. Indeed, notice that each element in the field $\mathcal{H}$ generated by X is X-definable, in other words it is the only point in $\mathcal{K}$ satisfying a suitable $L(X)$-formula. Consequently there is no great loss of generality, when discussing the types over X, in replacing X by H and hence in assuming that $X = H$ is the domain of a subfield $\mathcal{H}$.

   Owing to quantifier elimination, every n-type p over H is fully determined by its formulas $f(\vec{v}) = 0$ where $f(\vec{x})$ ranges over polynomials in $H[\vec{x}]$. Indeed every L_H-formula is equivalent in ACF to a Boolean combination of equations over $\mathcal{H}$. Notice that the polynomials $f(\vec{x}) \in H[\vec{x}]$ for which $f(\vec{v}) = 0 \in p$ form a proper prime ideal in $\mathcal{H}[\vec{x}]$. This is easy to check. Just take a realization $\vec{a}$ of p in a suitable extension, and notice that, for $f(\vec{x})$, $g(\vec{x})$, $h(\vec{x})$ in $H[\vec{x}]$, if $f(\vec{a}) = g(\vec{a}) = 0$, then $f(\vec{a}) + g(\vec{a}) = f(\vec{a}) \cdot h(\vec{a}) = 0$; moreover, if $\vec{a}$ is a root of $f(\vec{x}) \cdot g(\vec{x})$, then it annihilates $f(\vec{x})$ or $g(\vec{x})$. Conversely, let I be a (proper) prime ideal in $\mathcal{H}[\vec{x}]$ and look at the set of formulas

$$\{f(\vec{v}) = 0 \,:\, f(\vec{x}) \in I\} \cup \{\neg(g(\vec{v}) = 0) \,:\, g(\vec{x}) \in H[\vec{x}] - I\}.$$

   This defines a (complete) n-type over H, the one of $I + \vec{x}$ viewed as an element of the extension of $\mathcal{H}$ provided by the field of quotients of the integral domain $\mathcal{H}[\vec{x}]/I$.

   In this way we obtain a connection (indeed a bijection) between n-types over H and prime ideals in $\mathcal{H}[\vec{x}]$. We shall provide a more detailed analysis of this point in Chapter 8. Now let us examine closely the case $n = 1$. When I is the ideal generated by some monic irreducible $f(x) \in H[x]$, then we obtain the type defined by the single equation $f(v) = 0$ (and its consequences). So the roots of $f(x)$ share a common 1-type over H. This type p is clearly realized in $\mathcal{K}$ because $\mathcal{K}$ is algebraically closed; actually $\mathcal{K}$ contains all the realizations of p. p is isolated by $f(v) = 0$, and algebraic, because $f(x)$ has only finitely many roots.

   Otherwise $I = 0$. Now the corresponding 1-type p concerns all the elements which are transcendental over H. When the transcendence degree of $\mathcal{K}$ over $\mathcal{H}$ is bigger than 0, then p has (infinitely many) realizations in $\mathcal{K}$, in particular it is not algebraic. Otherwise, if $\mathcal{K}$ is just the algebraic closure of $\mathcal{H}$, then there is no room to satisfy p in $\mathcal{K}$. One easily sees that p is not even isolated.

3. Now take the language $L = \{+, \cdot, -, 0, 1, \leq\}$ for ordered fields and a real closed field $\mathcal{K}$. We discuss directly the types over K, for simplicity. Owing to quantifier elimination, every n-type over K is fully determined by its equations and disequations $f(\vec{v}) = 0, \quad g(\vec{v}) \geq 0$, where $f(\vec{x})$ and $g(\vec{x})$ range over polynomials in n unknowns with coefficients in K. In particular, when we look at a 1-type p over K, we have to consider polynomials in $K[x]$. But then, as explained in Chapter 2, § 5, p is completely determined by its formulas $v \leq a$, $b \leq v$ (and negations), with a and b in K. So a nonalgebraic p is given by the cut it defines in K.

4. Let $L = \{\leq\}$, consider the L-structure $\mathcal{A} = (\mathbf{R}, \leq)$ and the subset $X = \mathbf{Q}$. If a, $a' \in \mathbf{R}$ and $a \neq a'$ (say $a < a'$), there is $r \in \mathbf{Q}$ such that $a < r$ and $r < a'$. So $v < r \in tp(a/\mathbf{Q})$, but $v < r \notin tp(a'/\mathbf{Q})$. In conclusion, there are at least $2^{\aleph_0}$-types over $\mathbf{Q}$ in $(\mathbf{R}, \leq)$.

In general, for a countable language, there are at most $2^{max\{\aleph_0, |X|\}}$ n-types over a set X for every positive integer n.

Now let us mention another relevant technical fact about types; it will be useful several times later. Let $\mathcal{A}$ be a model of T, $X \subseteq A$, f be an elementary function from X into a model $\mathcal{B}$ of T (for instance, let f be the restriction to X of some isomorphism between $\mathcal{A}$ and $\mathcal{B}$). For $p \in S(X, \mathcal{A})$, put

$$f(p) = \{\varphi(v, f(\vec{x})) : \varphi(v, \vec{x}) \in p \ \ (\text{and}\, \vec{x}\, \text{in}\, X)\}.$$

Then $f(p)$ is a type over $f(X)$ in $\mathcal{B}$. Indeed f determines in this way a homeomorphism between $S(X, \mathcal{A})$ and $S(f(X), \mathcal{B})$. In particular $f(p)$ is isolated, or algebraic, exactly when p is.
To conclude, it is worth underlining that, just as the algebra $\mathcal{B}_n(X, \mathcal{A})$ of X-definable subsets of A^n, similarly the space $S_n(X, \mathcal{A})$ of n-types over X does not depend directly on the model $\mathcal{A}$ where X is elementarily embedded, but only on the theory of $\mathcal{A}_X$.

5.4 Saturated models

Let T be a complete theory with infinite models in a countable language L. A model $\mathcal{A}$ of T may not realize all the 1-types over a subset X: a trivial counterexample is provided, when $X = A$, by the consistent 1-type $\{\neg(v = a) : a \in A\}$, which cannot satisfied by any element in A. The saturated models of T are, roughly speaking, those realizing as many types as possible over their subsets. In particular, for an infinite cardinal λ, the λ-*saturated* models of T are those able to satisfy any 1-type over an arbitrary subset of power less than λ.

Definition 5.4.1 *Let λ be an infinite cardinal. A model $\mathcal{B}$ of T is said to be* **λ-saturated** *if and only if, for every subset X of B or power $< \lambda$, $\mathcal{B}$ realize every 1-type over X.*

Of course this definition makes sense when $\lambda \leq |B|$ (as the previous counterexample shows); it is also clear that, if $\lambda \geq \mu \geq \aleph_0$ are cardinal numbers and $\mathcal{B}$ is λ-saturated, then $\mathcal{B}$ is μ-saturated, too. The first question one may raise at this point just concerns the existence of λ-saturated models. The answer is positive, as the next theorem shows.

Theorem 5.4.2 *Let λ be an infinite cardinal. Then every model $\mathcal{A}$ of T has a λ-saturated elementary extension.*

Proof. We provide a comparatively simple argument working when $\lambda = \aleph_0$. For an uncountable λ this approach does not work any more, and one needs a more complicated proof of a quite different style (but the result remains true). So take a model $\mathcal{A}$ of T, we are looking for an elementary extension $\mathcal{A}' > \mathcal{A}$ realizing any 1-type over a finite subset of A'. We do know that every 1-type over any (finite or infinite) subset of A is satisfied in some elementary extension of A. Accordingly, well order the set P of 1-types over finite subsets of A, $P = \{p_\nu : \nu < \alpha\}$ and associate with any $\nu < \alpha$ an elementary extension $\mathcal{A}(\nu)$ of $\mathcal{A}$ realizing p. Do this in such a way that, if $\nu < \mu < \alpha$ then $\mathcal{A}(\nu)$ is an elementary substructure of $\mathcal{A}(\mu)$ (the reader may check the details as an exercise). Now use the Elementary Chain Theorem (1.3.19) and deduce that $\mathcal{A}^\star = \bigcup_{\nu<\alpha} \mathcal{A}(\nu)$ is an elementary extension of $\mathcal{A}$, and even of $\mathcal{A}(\nu)$ for any $\nu < \alpha$; it is also clear that $\mathcal{A}^\star$ realize any type $p \in P$ and so every 1-type over a finite subset of A. But this does not imply that $\mathcal{A}^\star$ is $\aleph_0$-saturated, namely realizes every 1-type over a finite subset of itself. However we can repeat the previous procedure and define, for every natural n, an elementary extension $\mathcal{A}_n$ of $\mathcal{A}$ in such a way that $\mathcal{A}_0$ is just $\mathcal{A}$ and, for every n, $\mathcal{A}_{n+1}$ is an elementary extension of $\mathcal{A}_n$ realizing any 1-type over an arbitrary finite subset of A_n. Using the Elementary Chain Theorem once again, we deduce that $\mathcal{A}' = \bigcup_{n\in\mathbf{N}} \mathcal{A}_n$ is an elementary extension of $\mathcal{A}$, and even of $\mathcal{A}_n$ for every n. Moreover $\mathcal{A}'$ is $\aleph_0$-saturated. In fact, let X be a finite subset of A'; there is some n for which $X \subseteq A_n$; consequently every 1-type over X is realized in $\mathcal{A}_{n+1}$ and, through it, in $\mathcal{A}'$. ♣

Now let us show some basic properties of λ-saturated models.

Theorem 5.4.3 (Weak Homogeneity Theorem) *Let $\mathcal{B}$ be a λ-saturated model of T. Let $\mathcal{A}$ be a model of T, X be a subset of A of power $< \lambda$ and f be an elementary function from X into $\mathcal{B}$. Then, for every $a \in A$ there is an elementary function of $X \cup \{a\}$ into $\mathcal{B}$ enlarging f. (A model $\mathcal{B}$ with this property is called* **weakly λ-homogeneous***).*

Proof. Let $p = tp(a/X)$, so $f(p)$ is a complete n-type over $f(X)$. As f is 1-1, $|f(X)| < \lambda$. As $\mathcal{B}$ is λ-saturated, $\mathcal{B}$ contains some realization b of $f(p)$; hence, for every $L(X)$-formula $\gamma(v, \vec{x})$,

$$\mathcal{A} \models \gamma(a, \vec{x}) \Leftrightarrow \gamma(v, \vec{x}) \in p \Leftrightarrow \gamma(v, f(\vec{x})) \in f(p) \Leftrightarrow \mathcal{B} \models \gamma(b, f(\vec{x})).$$

So enlarge f to a function g of $X \cup \{a\}$ into B putting $g(a) = b$. Clearly g is what we are looking for. ♣

By definition, a λ-saturated model $\mathcal{B}$ of T realizes any 1-type over a subset of power $< \lambda$. But actually $\mathcal{B}$ satisfies every type (even with more than 1 variable) over such a subset. Let us see why.

Corollary 5.4.4 *Let λ be an infinite cardinal, $\mathcal{B}$ be a λ-saturated model of T, X be a subset of B of power $< \lambda$, n be a positive integer. Then every n-type p over X is realized in $\mathcal{B}$.*

Proof. We know that p is realized in some model $\mathcal{A}_X$ of $\mathrm{Th}(\mathcal{B}_X)$, say by $\vec{a} = (a_1, \ldots, a_n) \in A^n$. As $\mathcal{A}_X \equiv \mathcal{B}_X$, the identity map of X is an elementary function from X into $\mathcal{B}$. Using the Weak Homogeneity Theorem over and over again, one extends this map to an elementary function g from $X \cup \{a_1, \ldots, a_n\}$ into $\mathcal{B}$. Let $\vec{b} = g(\vec{a})$. For every $L(X)$-formula $\varphi(\vec{v}, \vec{x})$,

$$\mathcal{A} \models \varphi(\vec{a}, \vec{x}) \Leftrightarrow \mathcal{B} \models \varphi(\vec{b}, \vec{x}).$$

Hence $tp(\vec{b}/X) = p$. ♣

Theorem 5.4.5 (Universality Theorem) *Let $\mathcal{B}$ be a λ-saturated model of T, $\mathcal{A}$ be a model of T of power $< \lambda$. Then there is an elementary embedding of $\mathcal{A}$ into $\mathcal{B}$ (and so $\mathcal{A}$ is an elementary substructure of $\mathcal{B}$ up to isomorphism). In this case $\mathcal{B}$ is called* **λ-universal**.

Proof. Well order $A = \{a_\nu : \nu < \alpha\}$ and, for $\nu < \alpha$, put $A_\nu = \{a_\mu : \mu < \alpha\}$. As $\mathcal{A} \equiv \mathcal{B}$, the empty map is an elementary function from $A_0 = \emptyset$ (viewed as a subset of A) into $\mathcal{B}$. Using the λ-saturation (actually the weak λ-homogeneity) of $\mathcal{B}$, extend this map to elementary functions from A_ν into $\mathcal{B}$ for every $\nu < \alpha$, whose domain progressively includes every element in A. One eventually gets an elementary function (and so an elementary embedding) of $\mathcal{A}$ into $\mathcal{B}$. ♣

Now we deal with the "most saturated" models of T; as observed before, they are the models saturated in their own power.

Definition 5.4.6 *A model $\mathcal{B}$ of T is said to be* **saturated** *if and only if $\mathcal{B}$ is $|B|$-saturated (and consequently realizes every 1-type over subsets of B of power $< |B|$).*

Such a model $\mathcal{B}$ is (also) weakly $|B|$-homogeneous and $|B|$-universal.
First let us observe that, for every infinite cardinal λ, there is at most one saturated model of T of power λ (up to isomorphism).

Theorem 5.4.7 (Uniqueness Theorem) *Let $\mathcal{B}$ and $\mathcal{B}'$ be saturated models of T of the same power. Then $\mathcal{B} \simeq \mathcal{B}'$.*

Proof. Recall the Weak Homogeneity Theorem: as both $\mathcal{B}$ and $\mathcal{B}'$ are saturated, if $X \subseteq B$, $X' \subseteq B'$, $|X| < |B|$ and f is an elementary function of X into $\mathcal{B}'$ with image X' (whence $|X'| = |X| < |B'|$ and f^{-1} is an elementary function from X' into $\mathcal{B}$), then

(i) for every $a \in B$, one can enlarge f to an elementary function g from $X \cup \{a\}$ into $\mathcal{B}'$,

(ii) for every $a' \in B'$, one can enlarge f^{-1} to an elementary function g from $X' \cup \{a'\}$ into $\mathcal{B}$.

Now observe that, as $\mathcal{B} \equiv \mathcal{B}'$, the empty map is an elementary function of $\emptyset \subseteq B$ into $\mathcal{B}'$ (and conversely). Start from this function, well order both B and B' and use alternatively (i) and (ii) in a suitable (possibly transfinite) induction procedure. One eventually gets an isomorphism between $\mathcal{B}$ and $\mathcal{B}'$.
Let us provide the details of this construction. Let λ denote the common power of B and B'; view λ as an initial ordinal; let $\{b_\nu : \nu < \lambda\}$, $\{b'_\nu : \nu < \lambda\}$. List B, B' respectively. For every $\nu \leq \lambda$, one builds two elementary functions f_ν, f'_ν, the former from a subset of B including all the b_μ's with $\mu < \lambda$ and having power $< \lambda$ into $\mathcal{B}'$, the latter from a subset of B' including

all the b'_μ's with $\mu < \lambda$ and having power $< \lambda$ into $\mathcal{B}$, in such a way that, for $\mu \leq \nu \leq \lambda$, f_ν enlarges f_μ and f'^{-1}_μ and f'_ν enlarges f'_μ and f_ν^{-1}, and eventually f_λ and f'_λ are two isomorphisms, the one inverse of the other one, between $\mathcal{B}$ and $\mathcal{B}'$. We proceed by induction on ν.
When $\nu = 0$, we know what to do: f_0 and f'_0 are just the empty function. For a limit ν, put $f_\nu = \cup_{\mu<\nu} f_\mu$ and $f'_\nu = \cup_{\mu<\nu} f'_\mu$. Finally suppose $\nu = \mu + 1$ successor. First build f_ν. Its domain is $Im\, f'_\mu \cup \{a_\nu\}$; using (i), enlarge $f'^{\,-1}_\mu$, hence f_μ, to include a_ν in the domain; f_ν is what we form in this way. Now construct f'_ν. Its domain is $Im\, f_\nu \cup \{b_\nu\}$; using (ii), enlarge f_ν^{-1} to include b_ν in the domain; let f'_ν be the resulting function. Clearly this machinery produces f_λ and f'_λ as required. ♣

As a consequence we obtain:

Corollary 5.4.8 (Strong Homogeneity Theorem) *Let $\mathcal{B}$ a saturated model of T, X be a subset of B of power $< |B|$, f be an elementary function of X into $\mathcal{B}$. Then f can be enlarged to an automorphism of $\mathcal{B}$ (and then $\mathcal{B}$ is called* **homogeneous***).*

Proof. In the language $L(X)$ consider the structures $\mathcal{B}_X$ and $\mathcal{B}_{f(X)}$. As f is an elementary function, $\mathcal{B}_X \equiv \mathcal{B}_{f(X)}$, and so $\mathcal{B}_X$ and $\mathcal{B}_{f(X)}$ are models of the same complete theory. As $|X| = |f(X)| < |B|$, $\mathcal{B}_X$ and $\mathcal{B}_{f(X)}$ are saturated also in $L(X)$. Finally $\mathcal{B}_X$ and $\mathcal{B}_{f(X)}$ have the same power. Accordingly there is some isomorphism (in $L(X)$) between $\mathcal{B}_X$ and $\mathcal{B}_{f(X)}$, whose restriction to L determines an automorphism of $\mathcal{B}$ enlarging f. ♣

As a particular case, consider $X \subseteq B$, $|X| < |B|$ and two tuples $\vec{a}$, $\vec{a}'$ in B^n having the same n-type over X. The function fixing X pointwise and mapping $\vec{a}$ into $\vec{a}'$ is elementary, hence can be enlarged to an automorphism of $\mathcal{B}$. As said, this automorphism acts identically on X and maps $\vec{a}$ into $\vec{a}'$.

Now we wonder for which cardinals λ a complete theory T may have a saturated model of power λ. This is a quite delicate and deep question, and the answer is not easy. Of course, stronger assumptions on T may sometimes ensure these existence results. For instance, one can see:

- if T is λ-categorical in some infinite cardinal λ, then the unique model of T of power λ is saturated (we will see why for a countable λ in the final section of Chapter 7, where we will also discuss the uncountable case; the examples at the end of the present section partly concern this point).

Furthermore the (complete) theories having a countable saturated model can be characterized in the following way.

- T has a saturated model of power $\aleph_0$ if and only if T has at most countably many n-types over $\emptyset$ for every positive integer n (the reader may check this as an exercise).

But what we can say from a general perspective, when dealing with arbitrary complete theories T? In this abstract framework, proving the existence of saturated models seeems related to some deep set theoretic assumptions, like the existence of large cardinals. In fact one shows:

Theorem 5.4.9 *Let $\lambda > \aleph_0$ be an inaccessible cardinal. Then T has a saturated model of power λ.*

The existence of an uncountable inaccessible cardinal is a quite delicate matter. But assume momentarily that such a saturated model Ω exists (for a given inaccessible $\lambda > \aleph_0$). Recall that Ω is unique up to isomorphism. Furthermore

- Ω is λ-universal: every model of T of power $< \lambda$ can be embedded as an elementary substructure in Ω;

- Ω is λ-homogeneous: if X is a subset of Ω of power $< \lambda$ and $\vec{a}, \vec{a}'$ are two tuples in Ω having the same type over X, then there is an automorphism of Ω fixing X pointwise and mapping $\vec{a}$ into $\vec{a}'$.

It is a general agreement (and habit) in Model Theory to assume that such a model Ω exists. This makes things easier and, on the other hand, is sufficiently plausible; in particular, one can check that everything is shown inside Ω, so assuming that Ω exists, can be proved (at the cost of some major complications) even avoiding any reference to Ω.
Which are the benefits of working in Ω? As we said, the cardinality of Ω is very large, and so one can reasonably suppose that all the models we expect to handle have a smaller power. But this implies that they actually are elementary substructures of Ω of a smaller size (up to isomorphism). Under this perspective, the subsets of models of T can be directly viewed as subsets of Ω with power $< |\Omega|$: we will call these subsets *small* subsets of Ω, just to tell them from the other subsets of Ω admitting its same inaccessible cardinality.
As already said, referring to small subsets of Ω instead of subsets of arbitrary models makes our life, and also our notation simpler. For instance, take a

small X and a positive integer n. When defining $\mathcal{B}_n(X, \mathcal{A})$ and $S_n(X, \mathcal{A})$, we had to explicitly refer to a model $\mathcal{A}$ of T (now an elementary substructure of Ω) where X is embedded. We also emphasized that $\mathcal{B}_n(X, \mathcal{A})$ and $S_n(X, \mathcal{A})$ do not depend directly on $\mathcal{A}$, but only on the theory of $\mathcal{A}_X$: the choice of $\mathcal{A}$ is quite arbitrary within these bounds. But then it is convenient to refer to $\mathcal{A} = \Omega$, as a universal model where X is embedded. This is what we will do from now on. Indeed we will write, when there is no danger of misunderstanding, $S_n(X)$ to mean $S_n(X, \Omega)$ for every small X; so $S(X)$ will denote the union of the spaces $S_n(X)$ when n ranges over positive integers.

At last, let us propose some algebraic examples concerning more or less saturated structures.

Examples 5.4.10 1. Which algebraically closed fields $\mathcal{K}$ are saturated in a given cardinality $\lambda \geq \aleph_0$? To answer, we may recall that, for any $p = 0$ or prime, the theory ACF_p is categorical in every uncountable power; so, according to what we said before, every uncountable algebraically closed field (in any characteristic) is saturated. To confirm this from the algebraic point of view and to discuss the existence of saturated models in the countable case, we can refer to Steinitz's analysis of algebraically closed fields $\mathcal{K}$ and recall that such a $\mathcal{K}$ is the algebraic closure of $\mathcal{K}_0(S)$, where $\mathcal{K}_0$ is the prime subfield and S is a transcendence basis of $\mathcal{K}$, so a maximal algebraically independent subset. The isomorphism type of $\mathcal{K}$ is fully determined by its characteristic and its *transcendence degree*, i. e. the power of S. Moreover, when S is infinite, this transcendence degree equals $|K|$. Now let $|K| \geq \lambda$ and take a subset H of K of power $< \lambda$. As we saw in the last section, H can be replaced by the subfield generated by H; this does not change its size, except when H is finite; however, even in this case, each point in this subfield is H-definable. Every algebraic 1-type over H is clearly realized in $\mathcal{K}$ because $\mathcal{K}$ is algebraically closed. So the point is: can we realize the remaining 1-type, the one of the elements which are transcendental over H, for any H?

If $\mathcal{K}$ has an infinite transcendence degree, then this degree is just $|K| \geq \lambda > |H|$, so it is strictly larger than the transcendence degree of the field generated by H. This means that we can realize our type inside $\mathcal{K}$ for every H.

Otherwise, let H be generated by a finite transcendence basis S of K. Clearly $|H| < \lambda$, but now there is no way to realize our type inside $\mathcal{K}$.

In conclusion, for $|K| \geq \lambda$, $\mathcal{K}$ is λ-saturated if and only if its transcendence degree is infinite. In particular every algebraically closed field satisfying this condition is saturated (in its own power). So every uncountable algebraically closed field is saturated (this applies also to the complex field), while the only countable saturated algebraically closed field (in a fixed characteristic) is the algebraic closure of $\mathcal{K}_0(S)$ where $\mathcal{K}_0$ is the prime subfield and S is a countable (infinite) algebraically independent set.

2. The real ordered field is not $\aleph_0$-saturated. This is perhaps remarkable, if one remembers that the reals are a complete ordered field (in the sense that every Cauchy sequence has a limit), and even the only complete ordered field up to isomorphism: in other words, given a Cauchy sequence $(r_n)_{n>0}$ in $\mathbf{R}$, the 1-type defined by

$$|v - r_n| < \frac{1}{n},$$

when n ranges over positive integers, has a (unique) realization in the real field.

However, look at the type of a positive infinitesimal nonzero element. This is defined by the cut

$$0 < v < \frac{1}{n},$$

for n as before. So it is a type over the empty set. Any finite portion is satisfied among the reals, so this infinitesimal element lives in some elementary extension of the real field. Nevertheless the type cannot be realized in $\mathbf{R}$.

Further examples will be discussed in the next section within modules.

5.5 A parenthesis: pure injective modules

Saturated models are very large, powerful and rich. Within algebraically closed fields, they remind, and actually coincide with the universal domains introduced by André Weil in his "*Foundations of Algebraic Geometry*" [176]; in fact, as we saw in the last section, they are just the algebraically closed fields with an infinite transcendence degree (over the prime subfield). Hence it is not surprising to realize that there do exist in other parts of Algebra some notions resembling saturation, perhaps in a weaker form. This is the case of *pure injectivity* within module theory.

So, fix a (countable) ring R (with identity), and consider (say left) R-modules. We want to recall in this section what a pure injective R-module is, and why these modules are so important. First we need introduce a particular class of embeddings concerning modules.

Definition 5.5.1 *Let $\mathcal{M}$ and $\mathcal{N}$ be R-modules. An R-module homomorphism f from $\mathcal{M}$ into $\mathcal{N}$ if called* **pure** *if, for every pp-formula $\varphi(\vec{v})$ in L_R and every $\vec{a}$ in M,*

$$(\star) \qquad \mathcal{M} \models \varphi(\vec{a}) \quad \Leftrightarrow \quad \mathcal{N} \models \varphi(f(\vec{a})).$$

Notice that a pure homomorphism is 1-1 and hence is an embedding. To see this, just apply the definition to the formula $v_1 = v_2$. Moreover the implication $\Rightarrow$ in $(\star)$ is trivial, hence the qualifying point in the definition of purity is $\Leftarrow$. When $\mathcal{M}$ is a submodule of $\mathcal{N}$, $\mathcal{M}$ is called *pure* in $\mathcal{N}$ if its inclusion embedding is pure; in this case, one says also that $\mathcal{N}$ is a pure extension of $\mathcal{M}$.
The algebraic content of purity is the following: if a linear system $A \cdot \vec{a} = B \cdot \vec{x}$ with parameters $\vec{a}$ from M admits a solution in the extension $\mathcal{N}$, then it admits a solution already in $\mathcal{M}$; here A and B are, of course, matrices with entries in R and suitable sizes.

Examples 5.5.2 1. If $\mathcal{M}$ is an elementary substructure of $\mathcal{N}$, then $\mathcal{M}$ is pure in $\mathcal{N}$. Indeed, $(\star)$ holds for arbitrary formulas.

2. If $\mathcal{M}$ is a direct summand of $\mathcal{N}$, then $\mathcal{M}$ is pure in $\mathcal{N}$. In fact, any solution in $\mathcal{N}$ of a linear system as before projects itself to a solution in $\mathcal{M}$.

3. Let $R = \mathbf{Z}$, so let us deal with abelian groups. We saw in Chapter 2 that, in this particular framework, *pp*-formulas reduce to torsion or divisibility conditions. Hence it is not difficult to realize that, over $\mathbf{Z}$, $\mathcal{M}$ is pure in $\mathcal{N}$ if and only if $rM = M \cap rN$ for every integer r, as usually required in the definition of purity in the handbooks of abelian group theory.

Incidentally, let us quote a noteworthy result of Sabbagh.

Theorem 5.5.3 *Let $\mathcal{M}$ and $\mathcal{N}$ be R-modules, $\mathcal{M}$ be a submodule of $\mathcal{N}$. Then $\mathcal{M}$ is an elementary substructure of $\mathcal{N}$ if and only if $\mathcal{M}$ is pure in $\mathcal{N}$ and $\mathcal{M}$ and $\mathcal{N}$ are elementarily equivalent.*

Now we can introduce pure injectivity: this notion restricts in some sense the usual injectivity requirement to pure embeddings.

Definition 5.5.4 *An R-module $\mathcal{M}$ is called pure injective p. i. (but someone prefers to say* algebraically compact*) if and only if one of the following equivalent conditions holds:*

(i) *For every choice of R-modules $\mathcal{N}$ and $\mathcal{N}'$, for which $\mathcal{N}$ is a pure submodule of $\mathcal{N}'$, every homomorphism f from $\mathcal{N}$ to $\mathcal{M}$ lifts to a homomorphism f' from $\mathcal{N}'$ to $\mathcal{M}$.*

(ii) *$\mathcal{M}$ is a direct summand of every pure extension.*

The equivalence between (i) and (ii) is proved in the references quoted at the end of the chapter. (i) and (ii) chacterize pure injectivity from the algebraic point of view. But there is a third equivalent definition, disclosing a model theoretic flavour and showing that pure injectivity is directly related to saturation.

Theorem 5.5.5 *An R-module $\mathcal{M}$ is pure injective if and only if, for every countable subset X of M and every (incomplete) 1-type p of pp-formulas $\varphi(v, \vec{a})$ with parameters $\vec{a}$ from X, when every finite portion of p has a realization in M, then the whole p can be satisfied in M.*

Hence pure injectivity is just a weak form of saturation, restricted to incomplete 1-types of *pp*-formulas (they are usually called *pp*-types). Consequently every $\aleph_1$-saturated R-module $\mathcal{M}$ is p. i.; this fact, and Theorem 5.4.2 clearly imply that any R-module enlarges to a p. i. elementary extension. But something much stronger holds.

Theorem 5.5.6 *Let $\mathcal{M}$ be an R-module. Then $\mathcal{M}$ has a p.i. pure extension $\overline{\mathcal{M}}$ which is minimal in the following sense: for every pure and p.i. extension $\mathcal{N}$ of $\mathcal{M}$, $\mathcal{N}$ contains $\overline{\mathcal{M}}$ as a pure submodule up to isomorphism. Moreover $\overline{\mathcal{M}}$ is unique up to isomorphism fixing M pointwise. Finally, $\overline{\mathcal{M}}$ is an elementary extension of $\mathcal{M}$.*

For a proof, see the references at the end of the chapter.
These existence and uniqueness results justify the specific symbol $\overline{\mathcal{M}}$ to denote this p.i. extension of $\mathcal{M}$ and, furthermore, a special name to dub it; actually, $\overline{\mathcal{M}}$ is called "the" *pure injective hull* of $\mathcal{M}$.

Up to elementary equivalence, p. i. R-modules represent the whole class of R-modules, hence their algebraic analysis can help a description of the possible completions of $_RT$. But what can we say about a possible classification of pure injective R-modules? A usual and general technique in studying a given class of modules is

(a) to look for a possible (and hopefully unique) direct sum decomposition of any module of the class into indecomposable objects,

and then

(b) to classify the indecomposable modules of the class up to isomorphism.

By the way, recall that an R-module $\mathcal{M}$ is called *indecomposable* if $\mathcal{M} \neq 0$ but there is no way to express $\mathcal{M}$ as a direct sum of two nonzero submodules. This procedure is not always successful within arbitrary classes of modules, but is sufficiently satisfactory when applied to p. i. modules. Let us explain why. First of all it is comparatively easy to check that pure injectivity is preserved by direct summands. Moreover (a) does work within p. i. modules, due to the following result of Fisher and Ziegler.

Theorem 5.5.7 (Fisher-Ziegler) *Let $\mathcal{M}$ be a p.i. R-module. Then $\mathcal{M}$ decomposes uniquely (up to isomorphism) as*

$$\overline{\oplus_i \mathcal{U}_i} \oplus E,$$

where E has no indecomposable direct summand, $\overline{}$ denotes p.i. hull and, for every i, $\mathcal{U}_i$ is an indecomposable p.i. R-module. Moreover $\mathcal{M}$ is elementarily equivalent to $\oplus_i \mathcal{U}_i$.

Incidentally, pure injectivity is not preserved under infinite direct sums. To summarize what we have seen so far:

- every R-module is an elementary substructure of its p.i. hull;
- every p.i. R-module is elementarily equivalent to a direct sum of indecomposable p.i.'s, and even isomorphic to the p.i. hull of this sum up to a summand with no indecomposable direct factors.

So what we have to do now is to study isomorphism classes, or elementary equivalence classes of indecomposable p.i. R-modules.
Classification up to isomorphism is sometimes easy. For instance, when R is a field, so when we are dealing with R-vectorspaces, the only indecomposable object is just R (as a vectorspace over itself) and is pure injective.

But over other rings the situation changes dramatically and in some cases one has to conclude that no classification is possible. This is what happens, for instance, when $R = K\langle x, y\rangle$ is the ring of polynomials over a field K in two non-commuting variables x and y, so when one considers K-vectorspaces with two additional linear operators, corresponding to the action of x, y respectively; in fact, this class of modules encodes in some way the word problem for groups, and there are some good reasons to believe that this forbids any classification of indecomposable p.i. objects. See [136] for a detailed discussion of this point. As a consequence, even modules over $K[x, y]$ (with two commuting unknowns x and y) and, in another direction, over $K\langle x_1, \ldots, x_n\rangle$ with $n > 2$ inherit the same "wild" situation, excluding any possible classification.
So the key point is to realize for which rings R indecomposable p.i. modules can be classified up to isomorphism. In this perspective we would like to quote a beautiful result of Ziegler, using model theory to equip in a natural way indecomposable p.i. modules over any fixed ring R with a topological space structure.
In fact, let $_RZg$ denote the set of (isomorphism classes of) indecomposable p.i. R-modules. For every choice of *pp*-formulas $\varphi(v)$ and $\psi(v)$ (in one free variable) in L_R, let (φ / ψ) be the set of the elements $\mathcal{U}$ in $_RZg$ such that $\varphi(\mathcal{U})$ properly includes $\varphi(\mathcal{U}) \cap \psi(\mathcal{U})$. Then the (φ / ψ)'s are a basis for a topology of $_RZg$, which is always compact and seldom Hausdorff. $_RZg$ with this topology is called the *(left) Ziegler spectrum* of R. Again, see [136] or directly [181] for more details. What we can say in the restricted framework of these pages is that the knowledge of the Ziegler spectrum of R is a sort of fixed course towards the solution of several significant model theoretic (and also algebraic) problems about R-modules.
In particular, a successful analysis of the Ziegler spectrum can provide some useful information on the elementary equivalence class of any R-module $\mathcal{M}$. Let us see very briefly and roughly why. We saw in Chapter 2 that the complete theory of $\mathcal{M}$ is fully determined by the values (modulo ∞) of $[\varphi(\mathcal{M}) : \psi(\mathcal{M})]$ where $\varphi(v)$ and $\psi(v)$ range over *pp*-formulas in one free variable in L_R. Now, up to elementary equivalence, $\mathcal{M}$ can be replaced by a direct sum of indecomposable p. i . R-modules $\oplus_i \mathcal{U}_i$, where

$$\overline{\mathcal{M}} = \overline{\oplus_i \mathcal{U}_i} \oplus E$$

is the canonical decomposition of the pure injective hull $\overline{\mathcal{M}}$ of $\mathcal{M}$ according to the Fisher-Ziegler Theorem. Consequently, for $\varphi(v)$ and $\psi(v)$ as before,

$$[\varphi(\mathcal{M}) : \psi(\mathcal{M})] = [\varphi(\oplus_i \mathcal{U}_i) : \psi(\oplus_i \mathcal{U}_i)]$$

modulo ∞ (in the sense that these values are either finite and equal or both infinite). Now it is easy to realize that

$$[\varphi(\oplus_i \mathcal{U}_i) : \psi(\oplus_i \mathcal{U}_i)] = \sum_i [\varphi(\mathcal{U}_i) : \psi(\mathcal{U}_i)]$$

modulo ∞. So, when in the Ziegler spectrum the points (i.e. the indecomposable p.i. R-modules) are classified as well as their basic open neighbourhoods, the elementary equivalence type of $\mathcal{M}$ can be determined by saying which $\mathcal{U} \in_R Zg$ are involved in the decomposition $\oplus_i \mathcal{U}_i$ and how many times they occur. For instance, if U is known to be isolated -say by (φ / ψ)-, then the value (modulo ∞) of $[\varphi(\mathcal{M}) : \psi(\mathcal{M})]$ can witness if U occurs in the decomposition, and how many times. Of course, for a non-isolated $\mathcal{U}$, a finer analysis is necessary. In conclusion, when the spectrum is known, looking at the points in $_RZg$ and at their open neighbourhoods, one can effectively list the complete extensions of $_RT$. Let us illustrate this by some examples.

Examples 5.5.8 1. Let R be the ring $\mathbf{Z}$ of integers. Recall that every *pp*-formula $\varphi(v)$ of $L_{\mathbf{Z}}$ is logically equivalent within $_{\mathbf{Z}}T$ to a conjunction of torsion or divisibility conditions

$$rv = 0, \quad p^k | p^l v$$

where p is a prime, $l < k$ are positive integers and r is an integer. An effective list of indecomposable p. i. $\mathbf{Z}$-modules is known, and includes (up to isomorphism) exactly the following objects:

- the finite modules $\mathbf{Z}/p^n\mathbf{Z}$,
- the Prüfer groups $\mathbf{Z}/p^\infty$,
- the p-adic integers $\overline{\mathbf{Z}_{(p)}}$,
- the additive group of rationals $\mathbf{Q}$

where p ranges over primes and n over positive integers; the p-adic integers are just the p. i. hull of the localization $\mathbf{Z}_{(p)}$ of $\mathbf{Z}$ at p. It is not prohibitive to realize that the previous examples are indecomposable and p. i.; for instance, the pure injectivity of the Prüfer groups and the rationals comes directly from their divisibility. But the point is to show that their list exhausts all the possible cases: see [181] for a discussion, and a complete proof of this.

Now we have to study the topology of the Ziegler spectrum $_{\mathbf{Z}}Zg$. Here is its Cantor-Bendixson analysis.

- The isolated points (those having rank 0) are the finite modules $\mathbf{Z}/p^n\mathbf{Z}$; in fact, each of them is the only element in the open set $((p^{n-1}|v \wedge pv = 0) \,/\, (p^n|v \wedge pv = 0))$.
- Now forget the $\mathbf{Z}/p^n\mathbf{Z}$'s; then a Prüfer group $\mathbf{Z}/p^\infty$ gets isolated by $((p^n|v \wedge pv = 0) \,/\, v = 0)$ for any positive n, while the p-adic integers are isolated by $(p^{n-1}|v \,/\, p^n|v)$ for any n; hence all these groups have rank 1.
- The rational group is the only remaining point, and so gets rank 2.

This analysis was pursued by Eklof-Fisher in the particular case when R is a principal ideal domain, and supports and clarifies the previous work of Wanda Szmielew (sketched in our Chapter 2) for abelian groups. In particular, as a consequence, it implies the decidability of the theory ${}_{\mathbf{Z}}T$. Ziegler's approach was inspired by these particular cases, but is fully general and covers any ring R.

2. Hence what has been said on $\mathbf{Z}$ actually enlarges to principal ideal domains and, partly, to Dedekind domains. For instance, it applies to the ring $\mathcal{K}[x]$ of polynomials over a field $\mathcal{K}$ with a single unknown: also in this case one can accomplish a quite satisfactory description of the Ziegler spectrum, essentially repeating the analysis for the integers with the necessary variants. Compare this and what was observed before when the number of unknowns increases.

5.6 Omitting types

Saturated models realize many types. But non isolated types are not easy to satisfy in arbitrary models. Let us see why. We consider a complete theory T, and we denote its universe by Ω. Recall that we met isolated types in 5.3: a type p over a set $X \subseteq \Omega$ is isolated when p is the only type over X containing some $L(X)$-formula $\varphi(v)$. It is an easy exercise to check what follows.

Fact 5.6.1 For $p \in S(x)$, the following propositions are equivalent.

1. p is isolated;
2. there is some $L(X)$-formula $\varphi(\vec{v}) \in p$ such that p is just the set of $L(X)$-formulas $\psi(\vec{v})$ for which $\varphi(\Omega^n) \subseteq \psi(\Omega^n)$;

3. there is some $L(X)$-formula $\varphi(\vec{v})$ such that $\varphi(\Omega^n) \neq \emptyset$ and p is just the set of the $L(X)$-formulas $\psi(\vec{v})$ for which $\varphi(\Omega^n) \subseteq \psi(\Omega^n)$.

Actually what distinguishes 2 and 3 is that in 2 we require the additional condition $\varphi(\vec{v}) \in p$. But it is easy to see that $\varphi(\vec{v}) \notin p$ forces $\neg\varphi(\vec{v}) \in p$, hence $\varphi(\Omega^n) \subseteq \neg\varphi(\Omega^n)$, and so $\varphi(\Omega^n) = \emptyset$, a contradiction.
Now an isolated type p over X is trivially realized in every model $\mathcal{M}$ of $\mathrm{Th}(\Omega_X)$. For, Ω does contain some tuple satisfying p, and in particular the formula $\varphi(\vec{v})$ isolating p. So $\exists\vec{v}\varphi(\vec{v}) \in Th(\Omega_X)$.
On the contrary a non-isolated type p is not always realized. This is the content of the following result.

Theorem 5.6.2 (Omitting Types) *Let T be a complete theory in a language L, p be a non-isolated type over $\emptyset$. Then there exists a (countable) model $\mathcal{M}$ of T omitting p (in the sense that $p(\mathcal{M}) = \emptyset$).*

Proof. For simplicity, assume $p \in S_1(\emptyset)$ (the reader can easily adapt the following argument to more variables, if he (she) likes). We use some usual techniques of classical Model Theory. In detail, first we extend our language L by countably many new constant symbols

$$c_0, c_1, \ldots, c_n, \ldots \ (n \text{ natural}).$$

List the sentences of the enlarged language $L' = L \cup \{c_n : n \in \mathrm{N}\}$

$$\varphi_0, \varphi_1, \ldots, \varphi_n, \ldots$$

Now we build an increasing sequence of consistent theories in L'

$$T_0 \subseteq T_1 \subseteq \ldots \subseteq T_n \ldots,$$

all enlarging T and satisfying the following conditions, for every n:

1. T_n is axiomatized by T_0 and a unique further L'-sentence θ_n;

2. either φ_n or $\neg\varphi_n$ is in T_{n+1};

3. whenever φ_n or $\neg\varphi_n$ is in T_{n+1} and has the form $\exists v\psi_n(v)$, then $\psi_n(c_m) \in T_{n+1}$ for some new constant c_m not already occuring in any φ_i, θ_i for $i \leq n$;

4. finally, there is some formula $\sigma_n(v) \in p$ for which $\neg\sigma_n(c_n) \in T_{n+1}$.

T_0 is just T, or, more precisely, the L'-theory axiomatized by T. Now, assume T_n given, and let us build T_{n+1}.
Let $c_{i_0}, \ldots, c_{i_{h-1}}, c$ be the constants in $L' - L$ occuring in θ_n; replace these constants by new variables $w_0, \ldots, w_{h-1}, v$ chosen out of θ_n; place before $\exists w_0, \ldots, \exists w_{h-1}$ and get a formula $\theta'_n(v)$ in L, with the only free variable v. Take a model $\mathcal{M}'_n$ of T_n; as $\theta_n \in T_n$, $\theta'_n(\mathcal{M}'_n)$ is not empty (it includes $c^{\mathcal{M}'_n}$). As p is not isolated, the condition 3 in Fact 5.6.1 ensures that, for some formula $\sigma_n(v) \in p$,

$$\exists v(\theta'_n(v) \wedge \neg\sigma_n(v)) \in T \subseteq T_n.$$

Put $\neg\sigma_n(c_n)$ in T_{n+1}. This guarantees 4. Now look at φ_n. If $T_n \cup \{\neg\sigma_n(c_n), \varphi_n\}$ has some model, let T_{n+1} include also φ_n. Otherwise $\neg\varphi_n$ is a consequence of $T_n \cup \{\neg\sigma_n(c_n)\}$. This gives 2. When φ_n (or $\neg\varphi_n$) is of the form $\exists v \varphi_n(v)$, pick the least m such that c_m does not occur in φ_i, θ_i for $i \leq n$, and put $\psi_n(c_m)$ in T_{n+1}. This ensures 3.
Let T_{n+1} be the theory obtained in this way. Clearly T_{n+1} is consistent and satisfies 1.
Now form $T' = \bigcup_n T_n$. T' is consistent. Take any model $\mathcal{M}'$ of T'. By 2, T' is complete, $T' = \mathrm{Th}(\mathcal{M}')$: the sentences $\exists v \psi(v)$ $\mathcal{M}'$ satisfies are just those occuring in some T_n. Owing to 3, we can apply the Tarski-Vaught Theorem and deduce that $\{c_n^{\mathcal{M}'} : n \text{ natural}\}$ is the domain of a (countable) elementary substructure of $\mathcal{M}'$, and hence a countable model of T'. Its restriction to L is a countable model of T, and omits p because, for every n, $c_n^{\mathcal{M}'}$ cannot satisfy $\sigma_n(v)$. ♣

5.7 The Morley rank, at last

Let T be a complete theory in a countable language. We remind the program outlined in 5.2: we aim at measuring the complexity of definable sets X in Ω (equivalently, of $\sim$-classes of formulas with parameters in Ω) and, more generally, of sets of formulas with parameters in Ω, including types over *small* subsets of Ω. A reasonable way to get this is to define some function assigning, if possible, to every definable, or formula, or type, an ordinal value, according to some reasonable conditions, like monotonicity. An axiomatic introduction to this complexity measures can be found in [8] or [131, 134]; these functions are usually called *ranks*. There are several possible ways to define a rank, according to some particular algebraic or model theoretic features of the theory T. Morley's rank was the very first,

and fundamental, example in this direction. Historically, it was inspired by the Cantor-Bendixson analysis of topological spaces S. Recall that this equips any point in S with a rank, whose value is either an ordinal number or ∞ (where one assumes that ∞ is greater than any ordinal); in particular the Cantor-Bendixson rank assigns the value 0 to the isolated points of S, the value 1 to the points in S that get isolated within the relative topology once the isolated points are forgotten, and so on. Of course one can apply this Cantor-Bendixson machinery to the topological space $S(C)$, where C is a small subset of Ω, and hence to the types over C. Using the Stone duality between topological Boolean spaces and Boolean algebras, one can define a Cantor-Bendixson rank also for the elements of the dual algebra of $S(C)$, and hence for C-definable sets. However, if one translates literally the Cantor-Bendixson analysis into our particular setting, then the same definable object may obtain several possible ranks, according to which basic set of parameters C we refer to; in particular, if we replace C by a larger C', then we can expect that, over C', C-definable sets get a stronger complexity and consequently a bigger Cantor-Bendixson rank: some examples of this phenomenon will be provided later in the present section.
Morley's rank adapts the Cantor-Bendixson approach to obviate this difficulty. The recipe is just to refer to our universe Ω and so to evaluate any definable within arbitrary Ω-definable sets. But it is time to give, at last, the exact definition of the Morley rank; we will introduce also a related notion (*Morley degree*).
We consider a complete theory T in a countable language L: Ω denotes, as usual, the universe of T. Let X be a definable subset of Ω^n (for some positive integer n), we want to define the *Morley rank* of X $RM(X)$. First let us say what

$$RM(X) \geq \alpha$$

means for every ordinal α. We proceed by induction on α.

Definition 5.7.1 *When $\alpha = 0$, put $RM(X) \geq \alpha$ if and only if $X \neq \emptyset$. When α is a limit ordinal, put $RM(X) \geq \alpha$ if and only if $RM(X) \geq \nu$ for every ordinal $\nu < \alpha$. Finally, when $\alpha = \nu + 1$ is a successor ordinal, put $RM(X) \geq \alpha$ if and only if there are infinitely many pairwise disjoint definable subsets X_i ($i \in \mathbf{N}$) of X such that $RM(X_i) \geq \nu$ for every $i \in \mathbf{N}$.*

It is easy to observe that, if α, β are ordinals and $\alpha \geq \beta$, then

$$RM(X) \geq \alpha \quad \text{implies} \quad RM(X) \geq \beta.$$

Accordingly it makes sense to put:

Definition 5.7.2 *1. If $X = \emptyset$, then $RM(X) = -1$.*

2. *Let $X \neq \emptyset$ and suppose that there is some ordinal α such that $RM(X) \not\geq \alpha$ (so $RM(X) \not\geq \beta$ for every ordinal $\beta \geq \alpha$); the least ordinal α_0 with this property is a successor; if $\alpha_0 = \nu + 1$, then put $RM(X) = \nu$.*

3. *Finally suppose $X \neq \emptyset$, $RM(X) \geq \alpha$ for every ordinal α. Then put $RM(X) = \infty$.*

(Of course we assume $-1 < \alpha < \infty$ for every ordinal α). When $\varphi(\vec{v})$ is a formula (possibly with parameters from Ω), define

$$RM(\varphi(\vec{v})) = RM(\varphi(\Omega^n))$$

(where n is the length of $\vec{v}$).

Examples 5.7.3 1. $RM(X) = -1$ if and only if $X = \emptyset$.

2. $RM(X) = 0$ if and only if X is finite and non-empty. For instance, in algebraically closed fields, the (definable) zero set of a given polynomial of degree ≥ 1 in one unknown is finite and non-empty, and hence has Morley rank 0.

3. $RM(X) = 1$ if and only if X is infinite, but cannot partition as the union of infinitely many pairwise disjoint infinite definable subsets. For example, in algebraically closed fields, the only definable infinite 1-ary sets are cofinite. This implies that their Morley rank is 1. Notice that the same is true for infinite vectorspaces over a countable field.

4. Let $T = DLO^-$ be the theory of dense linear orders without endpoints. Fix $a < b$ in Ω and consider the interval $I =]a, b[= \{x \in \Omega : a < x < b\}$. I is definable, and $RM(I) = \infty$. In fact, for every a, b as before and ordinal α, $RM(]a, b[) \geq \alpha$. This can be easily checked by induction on α. The cases $\alpha = 0$ and α limit are trivial. So suppose $\alpha = \nu + 1$ for some ν. As the order $\leq$ is dense, we can choose $c \in]a, b[$, and observe that $]a, b[$ includes the disjoint intervals $]a, c[$, $]c, b[$, both having Morley rank $\geq \nu$, because of the induction hypothesis. Repeating this procedure, one finds infinitely many pairwise disjoint open intervals in $]a, b[$, all of Morley rank $\geq \nu$. Hence $RM(]a, b[) \geq \nu + 1$. Notice that even $[a, b[$, $]a, b]$, $[a, b]$ (for $a < b$) have Morley rank ∞.

Let us point out now some simple properties of RM:

Proposition 5.7.4 *Let X, Y be two definable subsets of Ω^n for some positive integer n.*

(i) *If $X \subseteq Y$, then $RM(X) \leq RM(Y)$.*

(ii) *For every automorphism f of Ω, $RM(X) = RM(f(X))$.*

(iii) *$RM(X \cup Y) = max\{RM(X), RM(Y)\}$.*

Proof. (i) Every definable subset of X is clearly a definable subset also of Y. So a simple induction argument shows that, for every ordinal α, if $RM(X) \geq \alpha$, then $RM(Y) \geq \alpha$.
(ii) It suffices to show $RM(X) \leq RM(f(X))$ (as the opposite relation $RM(X) \geq RM(f(X))$ can be obtained just by reversing the roles of X and $f(X)$ and replacing f by f^{-1}). In other words, we have to check that, for every ordinal α, $RM(X) \geq \alpha$ implies $RM(f(X)) \geq \alpha$. Proceed again by induction on α (and observe that, if Z is a definable subset of X, then $f(Z)$ is a definable subset of $f(X)$).
(iii) $\geq$ follows from (i). To check $\leq$, it suffices to prove that, for every ordinal α, $RM(X \cup Y) \geq \alpha$ implies either $RM(X) \geq \alpha$ or $RM(Y) \geq \alpha$. As before, we can proceed by induction on α. If $\alpha = 0$ and $RM(X \cup Y) \geq \alpha$, then $X \cup Y \neq \emptyset$ and consequently either X or Y is not empty. Now suppose α limit and $RM(X \cup Y) \geq \alpha$, namely $RM(X \cup Y) \geq \nu$ for every ordinal $\nu < \alpha$. If there exists some $\nu < \alpha$ such that $RM(X) \geq \mu$ for every ordinal μ satisfying $\nu \leq \mu < \alpha$, then $RM(X) \geq \alpha$. Otherwise, for every ordinal $\nu < \alpha$, there is some $\mu \geq \nu$ such that $\mu < \alpha$ and $RM(X) \not\geq \mu$. By induction, $RM(Y) \geq \mu$, whence $RM(Y) \geq \nu$. It follows $RM(Y) \geq \alpha$. At last, take a successor $\alpha = \nu + 1$ and suppose $RM(X \cup Y) \geq \alpha$. Then there are infinitely many pairwise disjoint infinite definable subsets Z_i ($i \in \mathrm{N}$) of $X \cup Y$, all having $RM \geq \nu$. Look at the sets

$$X_i = X \cap Z_i, \quad Y_i = Y \cap Z_i \qquad (i \in \mathrm{N}).$$

For every i, either $RM(X_i) \geq \nu$ or $RM(Y_i) \geq \nu$. Accordingly $RM(X_i) \geq \nu$, or $RM(Y_i) \geq \nu$ for infinitely many i's. In the former case $RM(X) \geq \alpha$; in the latter $RM(Y) \geq \alpha$. ♣

Now we are going to associate with any definable subset $X \subseteq \Omega^n$ (whose Morley rank is an ordinal) a positive integer: the *Morley degree* of X $GM(X)$.

Proposition 5.7.5 *Let $X \subseteq \Omega^n$ be a definable set whose Morley rank is an ordinal α. Then there is a maximal positive integer d such that X partitions as the union of d pairwise disjoint definable subsets of Morley rank α.*

Proof. Suppose not. Then, for every positive integer d, one can partition X as the union of d definable subsets of Morley rank α

$$X(d, 0), \ldots, X(d, d-1).$$

We can assume that, for every d and $i \leq d$, there is some $j < d$ such that $X(d+1, i) \subseteq X(d, j)$. In fact, $X(d+1, i)$ decomposes as the disjoint union of its subsets $X(d+1, i) \cap X(d, j)$ for $j < d$. Owing to Proposition 5.7.4, (iii), at least one of these subsets has the same Morley rank α as $X(d+1, i)$. By Proposition 5.7.4, (i), all these subsets have Morley rank $\leq \alpha$. Accordingly we can replace $X(d+1, i)$ by a subset $X(d+1, i) \cap X(d, j)$ (to which one possibly adds other subsets $X(d, j)$ of Morley rank $\leq \alpha$). Now a combinatorial argument (König's Lemma) ensures that there are positive integers

$$d_0 < d_1 < \ldots < d_m < \ldots \qquad (m \in \mathbf{N})$$

and natural numbers

$$i_0, i_1, \ldots, i_m, \ldots \qquad (m \in \mathbf{N})$$

such that, $\forall m \in \mathbf{N}$, $i_m < d_m$ and

$$X(d_m, i_m) \supset X(d_{m+1}, i_{m+1}), \quad RM(X(d_m, i_m) - X(d_{m+1}, i_{m+1})) = \alpha.$$

We obtain in this way infinitely many pairwise disjoint definable subsets of X

$$X(d_m, i_m) - X(d_{m+1}, i_{m+1})$$

(where m ranges over $\mathbf{N}$), all having Morley rank α. Consequently $RM(X) \geq \alpha + 1$, and this is a contradiction. ♣

Owing to this proposition, we can at last introduce the Morley degree as follows.

Definition 5.7.6 *Let X be a definable subset of Ω^n (for some positive integer n) whose Morley rank is an ordinal α. The* Morley degree *of X (denoted $GM(X)$) is the maximal positive integer d such that X can partition as the union of d definable subsets of Morley rank α.*

Remarks 5.7.7 1. If X is finite and non-empty (in other words $RM(X) = 0$), $GM(X)$ is just the size $|X|$ of X. In particular, when X is the zero set of a polynomial $f(x)$ of positive degree in one unknown over an algebraically closed field $\mathcal{K}$, then the Morley degree of X is just the number of roots of $f(x)$ in K.

2. A definable X having Morley rank 1 and Morley degree 1 (so an infinite definable X admitting no infinite coinfinite definable subset) is called *strongly minimal.* A theory T is strongly minimal if its universe (so the formula $v = v$) is: we met this notion in Chapter 2, and we shall deal again with it in the next sections. Recall that algebraically closed fields, as well as infinite vectorspaces over a countable field, admit a strongly minimal theory.

3. Let X, Y be two definable disjoint subsets of Ω^n such that both $RM(X)$ and $RM(Y)$ are ordinals, and $RM(X) \leq RM(Y)$. Then $GM(X \cup Y)$ equals

   $GM(X) + GM(Y)$ if $RM(X) = RM(Y)$,
   $GM(Y)$ otherwise

   (the reader may check this as an exercise).

When $\varphi(\vec{v})$ is a formula (possibly with parameters from Ω) and $RM(\varphi(\Omega^n))$ is an ordinal, we put $GM(\varphi(\vec{v})) = GM(\varphi(\Omega^n))$.

Now we want to discuss the following problem. Let X be a definable subset of Ω^n (for some positive integer n), α denote the Morley rank of X and d its Morley degree (if any): both α and d can be calculated by looking at the definable subsets of X. But suppose that X is M-definable for some model $\mathcal{M}$ of T. Then we wonder whether the values of α and d are already witnessed by the M-definable subsets of X, in other words whether α and d remain the same after replacing "definable" by "M-definable" in the definition of RM and GM.

Examples 5.7.8 1. We know that $RM(X) = 0$ if and only if X is finite; furthermore, in this case, $GM(X)$ is just the size $|X|$ of X. Suppose $X = \varphi(\Omega^n, \vec{a})$ for a suitable tuple $\vec{a}$ of parameters of a model $\mathcal{M}$ of T. As $\mathcal{M}$ is an elementary substructure of Ω, $X = \varphi(\mathcal{M}^n, \vec{a})$. So the elements of X in M can witness $RM(X) = 0$, $GM(X) = |X|$.

2. $RM(X) = 1$ if and only if X is infinite, but X cannot contain infinitely many pairwise disjoint infinite definable subsets. As before, suppose

$X = \varphi(\Omega^n, \vec{a})$ for $\vec{a}$ in $\mathcal{M} \models T$. As $\mathcal{M} < \Omega$, $\varphi(\mathcal{M}^n, \vec{a})$ is infinite, so $\mathcal{M}$ can witness $RM(X) \geq 1$. It is also clear that $\varphi(\mathcal{M}^n, \vec{a})$ cannot contain infinitely many pairwise disjoint infinite M-definable sets.

3. Now consider a structure $\mathcal{M}_0 = (M_0, E)$ where E is an equivalence relation having exactly one equivalence class of size k for every positive integer k, but no infinite class. Let T denote the theory of $\mathcal{M}_0$. One easily sees that $\mathcal{M}_0$ is elementarily embeddable in every model of T. But a simple application of Compactness Theorem shows that there exist some other models of T containing also infinite equivalence classes, and indeed, for every positive integer t, T has a countable model $\mathcal{M}_t$ with exactly t infinite classes, while the universe Ω has infinitely many infinite classes. Now consider the formula $v = v$. The set it defines is just Ω, and consequently has Morley rank ≥ 2, because it partitions into infinitely many infinite pairwise disjoint definable subsets (the infinite classes of E). Indeed one sees that Ω has Morley rank 2 and Morley degree 1. On the other side, there is only one non-algebraic 1-type over M_0 as, for every a and b in $\Omega - M_0$, one can find an M_0-automorphism of Ω mapping a in b. In particular, every M_0-definable set is finite or cofinite according to whether it excludes or includes the elements in $\Omega - M_0$. Hence Ω cannot partition in two infinite M_0-definable subsets, and $\mathcal{M}_0$ cannot witness $RM(\Omega) \geq 2$.

So we can wonder which models $\mathcal{M}$ of a given T can witness the Morley rank and degree of every M-definable set. As a partial answer, let us show that $\aleph_0$-saturated models have this feature.

Proposition 5.7.9 *Let T be a complete theory, $\mathcal{M}$ be an $\aleph_0$-saturated model of T. Then $\mathcal{M}$ can witness the Morley rank and (if it exists) the Morley degree of any M-definable set X.*

Proof. Put $X = \varphi(\Omega^n, \vec{a})$ for some suitable positive integer n and $\vec{a} \in M$. First let us deal with $RM(X)$. It suffices to show that, for every ordinal ν, if $RM(X) \geq \nu + 1$, then there are infinitely many pairwise disjoint M-definable subsets of X, all having Morley rank $\geq \nu$. On the other side we do know that there are infinitely many pairwise disjoint subsets $X_i \subseteq X$ ($i \in \mathbb{N}$), all having $RM \geq \nu$. For every $i \in \mathbb{N}$, let $X_i = \varphi_i(\Omega^n, \vec{a_i})$ with $\vec{a_i} \in \Omega$. As $\mathcal{M}$ is $\aleph_0$-saturated, for every natural k there are $\vec{b_0}, \ldots, \vec{b_k}$ in M such that $tp(\vec{a_0}, \ldots, \vec{a_k} / \vec{a}) = tp(\vec{b_0}, \ldots, \vec{b_k} / \vec{a})$. Consequently, for every $i \leq k$, $\varphi_i(\Omega^n, \vec{b_i}) \subseteq \varphi(\Omega^n, \vec{a})$ because $\forall \vec{w}(\varphi_i(\vec{w}, \vec{v_i}) \rightarrow \varphi(\vec{w}, \vec{a}))$ lies in the type of $(\vec{a_0}, \ldots, \vec{a_k})$ over $\vec{a}$. Moreover, for $i < j \leq k$, $\varphi_i(\Omega^n, \vec{b_i}) \cap \varphi(\Omega^n, \vec{b_j}) =$

$\emptyset$ because $\neg\exists\vec{w}\,(\varphi_i(\vec{w},\vec{v_i})\ \wedge\ \varphi_j(\vec{w},\vec{v_j}))$ is in $tp(\vec{a_0},\ldots,\vec{a_k}\,/\vec{a})$. Notice also that, $\forall i\leq k$, $\vec{a_i}$, $\vec{b_i}$ correspond to each other by some automorphism of Ω fixing $\vec{a}$ pointwise, and consequently $\varphi_i(\Omega^n,\vec{a_i})$ and $\varphi_i(\Omega^n,\vec{b_i})$ have the same $RM\geq\nu$. Finally, for every $i\leq k$, $\varphi(\Omega^n,\vec{b_i})$ is M-definable because $\vec{b_i}$ is in M. In conclusion, for every natural k, X contains at least $k+1$ M-definable pairwise disjoint subsets of Morley rank $\geq\nu$. So, the combinatorial argument in Proposition 5.7.5 can be repeated and shows our claim.
Now let $RM(X)=\alpha$ where α is an ordinal, $d=GM(X)$. We have to show that X can partition as the union of d M-definable subsets of Morley rank α. Again we know that X does admit such a decomposition in definable subsets

$$\varphi_0(\Omega^n,\vec{a_0}),\ldots,\varphi_{d-1}(\Omega^n,\vec{a}_{d-1})$$

with $\vec{a_0},\ldots,\vec{a_{d-1}}$ in Ω. On the other hand $tp(\vec{a_0},\ldots,\vec{a_{d-1}}\,/\vec{a})$ can be realized in $\mathcal{M}$, say by $\vec{b_0},\ldots,\vec{b}_{d-1}$, because $\mathcal{M}$ is $\aleph_0$-saturated. By proceeding as before, one sees that

$$\varphi_0(\Omega^n,\vec{b_0}),\ldots,\varphi_{d-1}(\Omega^n,\vec{b}_{d-1})$$

provide a partition as required. ♣

Now we want to define the Morley rank and degree of a complete type p over a small subset A of Ω: the Morley rank of p $RM(p)$ is an ordinal or ∞, while the Morley degree $GM(p)$ is defined only when $RM(p)$ is an ordinal, and is a positive integer. The reader can easily observe that this assignment of a Morley rank and degree can be easily extended to arbitrary set of formulas over A (although, in this enlarged framework, the Morley rank may assume also the value -1 -for inconsistent set of formulas-).

Definition 5.7.10 *Let A be a small subset of Ω, $p\in S(A)$. The Morley rank of p $RM(p)$ is the least Morley rank of a formula in p. If $RM(p)$ is an ordinal α, $GM(p)$ (the Morley degree of p) is the least Morley degree of a formula in p of Morley rank α.*

Accordingly, with any $p\in S(A)$ it is associated a formula $\varphi(\vec{v})\in p$ such that $RM(\varphi(\vec{v}))=RM(p)$ and, when $RM(p)$ is an ordinal, $GM(\varphi(\vec{v}))=GM(p)$. Of course, one may wonder whether this formula $\varphi(\vec{v})$ is unique. To clarify this question, first notice:

Proposition 5.7.11 *Let $p\in S(A)$ satisfy $RM(p)<\infty$, $\varphi(\vec{v})$ be a formula in p having the same Morley rank and degree as p. Let $\varphi'(\vec{v})$ be any formula with parameters from A. Then $\varphi'(\vec{v})\in p$ if and only if $RM(\varphi(\Omega^n)-\varphi'(\Omega^n))<RM(p)$.*

Proof. First suppose $\varphi'(\vec{v}) \in p$. Then $\varphi(\vec{v}) \wedge \varphi'(\vec{v}) \in p$; as $RM(\varphi(\vec{v}) \wedge \varphi'(\vec{v})) \leq RM(\varphi(\vec{v}))$, $\varphi(\vec{v}) \wedge \varphi'(\vec{v})$ have the same Morley rank and degree as $\varphi(\vec{v})$, hence as p. Consequently $RM(\varphi(\vec{v}) \wedge \neg\varphi'(\vec{v})) < RM(p)$.
Conversely, suppose $RM(\varphi(\Omega^n) - \varphi'(\Omega^n)) < RM(p)$, so $\varphi(\vec{v}) \wedge \neg\varphi'(\vec{v})$ cannot belong to p. In particular $\neg\varphi'(\vec{v}) \notin p$, and so $\varphi'(\vec{v}) \in p$. ♣

At this point one deduces:

Corollary 5.7.12 *Let $p \in S(A)$ satisfy $RM(p) < \infty$, $\varphi(\vec{v})$ be a formula in p having the same Morley rank and degree as p. Let $\varphi'(\vec{v})$ be a formula of $L(A)$ also having the same Morley rank and degree as p. Then $\varphi'(\vec{v}) \in p$ if and only if $RM(\varphi(\Omega^n) \triangle \varphi'(\Omega^n)) < RM(p)$.*

Here $\triangle$ denotes symmetric difference: for X, X' sets, $X \triangle X' = (X - X') \cup (X' - X)$.

Proof. Let $\varphi'(\vec{v}) \in p$. Owing to Proposition 5.7.11, $RM(\varphi(\Omega^n) - \varphi'(\Omega^n)) < RM(p)$. But now we can reverse the roles of $\varphi(\vec{v})$ and $\varphi'(\vec{v})$, and so deduce $RM(\varphi(\Omega^n) - \varphi'(\Omega^n)) < RM(p)$. By Proposition 5.7.4, (iii), $RM(\varphi(\Omega^n) \triangle \varphi'(\Omega^n))$ equals $max\{RM(\varphi(\Omega^n) - \varphi'(\Omega^n)), RM(\varphi'(\Omega^n) - \varphi(\Omega^n))\}$, and hence is $< RM(p)$. Conversely suppose $RM(\varphi(\Omega^n) \triangle \varphi'(\Omega^n)) < RM(p)$. Then $RM(\varphi(\Omega^n) - \varphi'(\Omega^n)) < RM(p)$ as well, and so $\varphi'(\vec{v}) \in p$. ♣

It is easy to realize that, for every ordinal α, the relation $\equiv_\alpha$ identifying two $L(A)$-formulas $\varphi(\vec{v})$, $\varphi'(\vec{v})$ if and only if

$$RM(\varphi(\Omega^n) \triangle \varphi'(\Omega^n)) < \alpha$$

is an equivalence relation. Accordingly, when $RM(p)$ is an ordinal α, a formula $\varphi(\vec{v}) \in p$ with the same Morley rank and degree as p is uniquely determined (in p) up to $\equiv_\alpha$.

Remarks 5.7.13

1. Let $A \subseteq B$ be small subsets of Ω, $p \in S(A)$, $q \in S(B)$, $p \subseteq q$. Then $RM(p) \geq RM(q)$ and, when $RM(p) = RM(q)$, $GM(p) \geq GM(q)$. For, all the formulas in p belong to q as well.

2. Let $p \in S(A)$. Then p is algebraic if and only if $RM(p) = 0$ (recall that p is algebraic if and only if there is a formula $\varphi(\vec{v}) \in p$ such that $\varphi(\Omega^n)$ is finite and non-empty, in other words has Morley rank 0). We saw that, for types in a suitable setting, algebraic just means isolated: in this way we recover in our framework a very basic and familiar property of the Cantor-Bendixson rank. For algebraically closed fields (of a fixed characteristic) the only nonalgebraic 1-type over a small

subset of Ω has Morley rank 1. The same is true for every strongly minimal theory.

So we have just introduced Morley rank and degree for types $p \in S(A)$ by referring to Morley rank and degree of formulas. Conversely, for every $L(A)$-formula $\varphi(\vec{v})$, we can recover the Morley rank and degree of $\varphi(\vec{v})$ (and of the corresponding A-definable set $\varphi(\Omega^n)$) from the Morley rank and degree of the types over A containing $\varphi(\vec{v})$ (provided that, of course, there is some type over A containing this formula, in other words $\varphi(\Omega^n) \neq \emptyset$).

Proposition 5.7.14 *Let A be a small subset of Ω, $\varphi(\vec{v})$ be an $L(A)$-formula (in the free variables $\vec{v} = (v_1, \ldots, v_n)$) such that $\varphi(\Omega^n) \neq \emptyset$. Then*

$$RM(\varphi(\Omega^n)) = max\{RM(p) : p \in S_n(A),\ \varphi(\vec{v}) \in p\}$$

and, when $RM(\varphi(\Omega^n))$ is an ordinal α,

$$GM(\varphi(\Omega^n)) = \sum_p GM(p)$$

where p ranges over the n-types over A containing $\varphi(\vec{v})$ and having rank α.

Proof. Clearly, if $p \in S_n(A)$ and $\varphi(\vec{v}) \in p$, then $RM(p) \leq RM(\varphi(\Omega^n))$. So we have preliminarily to show that there exists at least one type $p \in S_n(A)$ containing $\varphi(\vec{v})$ and having its rank.
First suppose $RM(\varphi(\Omega^n)) = \infty$. By compactness the set of $L(A)$-formulas

$$p_0 = \{\varphi(\vec{v})\} \cup \{\neg\vartheta(\vec{v}) : RM(\vartheta(\Omega^n)) < \infty\}$$

is consistent. In fact, let $\vartheta_0(\vec{v}), \ldots, \vartheta_s(\vec{v})$ be $L(A)$-formulas of rank $< \infty$; by Proposition 5.7.4, (iii), also $\bigvee_{i \leq s} \vartheta_i(\vec{v})$ have rank $< \infty$. Moreover $\varphi(\Omega^n) \not\subseteq \bigcup_{i \leq s} \vartheta_i(\Omega^n)$, otherwise even $\varphi(\vec{v})$ has Morley rank $< \infty$. Accordingly $\varphi(\Omega^n) \cap (\bigcap_{i \leq s} \neg\vartheta_i(\Omega^n)) \neq \emptyset$. So extend p_0 to a type $p \in S_n(A)$; $\varphi(\vec{v}) \in p$, and p has rank ∞.
Now suppose $RM(\varphi(\Omega^n)) = \alpha$ where α is an ordinal. Decompose $\varphi(\vec{v})$ as $\bigvee_{i \leq s} \varphi_i(\vec{v})$ where $\varphi_0(\vec{v}), \ldots, \varphi_s(\vec{v})$ are $L(A)$-formulas of Morley rank α and minimal Morley degree and, for $i < j < s$, $RM(\varphi_i(\Omega^n) \cap \varphi_j(\Omega^n)) < \alpha$. The formulas $\varphi_0(\vec{v}), \ldots, \varphi_s(\vec{v})$ occurring in this decomposition are uniquely determined up to $\equiv_\alpha$, as they have minimal degree. Furthermore the Morley degree of $\varphi(\vec{v})$ is the sum of the degrees of $\varphi_i(\vec{v})$ for $i \leq s$. Now fix $i \leq s$ and consider

$$q_i = \{\varphi_i(\vec{v})\} \cup \{\neg\vartheta(\vec{v}) : \vartheta(\vec{v})\ L(A)\text{–formula},\ RM(\varphi_i(\vec{v}) \wedge \vartheta(\vec{v})) < \alpha\}.$$

By proceeding as in the previous case, one checks that q_i is consistent, and so enlarges to a type $p_i \in S_n(A)$. Furthermore this type p_i is unique: in fact, if $\vartheta(\vec{v})$ is an $L(A)$-formula and $RM(\varphi_i(\vec{v}) \wedge \vartheta(\vec{v})) = \alpha$, then, owing to the choice of $\varphi(\vec{v})$ (and the minimality of its degree over A), it follows $RM(\varphi(\vec{v}) \wedge \neg\vartheta(\vec{v})) < \alpha$, whence $\vartheta(\vec{v})$ is in p_i. Clearly $RM(p_i) = \alpha$. Using the minimality of the degree of $\varphi_i(\vec{v})$ once again, one deduces $GM(p_i) = GM(\varphi_i(\vec{v}))$. Finally $\varphi_i(\vec{v}) \in p_i$, and so $\varphi(\vec{v}) \in p_i$.
So we have found types $p_0, \ldots, p_s \in S_n(A)$ as claimed. To accomplish the proof (in particular, to show the equality about degrees) it suffices to observe that the only n-types p over A containing $\varphi(\vec{v})$ and having rank α are just $p_0, \ldots, p_s$. In fact, if $\varphi(\vec{v}) \in p$, then there is a unique $i \leq s$ such that $\varphi_i(\vec{v}) \in p$. As p has rank α, it follows $q_i \subseteq p$, whence $p_i = p$. ♣

5.8 Strongly minimal sets

As said before, a definable set is *strongly minimal* if it is infinite, but admits no partition into 2 disjoint infinite definable subsets in Ω. A theory is called *strongly minimal* if the domain of its universe is.
Recall also from Chapter 2 that a structure $\mathcal{A}$ is said to be *strongly minimal* if its theory is, and *minimal* when its domain A admits no partition into 2 disjoint A-definable infinite subsets. Of course, strongly minimal structures are minimal, but the converse is not true (see, again, Example 2.7.2).
Clearly strongly minimal sets are the simplest infinite definable sets and hence provide a matter worthy of some interest. Incidentally, due to the new approach to structures provided in 5.2, any strongly minimal set, and, more generally, any definable D in Ω can be naturally regarded as a structure in its own right. It suffices to assume as definable sets in D the traces in D of the definable sets in Ω, hence $D^n \cap X$ for every definable $X \subseteq \Omega^n$. It is easy to check that the resulting collection satisfies the condition in Theorem 5.2.1. When D is strongly minimal, the resulting new structure is strongly minimal, too.
Accordingly, one can examine strongly minimal structures and theories instead of strongly minimal sets. We already met some examples in this area. In fact we know that infinite vectorspaces over a countable fields, as well as algebraically closed fields (in a fixed characteristic), and even mere infinite sets, admit a strongly minimal complete theory. Let us analyse closerly these examples.

Examples 5.8.1 1. Let $\mathcal{K}$ be a countable field and consider the theory ${}_{\mathcal{K}}T'$ of infinite vectorspaces over $\mathcal{K}$. Recall that each of them is fully determined, up to isomorphism, by a cardinal number (its *dimension* over $\mathcal{K}$). In view of a possible generalization, let us remind very quickly why. So let $\mathcal{V}$ be any vectorspace over $\mathcal{K}$. For $a \in V$, $X \subseteq V$ define

$$a \prec X \quad (a \text{ linearly dependent on } X)$$

if and only if a is in the subspace $\langle X \rangle$ of V spanned by X (in other words, a is a finite linear combination of elements in $X \cup \{0\}$). One introduces in this way a subset $\prec$ of $V \times \mathcal{P}(V)$ (the *linear dependence relation*). Moreover one sees that $\prec$ satisfies the following properties: for a, b in V and $X, Y \subseteq V$,

(D1) if $a \in X$, then $a \prec X$;

(D2) if $a \prec X$, then there is some finite $X_0 \subseteq X$ such that $a \prec X_0$;

(D3) if $a \prec X$ and, for every $x \in X$, $x \prec Y$, then $a \prec Y$;

(D4) if $a \prec X \cup \{b\}$ but $a \not\prec X$, then $b \prec X \cup \{a\}$.

Checking (D1) and (D2) is trivial, while (D3), (D4) just need some simple calculations.

Now define a subset B of V *linearly independent* if and only if, for every $b \in B$, $b \not\prec B - \{b\}$; and say that $B \subseteq V$ is a *basis* of $\mathcal{V}$ if and only if B is linearly independent and, for every $a \in V$, $a \prec B$.

Using (D1) - (D4) and Zorn's Lemma (with no specific reference to the algebraic framework of vectorspaces), one shows that $\mathcal{V}$ admits some basis B and that two bases of $\mathcal{V}$ have the same power. Accordingly one defines the *dimension* of $\mathcal{V}$ over $\mathcal{K}$ as the power of any basis of $\mathcal{V}$.

Finally one proves that two vectorspaces over $\mathcal{K}$ are isomorphic if and only if they have the same dimension. This time, the proof needs also the following fact: if $\mathcal{V}$ is a vectorspace over $\mathcal{K}$, $X \subseteq V$, a, $b \in V$ and both a and b are linearly independent of X, then there is an automorphism of $\mathcal{V}$ fixing X pointwise and mapping a into b (in other words, a and b have the same type over X).

2. Let $p = 0$ or p be prime, consider now the theory ACF_p of algebraically closed fields of characteristic p. Also in this case it is known that every model $\mathcal{K}$ of ACF_p is fully determined up to isomorphism by a cardinal number: its transcendence degree (over the prime subfield). Under

this point of view, the setting is quite similar to what we saw in the previous example. In fact, for every algebraically closed field $\mathcal{K}$ of characteristic p, one introduces a subset $\prec$ of $K \times \mathcal{P}(K)$ (now called the *algebraic dependence relation*) in the following way: for $a \in K$ and $X \subseteq K$, one puts

$$a \prec X \quad (a \textit{ algebraically dependent } \text{on } X)$$

if and only if a is algebraic over the subfield generated by X. One checks that (D1) - (D4) still hold for $a, b \in K$ and $X, Y \subseteq K$ (although the proofs, especially that of (D3) and (D4), require now some major algebraic difficulties).

Anyhow, as in the previous example, we can define a subset $B \subseteq K$ *algebraically independent* if and only if, for every $b \in B$, $b \not\prec B - \{b\}$; and we can say that B is a *transcendence basis* of $\mathcal{K}$ if B is algebraically independent and, for every $a \in K$, $a \prec B$. The existence of a transcendence basis of $\mathcal{K}$ and the fact that two transcendence bases of $\mathcal{K}$ have the same power are shown exactly as in the previous examples, despite the different algebraic framework, by using only (D1) - (D4) (and Zorn's Lemma). The cardinality of a transcendence basis of $\mathcal{K}$ is called the *transcendence degree* of $\mathcal{K}$. As in the previous examples, one shows that an algebraically closed field $\mathcal{K}$ of characteristic p is fully determined up to isomorphism just by its transcendence degree (so by a cardinal number). The reason is, again, the fact that, if a, $b \in K$, $X \subseteq K$ and a, $b \not\prec X$ (so both a and b are transcendental over the subfield generated by X), then there is an automorphism of $\mathcal{K}$ fixing X pointwise and mapping a into b (whence a and b have the same type over X).

3. Let us propose a more trivial example. Let $L = \emptyset$, $\mathcal{M}$ be an infinite set. For $a \in M$, $X \subseteq M$, put

$$a \prec X \quad \Leftrightarrow \quad a \in X;$$

this again defines a subset $\prec$ of $M \times \mathcal{P}(M)$ clearly satisfying (D1) - (D4). In this case the only possible basis of M is just M, and actually the power of M fully determines the isomorphism class of M among L-structures (i. e. nonempty sets).

Before continuing our analysis of strongly minimal theories and sets, in order to exclude any possible misunderstandings and to prepare the next Chapter

7, let us point out a key remark: we cannot expect that, for any complete theory T, every model of T is determined up to isomorphism by a single cardinal invariant. Here is a simple "counterexample".

Example 5.8.2 Let T be the theory of the structures (A, B) where $B \subseteq A$ and both B and $A - B$ are infinite. Notice that T is complete, for example because it satisfies the assumptions of Vaught's Theorem: it has no finite models and is $\aleph_0$-categorical, as the only countable model (up to isomorphism) must satisfy $|B| = |A - B| = \aleph_0$. In order to characterize a model (A, B) of T up to isomorphism, one needs an **ordered pair** $(|B|, |A - B|)$ of cardinal numbers.

Of course this example can be easily generalized. Just think of a structure $\mathcal{A} = (A, (B_n)_{n<N})$ where $2 \leq N \leq \omega$ and the B_n's are pairwise disjoint infinite subsets of A. In this extended framework the isomorphism type of $\mathcal{A}$ is given by the ordered sequence of cardinals

$$\left((|B_n|)_{n<N}, \; |A - \bigcup_{n<N} B_n|\right);$$

notice that this sequence is infinite when $N = \omega$.
Now let us restrict again our attention to strongly minimal theories. We wonder if the similar behaviour observed in Examples 1, 2 and 3 in 5.8.1 for infinite $\mathcal{K}$-vectorspaces, algebraically closed fields and infinite sets is founded on some common basis. To clarify this point, let us give a further glance at these examples.

Examples 5.8.3 1. Let $\mathcal{V}$ be a vectorspace over $\mathcal{K}$, $a \in V$, $X \subseteq V$. Recall that $a \prec X$ means that there are $n \in \mathbf{N}$, $k_0, \ldots, k_n \in K$, $x_0, \ldots, x_n \in X$ such that $a = \sum_{i \leq n} k_i x_i$; in other words a is the only element of V satisfying the $L(X)$-formula $v = \sum_{i \leq n} k_i x_i$.

2. Let $\mathcal{K} \models ACF_p$, $a \in K$, $X \subseteq K$. Now $a \prec X$ if and only if there is a polynomial $f(t) \in K[t] - \{0\}$ with coefficients in the subfield generated by X such that $f(a) = 0$. There is no loss of generality in assuming that the coefficients of $f(t)$ are also in the subring generated by X (so in the L-substructure generated by X). Accordingly every coefficient of $f(t)$ gets X-definable, and "$f(v) = 0$" can be written as an $L(X)$-formula. Of course there are only finitely many elements satisfying this formula, and a is one of them.

3. Let A be an infinite set, $a \in A$, $X \subseteq A$. Now $a \prec X$ simply means $a \in X$, hence that a is the unique element in A satisfying the $L(X)$-formula $v = a$ (a is in X!).

5.9 Algebraic closure and definable closure

What we observed at the end of the previous section suggests the following definitions. Let $\mathcal{M}$ be a structure in a language L, $X \subseteq M$.

Definition 5.9.1 *1. The* **algebraic closure** *of X (denoted by $acl(X)$) is the union of all the finite X-definable subsets of M (and so is the set of the elements $a \in M$ such that, for some $L(X)$-formula $\varphi(v)$, $a \in \varphi(\mathcal{M})$ and $\varphi(\mathcal{M})$ is finite).*

2. The **definable closure** *of X (denoted by $dcl(X)$) is the set of all the X-definable elements of M (and hence of the elements of M such that, for some $L(X)$-formula $\varphi(v)$, $\varphi(\mathcal{M}) = \{a\}$).*

Clearly $dcl(X) \subseteq acl(X)$ for every X and $\mathcal{M}$. Sometimes $dcl(X) = acl(X)$. For instance, this is the case when $\mathcal{M}$ expands a linear order $\leq$ and X is any subset of M.
In fact, let $a \in acl(X)$, $\varphi(v)$ be an $L(X)$-formula such that $\varphi(\mathcal{M})$ is finite and includes a. More precisely, let $a_0 < \ldots < a_r$ be the elements of $\varphi(\mathcal{M})$, and let $a = a_i$ for a unique $i \leq r$. Then a is the only element in M satisfying the formula

$$\varphi(v) \wedge \exists !^i\, w\, (\varphi(w) \wedge w < v).$$

However there do exist some structures $\mathcal{M}$ and subsets $X \subseteq M$ such that $dcl(X) \neq acl(X)$, as we will see in the next lines.

Remarks 5.9.2 1. Let $\mathcal{M}$, $\mathcal{N}$ be structures for L such that $\mathcal{M}$ is an elementary substructure of $\mathcal{N}$, $X \subseteq M$ (hence $X \subseteq N$). Of course, we might expect to have to form two algebraic closures of X, the former in $\mathcal{M}$ and the latter in $\mathcal{N}$. However these two sets coincide. Let us see why. First notice that, for every $L(X)$-formula $\varphi(v)$, $\varphi(\mathcal{M})$ is finite if and only if $\varphi(\mathcal{N})$ is, and these sets have the same power. In fact, if $|\varphi(\mathcal{M})| = k$, then $\mathcal{M} \models \exists !k\; v\; (\varphi(v))$; as $\mathcal{M} < \mathcal{N}$, $\mathcal{N} \models \exists !k\; v\; (\varphi(v))$, whence $|\varphi(\mathcal{N})| = k$. Of course this argument holds even in the other direction, from $\mathcal{N}$ to $\mathcal{M}$. Moreover, using $\mathcal{M} < \mathcal{N}$ once again, one sees that every element of a finite $\varphi(\mathcal{M})$ satisfies $\varphi(v)$ in $\mathcal{N}$ as well. Consequently, for every $L(X)$-formula $\varphi(v)$ such that $\varphi(\mathcal{M})$ is finite,

$\varphi(\mathcal{M}) = \varphi(\mathcal{N})$. In conclusion, the algebraic closure of X is the same in $\mathcal{M}$ and in $\mathcal{N}$.

Clearly the same can be said about the definable closure. Consequently, for a complete theory T with universe Ω and for a small $X \subseteq \Omega$, $acl(X)$ and $dcl(X)$ do not depend on the model of T containing X one refers to.

2. Let T be a complete theory, X be a small subset of Ω, $a \in \Omega$. Then

   $a \in acl(X) \Leftrightarrow$ there are at most finitely many elements $f(a)$ when f ranges over the automorphisms of Ω fixing X pointwise.

   In fact, take $a \in acl(X)$. Let $\varphi(v)$ be an $L(X)$-formula such that $a \in \varphi(\Omega)$ and $\varphi(\Omega)$ is finite. For every automorphism f of Ω fixing X pointwise, $f(a) \in \varphi(\Omega)$ as well. Accordingly the images of a under the automorphisms of Ω fixing X pointwise form a finite set.

   Conversely, let $a \notin acl(X)$, so, for every $L(X)$-formula $\varphi(v)$, if $a \in \varphi(\Omega)$, then $\varphi(\Omega)$ is infinite. Let $a_0, \ldots, a_n$ be different elements, all realizing $tp(a/X)$. For every formula $\varphi(v) \in tp(a/X)$, there is some $b \in \varphi(\Omega)$, $b \neq a_0, \ldots, a_n$. As $tp(a/X)$ is closed under finite conjunction, $tp(a/X) \cup \{\neg(v = a_i) : i \leq n\}$ is consistent and can be enlarged to a complete type over $X \cup \{a_0, \ldots, a_n\}$: any element a_{n+1} realizing this type satisfies $a_{n+1} \models tp(a/X)$, $a_{n+1} \neq a_0, \ldots, a_n$. So there are infinitely many realizations of $tp(a/X)$ in Ω, and each of them is the image of a under some automorphism of Ω fixing X pointwise.

   A similar argument shows

   $a \in dcl(X) \Leftrightarrow f(a) = a$ for every automorphism f of Ω fixing X pointwise.

Now let us come back to the Examples 5.8.1 of the last section. We want to examine *acl* and *dcl* in their frameworks.

Examples 5.9.3 1. Let $\mathcal{V}$ be a vectorspace over a countable field $\mathcal{K}$, $X \subseteq V$. Then $dcl(X) = acl(X)$ coincide with the subspace $\langle X \rangle$ of $\mathcal{V}$ spanned by X.

In fact, we know that $dcl(X) \subseteq acl(X)$ in general, and we have already seen that $\langle X \rangle \subseteq dcl(X)$. Hence it suffices to prove $acl(x) \subseteq \langle X \rangle$. Notice that, if a and b are two elements of Ω, and both of them are linearly independent of X, then there is some automorphism of Ω fixing X pointwise and mapping a into b. So, for $a \notin \langle X \rangle$, every element in $\Omega - \langle X \rangle$ is the image of a under some automorphism of Ω acting identically on X; accordingly $a \notin acl(X)$.

2. Now consider an algebraically closed field $\mathcal{K}$ (of some given characteristic p). If $X \subseteq K$, then $acl(X)$ coincide with the *algebraic closure* (in the usual field theoretic sense) of the subfield of $\mathcal{K}$ generated by X.

   In fact, we already saw that $\supseteq$ holds. Conversely, if $a, b \in \Omega$ are transcendental over the subfield generated by X, then there is some automorphism of Ω fixing X pointwise and mapping a into b. Just as in 1, this excludes that a, or b, or any element transcendental over the subfield generated by X can belong to $acl(X)$.

   This time, $dcl(X) \neq acl(X)$. The reason is very simple: in fact, if $a, b \in K$ are two different roots of the same irreducible polynomial with coefficients in the subring generated by X, then one can build an automorphism of $\mathcal{K}$ fixing X pointwise and mapping a into b. Hence $a \in acl(X)$, but $a \notin dcl(X)$.

   Indeed one can check that, if $\mathcal{K}$ has characteristic 0, then $dcl(X)$ is just the subfield of $\mathcal{K}$ generated by X, while, if $\mathcal{K}$ has a prime characteristic p, then $dcl(X)$ coincides with the closure of the subfield of $\mathcal{K}$ generated by X under the inverse function of the Frobenius morphism Fr (the one taking any $a \in K$ to $Fr(a) = a^p$).

3. Finally let M be an infinite set, $X \subseteq M$. Now $acl(X) = dcl(X) = X$.

   In fact $acl(X) \supseteq dcl(X) \supseteq X$ is clear. So it suffices to show $acl(X) \subseteq X$. Owing to the Remark 5.9.2.1, we can assume $|X| < |M|$. Any two elements of $M - X$ correspond to each other by a permutation of M fixing X pointwise. Hence any X-definable subset of M overlapping $M - X$ includes $M - X$, and so is infinite. In conclusion, $acl(X) \subseteq X$, as claimed.

Anyhow, in these examples, we recognize a common feature: in fact, for every small subset X of Ω and $a \in \Omega$,

$$a \prec X \quad \Leftrightarrow \quad a \in acl(X).$$

But, before examining closerly this point, let us propose a couple of further examples.

Examples 5.9.4 1. Let $T = DLO^-$, X be a small subset of Ω. As T is concerned with linear orders, $acl(X) = dcl(X)$. We claim $acl(X) = X$.

In fact, take $a \in \Omega - X$. Owing to Theorem 2.3.1, every element $b \in \Omega - X$ such that

$$b < x \quad \Leftrightarrow \quad a < x$$

for every $x \in X$ satisfies the same $L(X)$-formulas as a. Hence, if $a \in \varphi(\Omega)$ for some $L(X)$-formula $\varphi(v)$, $\varphi(\Omega)$ contains all these elements b. By Compactness, there are infinitely many b's; hence $\varphi(\Omega)$ must be infinite. In conclusion $a \notin acl(X)$.

2. Let $T = RCF$, X be a small subset of Ω. As before $acl(X) = dcl(X)$. But now $acl(X)$ equals the *real closure* of the subfield generated by X, that is the least real closed field including X (inside Ω). This can be equivalently introduced as the ordered field of the elements in Ω algebraic over (the subfield generated by) X. So it is obvious that $acl(X)$ includes it. Conversely let $a \in acl(X) = dcl(X)$, accordingly a is the only element in Ω satisfying a suitable $L(X)$-formula $\varphi(v)$. Owing the quantifier elimination, we can assume that $\varphi(v)$ is a disjunction of conjunctions -and indeed a single conjunction- of equations $f(v) = 0$ or disequations $g(v) > 0$ where $f(x)$ and $g(x)$ are nonzero polynomials with coefficients in the subfield generated by X. As a must be the only element satisfying these conditions, at least one equation occurs, whence a is algebraic over the subfield generated by X, as claimed.

The reader may check the details of the last example, and also calculate $acl(X)$, $dcl(X)$ in other familiar cases, for instance when X is a small subset of the universe Ω of dLO^+.

Now let us come back to a general framework. Accordingly, let $\mathcal{A}$ be a structure for a given language L. For $a \in A$, $X \subseteq A$, put

$$a \prec X \quad \Leftrightarrow \quad a \in acl(X).$$

We wonder whether this relation $\prec$ satisfies the conditions (D1) - (D4) in 5.8 (just as in the three main examples before). (D1) - (D3) still hold in this general setting. Let us see why.

(D1) For every $a \in A$ and $X \subseteq A$, if $a \in X$, then $a \prec X$.

In fact a is the only element satisfying $v = a$ (and so is even in $dcl(X)$).

(D2) For every $a \in A$ and $X \subseteq A$, if $a \in acl(X)$, then there is a finite subset X_0 of X such that $a \in acl(X_0)$.

In fact take $a \in acl(X)$; there is some $L(X)$-formula $\varphi(v)$ such that $\varphi(\mathcal{A})$ is finite and includes a. Let X_0 be the set of the parameters from X occurring in $\varphi(v)$. Clearly X_0 is what we are looking for.

(D3) For every $X \subseteq A$, $acl(acl(X)) \subseteq acl(X)$.

Let $a \in acl(acl(X))$, $\varphi(v, \vec{x})$ be a formula with parameters $\vec{x} = (x_0, \ldots, x_m)$ in $acl(X)$ such that $a \in \varphi(\mathcal{A}, \vec{x})$ and $\varphi(\mathcal{A}, \vec{x})$ is finite. Put $|\varphi(\mathcal{A}, \vec{x})| = k$. For every $i \leq m$, there is a formula $\vartheta_i(v, \vec{y_i})$ of $L(X)$ such that $\vartheta_i(\mathcal{A}, \vec{y_i})$ is finite and includes x_i. Put $\vec{y} = (\vec{y_0}, \ldots, \vec{y_m})$ and consider the formula

$$\vartheta(v, \vec{y}) : \exists w_0 \ldots \exists w_m \left(\bigwedge_{i \leq m} \vartheta_i(w_i, \vec{y_i}) \wedge \exists!^k z\, \varphi(z, \vec{w}) \wedge \varphi(v, \vec{w}) \right)$$

where $\vec{w}$ abbreviates the tuple $(w_0, \ldots, w_m)$. Clearly $a \in \vartheta(\mathcal{A}, \vec{y})$. Furthermore $\vartheta(\mathcal{A}, \vec{y})$ is the union of the sets $\varphi(\mathcal{A}, \vec{x'})$ when $\vec{x'} = (x'_0, \ldots, x'_m)$ satisfies $x'_i \in \vartheta_i(\mathcal{A}, \vec{y_i})$ and $|\varphi(\mathcal{A}, \vec{x'})| = k$. Accordingly $\vartheta(\mathcal{A}, \vec{y})$ is a finite union of sets of size k, an hence is finite.

Now let us wonder if even (D4) holds in any structure $\mathcal{A}$. Recall what (D4) states.

(D4) For every $a, b \in A$ and $X \subseteq A$, if $a \in acl(X \cup \{b\}) - acl(X)$, then $b \in acl(X \cup \{a\})$.

Example 5.9.5 Let $\mathcal{A} = (\mathbf{N}^2, f)$ where f is the 1-ary function such that, for every $x, y \in \mathbf{N}$, $f(x, y) = (x, 0)$. Put

$$a = (0, 0), \quad b = (0, 1).$$

Then a is the only element in $\mathbf{N}^2$ satisfying $v = f(b)$, whence $a \in acl(b)$. Moreover $a \notin acl(\emptyset)$ because any element $(x, 0)$ with $x \in \mathbf{N}$ is the image of a in some automorphism of $\mathbf{N}^2$. But $b \notin acl(a)$ because any element $(0, y)$ with $y \in \mathbf{N} - \{0\}$ is the image of $b = (0, 1)$ under some automorphism fixing $a = (0, 0)$.

Accordingly (D4) fails in general. However the following proposition holds.

Proposition 5.9.6 *Let $\mathcal{A}$ be a minimal structure. Then (D4) holds in $\mathcal{A}$: if $a, b \in A$, $X \subseteq A$ and $a \in acl(X \cup \{b\}) - acl(X)$, then $b \in acl(X \cup \{a\})$.*

Recall that $\mathcal{A}$ is minimal if and only if any subset of A definable in $\mathcal{A}$ is either finite or cofinite. Algebraically closed fields, vectorspaces (over countable fields) and pure infinite sets are minimal structures. On the contrary, the structure $\mathcal{A}$ of the last example is not minimal: for, the formula $f(v) = (0, 0)$ defines an infinite coinfinite set.

Proof. As the minimality of $\mathcal{A}$ is preserved under adding or forgetting parameters from A, there is no loss of generality in assuming $X = \emptyset$ (otherwise replace $\mathcal{A}$ by $\mathcal{A}_X$). Suppose towards a contradiction that there are a and b in A such that

$$a \in acl(b) - acl(\emptyset), \quad b \notin acl(a).$$

As $a \in acl(b)$, there is an L-formula $\varphi(v, w)$ for which $\varphi(\mathcal{A}, b)$ is finite and includes a; let k be the size of $\varphi(\mathcal{A}, b)$. Accordingly b satisfies the formula

$$\vartheta(v, a) : \varphi(a, v) \wedge \exists !^k z \, \varphi(z, v).$$

As $b \notin acl(a)$, $\vartheta(\mathcal{A}, a)$ must be infinite, hence cofinite. Let m denote the (finite) cardinality of $A - \vartheta(\mathcal{A}, a)$; so a satisfies

$$\gamma(v) : \exists !^m t \, \neg \vartheta(t, v).$$

As $a \notin acl(\emptyset)$, also $\gamma(\mathcal{A})$ must be infinite. Pick $k + 1$ pairwise different elements $a_0, \ldots, a_k$ in $\gamma(\mathcal{A})$. For every $i \leq k$, there are exactly m elements in $\neg\vartheta(\mathcal{A}, a_i)$. Exclude all these elements for $i \leq k$. We can still find some $b' \in A$ satisfying

$$\mathcal{A} \models \bigwedge_{i \leq k} \vartheta(b', a_i),$$

hence

$$\mathcal{A} \models \bigwedge_{i \leq k} \varphi(a_i, b') \wedge \exists !^k z \, \varphi(z, b').$$

Consequently $a_0, \ldots, a_k \in \varphi(\mathcal{A}, b')$, contradicting $|\varphi(\mathcal{A}, b')| = k$. This accomplishes the proof. ♣

Hence, when T is a strongly minimal theory, $\prec$ satisfies (D1)-(D4) in every model $\mathcal{A}$ of T. This allows to generalize the notions already introduced in the main examples: for $B \subseteq A$, one can say that

- B is *independent* if and only if, for every $b \in B$, $b \notin acl(B - \{b\})$,
- B is a *basis* of $\mathcal{A}$ if and only if B is independent and $A = acl(B)$.

Using Zorn's Lemma, one can still deduce the existence of a basis of $\mathcal{A}$, and the fact that two bases have the same cardinality; in this way $\mathcal{A}$ is naturally associated with a cardinal number, that is the power of any basis. This is called the *dimension* of $\mathcal{A}$. Of course, all this directly depends on (D1)-(D4). However the fact that this cardinal determines $\mathcal{A}$ up to isomorphism refers to another basic property of strongly minimal theories, ensuring that, for

every $X \subseteq A$, in particular for $X = \emptyset$, there is a unique non-algebraic 1-type p_X over X: the one containing every formula $\varphi(v)$ such that $\varphi(\Omega)$ is cofinite. In this perspective a basis B of $\mathcal{A}$ is actually a subset of $p_\emptyset(\mathcal{A})$, and indeed any element b in B realizes the unique non-algebraic 1-type over $B - \{b\}$. Consequently, if $\mathcal{A}, \mathcal{A}' \models T$ have the same dimension and B, B' are two bases of $\mathcal{A}, \mathcal{A}'$ respectively, then there does exist some bijection between B and B', but (which is more relevant) such a bijection is even an elementary function. Furthemore, as $A = acl(B)$ and $A' = acl(B')$, it can be extended with a little patience to an isomorphism from $\mathcal{A}$ onto $\mathcal{A}'$.
In conclusion, the models of a strongly minimal theory can be classified quite satisfactorily up to isomorphism in the way we have just described, and the main ingredients of this classification result are:

(a) $\prec$ satisfies (D1)-(D4);

(b) there is a unique non-algebraic 1-type over any small subset X of Ω.

To underline once again the relevance of (b), let us mention the behaviour of *o-minimal* theories. Recall that they are the complete theories of the infinite structures $\mathcal{A}$ expanding linear orders in such a way that the only definable subsets of A are the finite unions of singletons and open intervals (possibly with infinite endpoints). We will see in the Chapter 7 that there are good reasons to agree that the models of these theories cannot be classified up to isomorphism. In spite of this, in an o-minimal structure $\mathcal{A}$ $\prec$ can satisfy (D1) - (D4), hence (a) holds (but (b) does not). As (D1) - (D3) are always true, we have to check (D4). Here is the proof.

Proposition 5.9.7 *Let $\mathcal{A}$ be an o-minimal structure expanding the linear order $(A, \leq)$, a, $b \in A$, $X \subseteq A$. If $a \in acl(X \cup \{b\}) - acl(X)$, then $b \in acl(X \cup \{a\})$.*

Proof. As the o-minimality of $\mathcal{A}$ is preserved under adding or forgetting parameters from A, there is no loss of generality in assuming $X = \emptyset$. Let a, $b \in A$ satisfy $a \in acl(b) - acl(\emptyset)$, we have to show $b \in acl(a)$. As acl and dcl coincide in linearly ordered structures, a is in $dcl(b)$ and so there is an L-formula $\varphi(v, w)$ such that a is the only element in $\varphi(\mathcal{A}, b)$. We can even assume that $\varphi(v, w)$ defines a partial function f such that $f(b) = a$. Let B denote the set of the elements $x \in A$ such that $f(x) = a$. B is $\{a\}$-definable because f is $\emptyset$-definable. By o-minimality, B is the union of finitely many intervals $I_0, \ldots, I_k$ (where each interval may be open, closed or semi-closed, come down to a singleton, or admit infinite endpoints). Take k minimal and assume $I_0 < \ldots < I_k$. Notice that all the endpoints of these intervals in A

are $\{a\}$-definable. Consequently, if b equals one of them, then $b \in acl(a)$. So suppose that b is in the interior of some interval among $I_0, \ldots, I_k$. Let I denote this interval, $d_1 < d_2$ be its endpoints (where d_1, d_2 may possibly equal $-\infty$, $+\infty$ respectively). If I is finite, then it is easy to deduce again $b \in acl(a)$. Accordingly assume I infinite. Notice that d_1 cannot equal $-\infty$, otherwise a is the only element in A satisfying $\exists w \forall z(z \leq w \to f(z) = v)$, whence $a \in acl(\emptyset)$. Similarly, $d_2 < +\infty$.
For a suitably large positive integer n, let D_n denote the ($\emptyset$-definable) set of the left endpoints $d \in A$ of some open interval J having size $\geq n$ and satisfying the following further assumptions: f is (defined and) constant in J, and one cannot enlarge J to the left keeping this condition. Associate any $d \in D_n$ with the right endpoint $g(d) \in A \cup \{+\infty\}$ of a maximal interval J. By o-minimality, $g(d)$ is well defined: in fact, the elements $> d$ lying in the domain of f and having the same value in f as immediately after d form a definable set, and so a finite union of intervals (in the broader sense said before). Moreover, for $d < d'$ in D_n, $]d, g(d)[\cap]d', g(d')[= \emptyset$. Recall that D_n is definable, so, by o-minimality, D_n must be finite. As $d_1 \in D_n$, $d_1 \in acl(\emptyset) = dcl(\emptyset)$. As a is the image of any element in $]d_1, d_2[$ by f, a is in $acl(\emptyset)$ as well, and this contradicts our hypotheses. So I cannot be infinite, and we are done. ♣

Before concluding this section and the whole chapter, we would like to underline two more points.
The former specifically concerns a minimal, or o-minimal L-structure $\mathcal{M}$. Both these properties (minimality and o-minimality) are preserved under adding parameters from a subset X of M, so under passing from L to $L(X)$ and from $\mathcal{M}$ to $\mathcal{M}_X$: $\mathcal{M}_X$ remains minimal, or o-minimal. Indeed, if the theory of $\mathcal{M}$ is strongly minimal, then $Th(\mathcal{M}_X)$ is: in fact its models are the structures $\mathcal{N}_{f(X)}$ where $\mathcal{N}$ is an L-structure elementarily equivalent to $\mathcal{M}$ -so a minimal structure-, and f is an elementary function from X into $\mathcal{N}$. The same happens for an o-minimal $Th(\mathcal{M})$. So, for $\mathcal{M}$ as before, we can introduce for every $X \subseteq M$ a new dependence relation $\prec_X$ in $M \times \mathcal{P}(M)$ by putting, for $a \in M$ e $S \subseteq M$, $a \prec_X S$ if and only if $a \in acl(S)$ in $\mathcal{M}_X$, in other words if and only if $a \in acl(X \cup S)$ in $\mathcal{M}$. Moreover $\prec_X$ still satisfies (D1)-(D4).
Secondly, when our relation $\prec$ on $M \times \mathcal{P}(M)$ satisfies (D1) - (D4), we can equip not only M, but also every subsets S of M with a dimension $dim\, S$: this can be formally introduced as the minimal power of a subset B of S such that $S \subseteq acl(B)$ (in fact, it is easy to check that such a B is independent). Of course, this also concerns the framework sketched a few lines ago: for

$X \subseteq M$, the dimension of X over X $dim(S/X)$ is just the dimension of S with respect to $\prec_X$.
A final remark: in order to pursue and possibly accomplish our analysis of strong minimality, we can wonder how general the examples we proposed (pure sets, vectorspaces and algebraically closed fields) are. Do there exist any "new" significant instances, or do they exhaust all the possible cases? In Chapter 7 we will discuss the relevance of this question (due to Zilber) and we will provide its solution (due to Hrushovski).

5.10 References

The papers where Morley developed his ideas and, in particular, introduced his rank are [116, 117, 119, 120] and [118]. The duality between Boolean algebras and Boolean spaces is in [52]. The proof of Theorem 5.4.2 in the general setting, for any cardinal λ can be found in Poizat's book [131], which also includes full details on saturated models. Inaccessible cardinals are described in [66] or [78].
Now let us deal with modules: both [181] and [136] treat the matters of Section 5.5 (pure injectivity, Ziegler topology, and so on). Tame and wild classes of modules are also discussed in [136]. The classification of pure injective objects among abelian groups, and more generally among modules over a Dedekind domain, is given in [40].
An abstract treatment of dependence relations, including the main examples in Section 5.8, was developed in [173]. The details on the real closure of an ordered field can be found in [65].
An axiomatic introduction to ranks, and a comparison of Morley rank with other rank notions, are in [8]: see also [131].

Chapter 6

ω -stability

6.1 Totally transcendental theories

We continue here our treatment of Morley's Theorem and related ideas. All throughout this section, T is a complete theory with no finite models in a countable language L, and Ω denotes a big saturated model of T. In the last chapter we defined Morley rank and Morley degree as a complexity measure for definable sets. In particular we studied the simplest infinite definable sets with respect to this measure, i. e. the strongly minimal sets, those having Morley rank 1 and Morley degree 1; we considered also strongly minimal theories, i. e. the complete theories whose universe is strongly minimal. More generally, one can examine the theories in which every non empty definable set gets a(n ordinal) Morley rank. These were called by Morley *totally transcendental.*

Definition 6.1.1 *T is called* **totally transcendental** *if and only if, for every non empty definable set $X \subseteq \Omega^n$ (with n a positive integer), $RM(X)$ is an ordinal number.*

Hence totally transcendental theories exclude the theory DLO^- of dense linear orderings without endpoints, as observed in Example 5.7.3,4, but include, for instance, the strongly minimal theories, which satisfy $RM(\Omega) = GM(\Omega) = 1$; indeed, as we will see within a few lines, strong minimality implies $RM(\Omega^n) = n$, $GM(\Omega^n) = 1$ for every n, and so $RM(X) \leq n$ for every definable set $X \subseteq \Omega^n$.
Totally transcendental theories can be characterized in the following alternative and equivalent way.

Definition 6.1.2 *T is ω-stable if and only if, for every countable $A \subseteq \Omega$, the space $S(A)$ of types over A is also countable.*

One can equivalently ask that, for every countable $A \subseteq \Omega$ and positive integer n, $S_n(A)$ is countable. Actually it suffices to require $S_1(A)$ countable. In fact let n be an integer bigger than 1, $p \in S_n(A)$, $\vec{a} = (a_1, \ldots, , a_n)$ realize p in Ω^n. Then p is fully determined by:

(i) the $(n-1)$-type of $(a_1, \ldots, a_{n-1})$ over A (equivalently, the orbit of $(a_1, \ldots, a_{n-1})$ with respect to the A-automorphisms of Ω);

(ii) the 1-type of a_n over $A \cup \{a_1, \ldots, a_{n-1}\}$ (hence over a countable set) up to A-automorphism.

So, when $S_1(A)$ is countable, a simple induction proves that $S_n(A)$ is countable for every positive integer n.
Observe also that T is ω-stable if and only if, for every countable model $\mathcal{M}$ of T, $S_1(M)$ is countable. In fact, for every countable $A \subseteq \Omega$, there exists some countable model $\mathcal{M}$ of T including A (just apply the Löwenheim-Skolem Theorem to the theory of Ω_A). Every 1-type over A extends to some (possibly non unique) 1-type over M, and conversely restricting to $L(A)$ a 1-type over M determines a 1-type over A. So $|S_1(A)| \leq |S_1(M)|$.
Now let us point out:

Proposition 6.1.3 *A strongly minimal theory T is ω-stable.*

Proof. Let A be a countable subset of Ω. The language $L(A)$ is countable, as well as the set of the $L(A)$-formulas $\varphi(v)$ such that $\varphi(\Omega)$ is finite. Accordingly $acl(A)$ is countable, and the algebraic 1-types over A also form a countable set. As there is only one non-algebraic 1-type over A, $S_1(A)$ is countable, too. ♣

Example 6.1.4 *DLO^-* is not ω-stable. In fact $(\mathbf{Q}, \leq)$ is countable, but it is easy to realize that there are $2^{\aleph_0}$ 1-types over $\mathbf{Q}$: just observe that, for every $r, s \in \mathbf{R}$ with $r < s$, there is some $a \in \mathbf{Q}$ such that $r < a < s$. Accordingly $tp(r/\mathbf{Q}) \neq tp(s/\mathbf{Q})$ as the former type contains $v < a$, and the latter does not.

As said before, ω-stability is just equivalent to total transcendency. Indeed the following theorem holds.

Theorem 6.1.5 *T is totally transcendental if and only if it is ω-stable.*

Proof. Let T be totally transcendental. Then every type p over a small set $A \subseteq \Omega$ is fully determined by an $L(A)$-formula $\varphi(\vec{v})$ in p having the same Morley rank and degree as p. Then there are at most as many types over A as $L(A)$-formulas, and in conclusion at most $|A| + \aleph_0$ types over A. In particular, when A is countable, $S(A)$ is countable, too.
Now assume T ω-stable. First we claim that there exists an ordinal α_T such that, for every positive integer n and definable set $X \subseteq \Omega^n$, if $RM(X) \geq \alpha_T$, then $RM(X) = \infty$.
In fact, let $\varphi(\vec{v}, \vec{w})$ be a formula of L, $\vec{a}$, $\vec{b}$ be two sequences in Ω having the same length as $\vec{w}$ and the same type over $\emptyset$; so there is an automorphism of Ω mapping $\vec{a}$ into $\vec{b}$, hence $\varphi(\Omega^n, \vec{a})$ onto $\varphi(\Omega^n, \vec{b})$. It follows that $\varphi(\Omega^n, \vec{a})$ and $\varphi(\Omega^n, \vec{b})$ have the same Morley rank. On the other side, $S(\emptyset)$ is countable, as well as the set of all possible formulas $\varphi(\vec{v}, \vec{w})$. So there are at most countably many Morley ranks $\neq \infty$ of definable sets in Ω, and we can take α_T as the least ordinal greater than all these values. By the way, one can even show that α_T is countable.
Now suppose towards a contradiction that T is not totally transcendental. Then there exists a definable set $X = X_\emptyset$ having Morley rank ∞, namely $\geq \alpha_T$. Hence $RM(X_\emptyset) \geq \alpha_T + 1$, and so there exist two disjoint definable subsets X_0 and X_1 of $X_\emptyset$ both having Morley rank $\geq \alpha_T$ and consequently ∞. Repeating this procedure one builds, for every finite ordered sequence σ of 0 and 1, hence for every element in $\{0, 1\}^{<\omega}$, a definable subset X_σ of $X_\emptyset$ of Morley rank ∞ such that, for every $\sigma \in \{0, 1\}^{<\omega}$ and $i \in \{0, 1\}$,

$$X_{\sigma i} \subseteq X_\sigma, \quad X_{\sigma 0} \cap X_{\sigma 1} = \emptyset.$$

The set A of the parameters needed to define the X_σ's when σ ranges over $\{0, 1\}^{<\omega}$ is countable. For every $\sigma \in \{0, 1\}^\omega$, the $L(A)$-formulas "v belongs to $X_{\sigma|_n}$" (with n a natural number) form a consistent type, which can be extended to a (complete) type $p_\sigma \in S_1(A)$. Of course different sequences $\sigma, \sigma' \in \{0, 1\}^{<\omega}$ yield different types $p_\sigma \neq p'_\sigma$. Accordingly there are at least $2^{\aleph_0}$ 1-types over A, and this contradicts the fact that T is ω-stable. ♣

A consequence of this theorem is that total transcendence (in fact ω-stability) is preserved by interpretability.

Corollary 6.1.6 *Let $\mathcal{A}$ be a structure for L, $\mathcal{A}'$ be a structure for a possibly different language L' such that $\mathcal{A}$, $\mathcal{A}'$ are infinite and $\mathcal{A}'$ is interpretable in $\mathcal{A}$. If $Th(\mathcal{A})$ is ω-stable, then $Th(\mathcal{A}')$ is.*

Proof. (Sketch) We know that, for a suitable choice of a positive integer n, a subset S of A^n definable in $\mathcal{A}$ and an equivalence relation E in A^n

definable in $\mathcal{A}$, A' is just the quotient set S/E and the whole L'-structure $\mathcal{A}'$ can be definably recovered by $\mathcal{A}$. Furthermore, if Ω is a big saturated elementary extension of $\mathcal{A}$, then the machinery defining $\mathcal{A}'$ inside $\mathcal{A}$ singles in Ω a big saturated elementary extension Ω' of $\mathcal{A}'$ (of the same power as Ω) and every set X' definable in Ω' is the quotient set of some X definable in Ω with respect to E. At this point a simple argument of ordinal induction shows that the Morley rank of X' in $Th(\mathcal{A}')$ is smaller or equal to the Morley rank of X in $Th(\mathcal{A})$. Consequently, when $RM(X)$ is an ordinal, $RM(X')$ is an ordinal, too. In conclusion, if $Th(\mathcal{A})$ is ω-stable, then $Th(\mathcal{A}')$ is. ♣

Now let us deal again with Morley rank and degree for a strongly minimal theory T. We already seen that

$$RM(\Omega) = 1, \qquad GM(\Omega) = 1.$$

Recall that there is a unique 1-type of Morley rank 1 over any small subset X of Ω, the one of the elements out of $acl(X)$. Furthermore, for a, S in Ω,

$$a \prec_X S \Leftrightarrow a \in acl(X \cup S)$$

determines a dependence relation satisfying (D1) - (D4).
More generally, one can see that, for every positive integer n and n-tuple $\vec{a}$ in Ω, $RM(tp(\vec{a}/X))$ equals the dimension of $\vec{a}$ with respect to $\prec_X$. In particular $RM(tp(\vec{a}/X))$ cannot exceed n; indeed there is a unique n-type over X of Morley rank n, the one of an n-tuple $\vec{a}$ $\prec_X$-independent.
Furthermore

$$RM(\Omega^n) = n, \qquad GM(\Omega^n) = 1$$

as already said at the beginning of this section. The proofs of these facts require patience rather than ingenuousness, and can be found in the references quoted at the end of the chapter.
Notice that, in the particular case of algebraically closed fields, the Morley rank of a tuple $\vec{a}$ over a subfield F just equals the trascendence degree of $F(\vec{a})$ over F.

6.2 ω-stable groups

A paradigmatical example of ω-stable theory is the theory ACF_p of algebraically closed fields of a fixed characteristic $p = 0$ or prime. But consequently every structure definable, or interpretable in an algebraically closed field $\mathcal{K}$ has an ω-stable theory. This is the case of the groups $GL(n, \mathcal{K})$ and, more generally, of the algebraic groups over $\mathcal{K}$, as we will see in Chapter 8. A common feature of these examples is the presence of a binary operation (the addition in $\mathcal{K}$, the usual row-by-column multiplication in $GL(n, \mathcal{K})$) making them a group. Accordingly we propose the following definitions.

Definition 6.2.1 *An* **ω-stable structure** *is a structure having an ω-stable theory; an* **ω-stable group** *is an ω-stable structure with a binary operation making it a group (and possibly other operations and relations).*

Then any algebraically closed field $\mathcal{K}$ is an ω-stable group (with respect to the sum operation), as well as any group $GL(n, \mathcal{K})$ and, more generally, any algebraic group over $\mathcal{K}$ (as we will see in Chapter 8). Other examples can be found among abelian groups; for instance, we know that the theory of non-zero vectorspaces over $\mathbf{Q}$ is strongly minimal and consequently ω-stable. Accordingly every non-zero $\mathbf{Q}$-vectorspace is an ω-stable group (with respect to addition), and so every divisible torsionfree abelian group $\mathcal{A}$ is an ω-stable group (for, $\mathcal{A}$ can be naturally equipped with a structure of $\mathbf{Q}$-vectorspace). The aim of this section is to begin the analysis of ω-stable groups. In fact, part of its results will be accomplished, and hence fully understood and appreciated, only in the next Chapters 7 and 8, where more powerful tools will help our study. However the present pages will develop several basic notions in this area. Most of them clearly owe to the theory of algebraic groups, and part of them refer to the study of finite groups. Of course, we are also interested in classifying ω-stable groups and in understanding their connections with algebraic groups, or finite groups, or abelian groups; but we will examine in a closer and deeper way these themes in Chapter 8.
First let us observe what follows. Let $\mathcal{G}$ be an ω-stable group, H be a definable subgroup, $a \in G$. Notice that, if X is a subset of H definable in $\mathcal{G}$, then

$$aX = \{ag : g \in X\} \text{ is a definable subset of the coset } aH$$

and

$$Xa = \{ga : g \in X\} \text{ is a definable subset of } Ha.$$

It follows that, for any a, both aH and Ha have the same Morley rank and degree as H. Consequently, if $H \subseteq K$ are definable subgroups of $\mathcal{G}$ and $H \neq K$, then

$$RM(H) \leq RM(K) \text{ and, if } = \text{ holds, } GM(H) < GM(K)$$

(indeed K is a union of cosets of H). Now we show a chain condition for definable subgroups of ω-stable groups.

Theorem 6.2.2 *Let $\mathcal{G}$ be an ω-stable group. For every natural n, let H_n be a definable subgroup of $\mathcal{G}$ such that, for every n, H_{n+1} is a subgroup of H_n. Then there is natural n such that, for every $m \geq n$ in $\mathbf{N}$, $H_m = H_n$.*

Proof. Otherwise, for every natural n, there exists $m \in \mathbf{N}$ such that $m > n$ and $H_m \neq H_n$. Hence $RM(H_m) \leq RM(H_n)$ and, when $RM(H_m) = RM(H_n)$, $GM(H_m) < GM(H_n)$. In particular, fixed n, there is some $s \in \mathbf{N}$ tale che $s > n$ e $RM(H_s) < RM(H_n)$. But in this way one forms a strictly decreasing infinite sequence of ordinals, and this is a contradiction. ♣

Corollary 6.2.3 *Let $\mathcal{G}$ be an ω-stable group and, for every natural n, let K_n be a definable subgroup of $\mathcal{G}$. Then $\bigcap_{n \in \mathbf{N}} K_n$ is definable.*

Proof. For every natural n, put $H_n = \bigcap_{i \leq n} K_i$. Then H_n is a definable subgroup of $\mathcal{G}$ and $H_{n+1} \subseteq H_n$ for every n. Consequently there exists $m \in \mathbf{N}$ such that, for every $i \leq m$, $H_i = H_m$. So $\bigcap_{n \in \mathbf{N}} K_n = \bigcap_{n \in \mathbf{N}} H_n = H_m = \bigcap_{n \leq m} K_n$ and $\bigcap_{n \leq m} K_n$ is definable. ♣

Corollary 6.2.4 *Let $\mathcal{G}$ be an ω-stable group, f be a definable group homomorphism from $\mathcal{G}$ into $\mathcal{G}$ having a finite kernel. Then the image $f(G)$ of G has a finite index in $\mathcal{G}$.*

Proof. Otherwise one can build a strictly decreasing infinite sequence of subgroups

$$G \supset f(G) \supset f^2(G) \ldots \supset f^n(G) \ldots$$

contradicting Theorem 6.2.2. In fact, as f is definable, $f^n(G)$ is a definable subgroup of $\mathcal{G}$ for every natural n. Furthermore, if $f(G)$ has an infinite index in $\mathcal{G}$, then, for every $n \in \mathbf{N}$, $f^{n+1}(G)$ has an infinite index in $f^n(G)$, and so $f^{n+1}(G) \neq f^n(G)$. Let us see why. The case $n = 0$ is just our assumption. So take $n > 0$, and suppose that the index of $f^n(G)$ in $f^{n-1}(G)$ is infinite. Notice that, for a and b in $f^n(G)$, $f(a) - f(b) \in f^n(G)$ if and only if $a - b \in f^{n-1}(G) + Kerf$. As $Kerf$ is finite and $f^n(G)$ has an infinite index in $f^{n-1}(G)$, there are infinitely many elements in $f^{n+1}(G)$ pairwise inequivalent modulo $f^n(G)$, and so $f^{n+1}(G)$ has infinite index in $f^n(G)$. ♣

Now take any structure $\mathcal{G}$ expanding some given group $(G, \cdot)$ (so $\mathcal{G}$ is not necessarily an ω-stable group). As said before, if X is a definable subset of di G, then also

$$aX = \{ag : g \in X\}, \quad Xa = \{ga : g \in X\}$$

are for every $a \in G$, as well as $X^{-1} = \{g^{-1} : g \in X\}$. Notice also that, for every type p over G and $a \in G$, we can form the sets ap, pa of the formulas $\varphi(v)$ of $L(G)$ such that $\varphi(av) \in p$, $\varphi(va) \in p$ respectively. It is easy to see that ap, pa are types over G. More precisely, if x denotes any realization of p, then ap is the type of $a^{-1}x$ over G and pa is the type of xa^{-1} over G.
The functions from $G \times S(G)$ to $S(G)$ mapping any ordered pair (a,p) of $G \times S(G)$ into ap and pa define two actions (a left action and a right action respectively) of G on $S(G)$. Correspondingly we can consider, for every $p \in S(G)$, the left stabilizer $\{a \in G : ap = p\}$ of p in G and the right stabilizer $\{a \in G : pa = p\}$ of p in G. It is well known that both are subgroups of $\mathcal{G}$.
In a similar way, for every type p over G, we can form the set p^{-1} of the formulas $\varphi(v)$ in $L(G)$ such that $\varphi(v^{-1}) \in p$. Even p^{-1} is a type over G; indeed, if x is any realization of p, then p^{-1} is just the type of x^{-1} over G.
Now let $\mathcal{G}$ be an ω-stable group. As ω-stability is preserved by $\equiv$, there is no loss of generality for our purposes in replacing $\mathcal{G}$ by a suitably saturated elementary extension, and so in assuming that $\mathcal{G}$ itself is saturated in some uncountable power (and so is the universe of its complete theory).
Let $X \subseteq G$ be definable. As observed before X has the same Morley rank and degree as aX and Xa for every $a \in G$, and also as X^{-1}. Consequently, for every type $p \in S(G)$, $RM(p) = RM(ap) = RM(pa)$ per ogni $a \in G$ and $RM(p) = RM(p^{-1})$. Moreover

Lemma 6.2.5 *Let $\mathcal{G}$ be an ω-stable group, p be a type over G. Then both the left and the right stabilizers of p in G are definable subgroups of $\mathcal{G}$.*

Proof. We treat the left case (the right one can be handled in a similar way). For every L-formula $\varphi(v, \vec{w})$ let $G(p, \varphi)$ denote the set of the elements $a \in G$ such that, for every $\vec{g}$ and h in G,

$$\varphi(hv, \vec{g}) \in p \quad \Leftrightarrow \quad \varphi(hv, \vec{g}) \in ap.$$

It is easy to check that $G(p, \varphi)$ is a subgroup of $\mathcal{G}$. In fact, let $a, b \in G(p, \varphi)$; then, for every h and $\vec{g}$ in G,

$$\varphi(hv, \vec{g}) \in p \Leftrightarrow \varphi(hv, \vec{g}) \in ap \Leftrightarrow \varphi(hav, \vec{g}) \in p \Leftrightarrow \varphi(hav, \vec{g}) \in bp \Leftrightarrow$$

$$\Leftrightarrow \varphi(habv, \vec{g}) \in p \Leftrightarrow \varphi(hv, \vec{g}) \in abp;$$

whence $ab \in G(p, \varphi)$. Furthermore

$$\varphi(hv, \vec{g}) \in p \Leftrightarrow \varphi(ha^{-1}av, \vec{g}) \in p \Leftrightarrow \varphi(ha^{-1}v, \vec{g}) \in ap \Leftrightarrow \varphi(ha^{-1}v, \vec{g}) \in p \Leftrightarrow$$

$$\Leftrightarrow \varphi(hv, \vec{g}) \in a^{-1}p,$$

whence $a^{-1} \in G(p, \varphi)$, too.
Now let us momentainly assume what we will actually show only in Chapter 7, Theorem 7.5.5: due to ω-stability, $G(p, \varphi)$ is definable in $\mathcal{G}$ because there are some formulas with parameters in G defining the sets of the tuples $(h, \vec{g})$ in G for which

$$\varphi(hv, \vec{g}) \in p, \quad \text{or} \quad \varphi(hv, \vec{g}) \in ap \quad (\text{that is } \varphi(hav, \vec{g}) \in p).$$

Clearly the left stabilizer of p is the intersection of the subgroups $G(p, \varphi)$ when $\varphi(v, \vec{w})$ ranges over L-formulas. Hence, by Corollary 6.2.3, the left stabilizer is definable. ♣

Let us extend again out interest to arbitrary expansions $\mathcal{G}$ of groups $(G, \cdot)$ (so we are momentarily forgetting the ω-stable assumption). A definable subgroup H of $\mathcal{G}$ is called *connected* (in $\mathcal{G}$) if and only if H has no proper subgroup definable (in $\mathcal{G}$) and of finite index. Clearly $\mathcal{G}$ has at most one definable connected subgroup of finite index (if H_0 and H_1 satisfy these conditions, then $H_0 \cap H_1$ must equal both H_0 and H_1, so H_0 and H_1 coincide). However $\mathcal{G}$ may lack such a subgroup. For instance, the additive group of integers $\mathbf{Z}$ has a definable subgroup $n\mathbf{Z}$ of finite index n for every natural $n \neq 0$. But a simple application of Theorem 6.2.2 ensures that every ω-stable $\mathcal{G}$ has a connected definable subgroup of finite index.

Theorem 6.2.6 *Let $\mathcal{G}$ be an ω-stable group. Then there exists a unique subgroup G^0 of $\mathcal{G}$ which is definable, connected and of finite index in $\mathcal{G}$. Every definable subgroup of finite index of $\mathcal{G}$ includes G^0.*

Proof. Suppose that $\mathcal{G}$ has no definable connected subgroup of finite index. Accordingly every definable subgroup H of $\mathcal{G}$ of finite index has in its turn a proper subgroup definable in $\mathcal{G}$ and of finite index (in H and consequently in G). Then one can build a strictly decreasing infinite sequence of definable subgroups of $\mathcal{G}$

$$G = H_0 \supset H_1 \supset \ldots \supset H_n \supset H_{n+1} \cdots,$$

each of finite index in its predecessor, and so in $\mathcal{G}$. But this contradicts Theorem 6.2.2. This shows the existence of a subgroup as required. Its uniqueness was already observed. Of course this unique subgroup does deserve a specific symbol (G^0, for instance). Finally notice that, if H is a definable subgroup of finite index and $H \not\supseteq G^0$, then $H \cap G^0$ contradicts the fact that G^0 is connected. ♣

G^0 is called the *connected component* of $\mathcal{G}$. G^0 is a normal subgroup of $\mathcal{G}$. In fact both the right and the left multiplication by a given element $a \in G$ are definable. Hence, for every $a \in G$, $a^{-1}G^0a$ is a definable subgroup of $\mathcal{G}$ of the same index as G^0, and consequently equals G^0. But we can say even more.

Proposition 6.2.7 *Let $\mathcal{G}$ be an ω-stable group. The connected component G^0 of $\mathcal{G}$ is an $\emptyset$-definable subgroup; in particular it is invariant under any automorphism of $\mathcal{G}$.*

Proof. G^0 is definable, and so there are a formula $\varphi(v, \vec{w})$ in the language of $\mathcal{G}$ and a tuple $\vec{a}$ in G such that $G^0 = \varphi(\mathcal{G}, \vec{a})$. Let k denote the index of G^0 in $\mathcal{G}$. The formula

$$\varphi^0(v) : \exists \vec{w}(\text{“}\varphi(\mathcal{G}, \vec{w}) \text{ is a subgroup of } \mathcal{G} \text{ of index } k\text{”} \wedge \varphi(v, \vec{w}))$$

defines a subgroup $\varphi^0(\mathcal{G})$ of index k of $\mathcal{G}$. Then $\varphi^0(\mathcal{G}) \supseteq G^0$. But G^0 and $\varphi^0(\mathcal{G})$ have the same index in $\mathcal{G}$, whence $\varphi^0(\mathcal{G}) = G^0$. ♣

The connected component greatly clarifies the role of Morley degree of an ω-stable group $\mathcal{G}$. In fact the following result holds.

Theorem 6.2.8 *Let $\mathcal{G}$ be an ω-stable group, and let G^0 be its connected component. Then G^0 has the same Morley rank as G, and Morley degree 1.*

The former claim is almost trivial. Let k still denote the index of G^0 in G. Of course G is the union of the k cosets of G^0 in G. Each coset is of the form aG^0 for some $a \in G$, and so has the same Morley rank as G^0. Hence $RM(G) = RM(G^0)$.
With respect to the latter claim, again it is easy to observe that, if $\mathcal{G}$ is an ω-stable group of Morley degree 1, then $\mathcal{G}$ has to be connected. But showing that in general, for any ω-stable G, G^0 has Morley degree 1 is not so simple. Indeed we have to prove that G^0 does not decompose as the union of two disjoint subsets of the same Morley rank, and so that there is a unique 1-type p over G containing the formula $\varphi^0(v)$ defining G^0 and having the same Morley rank as G. This needs some further more sophisticated ideas and tools (to be introduced in Chapter 7). So we postpone the full details of Theorem 6.2.8 to Theorem 7.5.10. Here we limit ourselves to a preliminary result, saying that, if p is a type over G as required before (so containing the formula $\varphi^0(v)$ defining G^0 and having the same Morley rank as G), then G^0 equals both the left and the right stabilizers of p.

Lemma 6.2.9 *Let $\mathcal{G}$ be an ω-stable group, G^0 be its connected component, p be any type containing the formula $\varphi^0(v)$ defining G^0 and having the same Morley rank as G. Then G^0 equals the left and the right stabilizer of p.*

Proof. It suffices to handle the left case. Let S denote the left stabilizer of p. Then S is a definable subgroup of $\mathcal{G}$. Moreover, for every a in G, ap has the same Morley rank as p; as there are only finitely many types over G having the same Morley rank as G, there are only finitely many different types ap when a ranges over G. On the other side, for a and b in G, $ap = bp$ holds if and only if a and b are in the same left coset of S in G. So S has finite index in G, whence $S \supseteq G^0$. Conversely take $a \in S$. For every realization x of p in Ω, ax satisfies p as well. In particular both x and ax lie in $\varphi^0(\Omega)$; hence $a = axx^{-1}$ satisfies $\varphi^0(v)$. As $a \in G$, $a \in \varphi^0(G) = G^0$. ♣

To summarize, if $\mathcal{G}$ is an ω-stable group (of Morley rank α), G^0 is its connected component and $1 = a_0, a_1, \ldots, a_{k-1}$ are a set of representatives of the cosets of G^0 in G, then $G = \bigcup_{i<k} a_i G^0$ where each $a_i G^0$ has Morley rank α and Morley degree 1. So G has Morley degree $k = [G : G^0]$, there are exactly k 1-types over G of Morley rank α, and each of them is isolated (among the 1-types of Morley rank α) by a formula $v \in a_i G^0$ (more precisely $\varphi^0(a_i^{-1}v)$) for some $i < k$. These types are usually called the *generic types* of $\mathcal{G}$.
More generally, when $\mathcal{G}$ is an ω-stable group and A is any small subset of the universe Ω of the theory of $\mathcal{G}$,

- a type $p \in S(A)$ is said to be *generic* if and only if it has the same Morley rank as G;

- an element x of Ω is said to be *generic* over A if and only if its type over A is.

Clearly there are only finitely many generic types over any small $A \subseteq \Omega$. Moreover, if x is generic over A, then also x^{-1}, ax and xa (for a in A) are. It is worthy underlining that every $g \in G$ can be written (in Ω) as the product of two elements generic over G: in fact, if $x \in \Omega$ is generic over G, then also gx^{-1} is, and $g = (gx^{-1})x$.
Let us conclude this section by proposing another fundamental tool in the analysis of ω-stable groups, in particular of ω-stable groups of finite Morley rank (i. e. of Morley rank $< \omega$): this is the so called *Zilber's Indecomposability Theorem*. We need the following definition.

Definition 6.2.10 *Let $\mathcal{G}$ be a structure expanding a group $(G, \cdot)$, X be a (non empty) definable subset of G. X is said to be (left)* **indecomposable**

if and only if, for every definable subgroup H di $\mathcal{G}$, either X is included in a unique left coset of H in G, or X overlaps infinitely many such cosets.

Right indecomposability is defined in a specular way. It is easy to see that, for every definable X, X is left indecomposable if and only if X^{-1} is right indecomposable. In particular, when $X = X^{-1}$, X is left indecomposable if and only if it is right indecomposable.
Now we can state and prove Zilber's Indecomposability Theorem.

Theorem 6.2.11 (Zilber) *Let $\mathcal{G}$ be an ω-stable group of finite Morley rank (where finite means $< \omega$). For every i in a set I of indexes, let X_i denote a definable left (right) indecomposable subset of G containing the identity element 1 of $\mathcal{G}$. Then the subgroup of $\mathcal{G}$ generated by $\bigcup_{i \in I} X_i$ is definable and connected in $\mathcal{G}$. Moreover only finitely many X_i's are sufficient to generate it.*

Proof. Assume for simplicity X_i left indecomposable for every i: the right case can be handled in a similar way. As G has a finite Morley rank, we can choose a subset B of G such that B is a finite product of sets X_i or X_i^{-1} with $i \in I$ (and hence is definable) and B has a maximal Morley rank. Accordingly, for every $i \in I$, $RM(X_i B) = RM(X_i^{-1} B) = RM(B)$; in fact $1 \in X_i \cap X_i^{-1}$ and hence $B \subseteq X_i B$, $B \subseteq X_i^{-1} B$. Let p be a type over G containing the formula "$v \in B$" defining B and having the same Morley rank as B. Moreover let S denote the (left) stabilizer of p. Recall that S is definable (Lemma 6.2.5). We claim that S is just the subgroup generated by the X_i's when i ranges over I, and that only finitely many X_i's are sufficient to generate S.
First notice that, for every $i \in I$, $X_i \subseteq S$. It suffices to show that X_i intersects at most finitely many left cosets of S in G; in fact, in this case, as X_i is left indecomposable, X_i is wholly included in a single left coset of S in G, which must coincide with S as $1 \in X_i$. Accordingly take $a \in X_i$; the formula "$v \in X_i^{-1} B$" defining $X_i^{-1} B$ is in ap. We know that $RM(ap) = RM(p) = RM(B)$ and that "$v \in X_i^{-1} B$" occurs in at most finitely many types having its rank. Hence there are only finitely many pairwise different types of the form ap with $a \in X$, otherwise $RM(X_i^{-1} B) > RM(B)$, which contradicts what we have observed before. Consequently there are only finitely many left cosets of S in G of the form aS with $a \in X$ (recall $aS = bS \Leftrightarrow b^{-1}a \in S \Leftrightarrow bp = ap$ for every a and b), as claimed. So S is a subgroup containing each X_i.
Now we show $S \subseteq BB^{-1}$. Let $a \in S$, $x \models p$. So even ax realizes p, and both x and ax satisfy the formula "$v \in B$". Consequently $a = axx^{-1}$ satisfies

"$v \in BB^{-1}$" in Ω and hence in $\mathcal{G}$. In particular S is generated by the X_i's (actually, by a finite subfamily of them).
At this point it remains to show that S is connected in $\mathcal{G}$. So take a subgroup H of S definable in $\mathcal{G}$ and having a finite index in S. Accordingly, for every $i \in I$, X_i overlaps only finitely many left cosets of H, whence $X_i \subseteq H$. But this forces $H = S$. ♣

6.3 ω-stable fields

The aim of this section is twofold. First we want to apply what we have seen about ω-stable theories and, more particularly, ω-stable groups to prove a beautiful theorem due to Macintyre and characterizing the fields having an ω-stable theory. We already know that they include the algebraically closed fields; but now we will show that no further examples arise, so the ω-stable complete theories of (pure) fields are just the theories ACF_p where p is 0 or a prime. Secondly, we will provide a new proof of the fact that algebraically closed fields eliminate the imaginaries; this alternative approach mainly refers to their ω-stability and, owing to this feature, applies to other ω-stable settings, including differentially closed fields of characteristic 0 (as we will see later in this chapter).
Now let us state Macintyre's Theorem.

Theorem 6.3.1 (Macintyre) *Let $\mathcal{K}$ be an infinite integral domain with identity 1, and let $\mathcal{K}$ have an ω-stable theory. Then $\mathcal{K}$ is an algebraically closed field.*

Proof. First let us see that $\mathcal{K}$ is a field. Take $a \in K$, $a \neq 0$. Let $\subset$ denote proper inclusion. If $aK \subset K$, then $a^{n+1}K \subset a^nK$ for every positive integer n because $\mathcal{K}$ is a domain and so, for $b \in K - aK$, $a^nb \in a^nK - a^{n+1}K$. So $\mathcal{K}$ is an ω-stable group (with respect to the sum operation) with a strictly decreasing infinite sequence of definable subgroups $K \supset aK \supset a^2K \supset \dots$, which contradicts Theorem 6.2.2. Accordingly $aK = K$, whence there is some $c \in K$ such that $ac = 1$. So $\mathcal{K}$ is a field. Now suppose towards a contradiction that $\mathcal{K}$ is not algebraically closed. Hence $\mathcal{K}$ has a Galois extension $\mathcal{F}$ of finite degree > 1. Consequently there is some intermediate field $\mathcal{L}$ extending $\mathcal{K}$ and included in $\mathcal{F}$ such that the Galois group of $\mathcal{F}$ over $\mathcal{L}$ is (cyclic) of prime order q. On the other side $\mathcal{L}$ is an extension of finite degree of $\mathcal{K}$, and hence is definable in $\mathcal{K}$: in fact, let d denote the dimension of $\mathcal{L}$ as a vectorspace over $\mathcal{K}$, then $\mathcal{L}$ can be viewed as $\mathcal{K}^d$ equipped with suitably defined field operations. Hence $\mathcal{L}$ has an ω-stable theory, and an extension $\mathcal{F}$ with a (cyclic) Galois group of prime order q. With no loss of generality replace $\mathcal{K}$ by $\mathcal{L}$ and hence assume that $\mathcal{K}$ itself has a Galois extension $\mathcal{F}$ of prime degree q. Field theory tells us that, in this setting, $\mathcal{F} = \mathcal{K}(\alpha)$ where the minimum polynomial α over $\mathcal{K}$ is either

$$x^q - a \text{ with } a \in \mathcal{K},\ q \neq car\, \mathcal{K}$$

or

$$x^q - x - a \text{ with } a \in \mathcal{K},\ q = car\, \mathcal{K}.$$

To get a contradiction, we show that every polynomial of this form must be reducible. For this purpose, first recall that $\mathcal{K}$ is an ω-stable group with respect to addition $+$. Moreover K coincides with its connected component (with respect to $+$). In fact, take $a \in K$ and look at the multiplication by a. This gives an automorphism of the additive group of $\mathcal{K}$, and indeed an automorphism definable in $\mathcal{K}$. Consequently the multiplication by a fixes the connected component K^0. So $aK^0 = K^0$ for every $a \in K$, in other words aK^0 is an ideal of $\mathcal{K}$. Hence $K^0 = K$ as K is infinite and so K^0 cannot equal $\{0\}$.

Now consider $K^\star = K - \{0\}$; $K^\star$ is an ω-stable group with respect to multiplication. As K is infinite, $K^\star$ has the same Morley rank and degree as $\mathcal{K}$, in particular $K^\star$ has degree 1, and consequently $K^\star$ equals its connected component. Notice that the function of $K^\star$ into $K^\star$ mapping any element $k \in K^\star$ into k^q is a definable group homomorphism having a finite kernel $\{a \in K : a^q = 1\}$. Owing to Corollary 6.2.4, the image of this function has a finite index in $K^\star$ and consequently equals $K^\star$ as $K^\star$ is connected. Moreover $0^q = 0$ and so $k \mapsto k^q$ defines a surjective function from K onto K. In particular, for every $a \in K$ the polynomial $x^q - a$ is reducible.

In the same way, when $q = car\, \mathcal{K}$, the function h of K into K mapping any element $k \in K$ into $k^q - k$ is a definable endomorphism of the additive group $\mathcal{K}$ and has a finite kernel $\{a \in K : a^q = a\}$. Accordingly its image has finite index in K, and so coincides with K because K is connected. Then h is surjective and, for every $a \in K$, even the polynomial $x^q - x - a$ is reducible. This yields the required contradiction. ♣

It is worth underlining once again that Macintyre's Theorem deals not only with infinite fields, but more generally with integral domains with identity; moreover, as algebraically closed fields are strongly minimal, it classifies even the strongly minimal examples in this framework.

Now let us deal with the second matter of this section: elimination of imaginaries.

Theorem 6.3.2 *Every algebraically closed field $\mathcal{K}$ eliminates uniformly the imaginaries.*

Proof. Our new approach needs the following non trivial algebraic preliminaries, regarding arbitrary fields $\mathcal{K}$ and ideals I in $\mathcal{K}[\vec{x}]$ (where $\vec{x}$ still abbreviates $(x_1, \ldots, x_n)$): their details can be found in the references quoted at the end of the chapter.

1. The first concerns the *fields of definition* of I: these are subfields of $\mathcal{K}$ containing the coefficients of the polynomials of some generating subset of I. One can see that there is a (countable) field of definition $\mathcal{K}(I)$ of I satisfying the following additional condition: for every automorphism σ of $\mathcal{K}$,

   σ (more exactly, its natural extension to $\mathcal{K}[\vec{x}]$) fixes I setwise

   if and only if

   σ acts identically on $K(I)$.

2. Incidentally one can notice that, for every automorphism σ of I, if $\sigma(I) \supseteq I$ then $\sigma(I) = I$.

3. Now consider prime ideals in $\mathcal{K}[\vec{x}]$. More particularly, take pairwise different prime ideals $I_0, \ldots, I_m$ of $\mathcal{K}[\vec{x}]$ in the same conjugacy orbit: so, for every $j < h \leq m$, there is some automorphism of $\mathcal{K}$ taking I_j to I_h. Correspondingly one can find a subfield $\mathcal{K}_0$ of $\mathcal{K}$ such that, for every automorphism σ of $\mathcal{K}$, σ permutes $I_0, \ldots, I_m$ if and only if σ fixes K_0 pointwise. It suffices to form $I = \bigcap_{j \leq m} I_j$ and choose as $\mathcal{K}_0$ the field of definition $\mathcal{K}(I)$ associated with I as in 1. The key point here is that an automorphism σ fixes I setwise if and only if it permutes the I_j's. This is a consequence of two facts: the former is that, owing to 2, $I = \bigcap_{j \leq m} I_j$ is an irredundant decomposition of the (radical) ideal I as an intersection of prime ideals, and the latter is that this irredundant decomposition is unique up to the order of the involved prime ideals.

Let us deal at last with *algebraically closed* fields $\mathcal{K}$ and with Model Theory: we want to show that $\mathcal{K}$ uniformly eliminates the imaginaries. As observed in Chapter 4, it is sufficient for our purposes to prove that any such $\mathcal{K}$ eliminates the imaginaries and so to find, for every $\emptyset$-definable equivalence relation $E = E(\vec{v}, \vec{w})$ in K^n and $\vec{a}$ in K^n,

$\star$ a formula $\varphi(\vec{v}, \vec{z})$,

$\star$ a unique sequence $\vec{b}$ in K

such that $E(\mathcal{K}^n, \vec{a}) = \varphi(\mathcal{K}^n, \vec{b})$ (here the common length of $\vec{z}$ and $\vec{b}$ may be $\neq n$). There is no loss of generality in assuming $\mathcal{K}$ $\aleph_1$-saturated: in fact, given $\vec{a}$ in K^n, if there exists a unique $\vec{b}$ in some $\aleph_1$-saturated extension $\mathcal{K}'$ of $\mathcal{K}$ such that $\mathcal{K}' \models \forall \vec{v}(E(\vec{v}, \vec{a}) \leftrightarrow \varphi(\vec{v}, \vec{b}))$, then this tuple $\vec{b}$ must belong to $dcl(\vec{a})$ and consequently to K.
Now we use the ω-stability -more precisely the total transcendence- of $\mathcal{K}$, as promised. In fact, given $\vec{a} \in K^n$ and the formula $E(\vec{v}, \vec{a})$, there are only finitely many n-types $p_0, \ldots, p_q$ over K containing $E(\vec{v}, \vec{a})$ and having the same Morley rank as $E(\vec{v}, \vec{a})$. Let $I_0, \ldots, I_q$ be the prime ideals of $\mathcal{K}[\vec{x}]$ corresponding to these types. Partition these ideals into conjugacy orbits, so into equivalence classes with respect to the relation linking two ideals exactly when they correspond to each other by some automorphism of $\mathcal{K}$. Take any orbit O. Up to rearranging the indexes, one can assume $O = \{I_0, \ldots, I_m\}$ for some $m \leq q$. Apply 3 and find a (countable) subfield $\mathcal{K}(O)$ of $\mathcal{K}$ such that, for every automorphism σ of $\mathcal{K}$, σ permutes $I_0, \ldots, I_m$ if and only if σ fixes $\mathcal{K}(O)$ pointwise. Let $\mathcal{K}_0$ be the countable subfield of $\mathcal{K}$ generated by the $\mathcal{K}(O)$'s when O ranges over the orbits of $I_0, \ldots, I_q$. We claim that, for every σ, σ fixes $E(\mathcal{K}^n, \vec{a})$ setwise if and only if σ acts identically on K_0.
In fact, if σ preserves $E(\mathcal{K}^n, \vec{a})$, then σ permutes the types $p_0, \ldots, p_q$ containing $E(\vec{v}, \vec{a})$ and having the same Morley rank as $E(\vec{v}, \vec{a})$; hence σ permutes the corresponding ideals $I_0, \ldots, I_q$ as well, and in particular any single orbit O. Hence σ fixes pointwise each $\mathcal{K}(O)$ and in conclusion $\mathcal{K}_0$.
Conversely, let σ act identically on K_0, and so on any $K(O)$. It follows that σ permutes the ideals in any orbit O and so the whole set $I_0, \ldots, I_q$. Equivalently σ permutes $p_0, \ldots, p_q$. But σ has to preserve E because E is $\emptyset$-definable, and so takes E-classes to E-classes. In particular σ fixes $E(\mathcal{K}^n, \vec{a})$ because the formula $E(\vec{v}, \vec{a})$ lies in all the types $p_0, \ldots, p_q$.
Therefore any automorphism of $\mathcal{K}$ fixing $\vec{a}$ pointwise, and so preserving $E(\vec{v}, \vec{a})$, fixes K_0 pointwise as well; in other words, $K_0 \subseteq dcl(\vec{a})$.
Now assemble all the $L(K_0)$-formulas $\psi(\vec{v})$ for which $E(\mathcal{K}^n, \vec{a}) \subseteq \psi(\mathcal{K}^n)$. They form a possibly incomplete type p over K_0. We claim

$$E(\mathcal{K}^n, \vec{a}) = \bigcap_{\psi(\vec{v}) \in p} \psi(\mathcal{K}^n).$$

Otherwise there is some $\vec{d} \in K^n$ satisfying p and $\neg E(\vec{d}, \vec{a})$. This remains true for every $\vec{d'} \in K^n$ having the same type as $\vec{d}$ over K_0; in fact, as $\mathcal{K}$ is

$\aleph_1$-saturated and K_0 is countable, $\vec{d}$, $\vec{d'}$ correspond to each other by some automorphism σ of $\mathcal{K}$ fixing K_0 pointwise, hence $E(\mathcal{K}^n, \vec{a})$ setwise. Then $tp(\vec{d}/K_0)$ implies $\neg E(\vec{v}, \vec{a})$ in $\mathcal{K}$. Use compactness and determine $\theta(\vec{v}) \in tp(\vec{d}/K_0)$ such that $\theta(\mathcal{K}^n) \subseteq \neg E(\mathcal{K}^n, \vec{a})$, equivalently $\neg\theta(\mathcal{K}^n) \supseteq E(\mathcal{K}^n, \vec{a})$. But this forces $\neg\theta(\vec{v}) \in p \subseteq tp(\vec{d}/K_0)$, a contradiction.
Consequently $E(\vec{v}, \vec{a})$ is a consequence of p. Again use compactness and find a formula $\varphi(\vec{v}, \vec{b})$ in p - so with $\vec{b}$ in K_0 - implying $E(\vec{v}, \vec{a})$, and hence equivalent to it. At this point it remains to show the uniqueness of $\vec{b}$. This can be obtained by some simple manipulations, slightly changing $\varphi(\vec{v}, \vec{b})$. In fact, take any $\vec{b'} \neq \vec{b}$ in K having the same type as $\vec{b}$ over $\emptyset$. There is some automorphism σ of $\mathcal{K}$ sending $\vec{b}$ to $\vec{b'}$, so σ does not act identically on K_0 because it moves $\vec{b'}$; accordingly σ does not fix setwise $E(\mathcal{K}^n, \vec{a}) = \varphi(\mathcal{K}^n, \vec{b})$, in other words $\varphi(\mathcal{K}^n, \vec{b}) \neq \varphi(\mathcal{K}^n, \vec{b'})$. This is true for every $\vec{b'} \neq \vec{b}$ realizing $tp(\vec{b}/\emptyset)$. By compactness, there is some formula $\eta(\vec{z})$ in $tp(\vec{b}/\emptyset)$ for which

$$\mathcal{K} \models \forall \vec{z}\,(\eta(\vec{z}) \wedge \neg(\vec{z} = \vec{b}) \longrightarrow \neg(\forall \vec{v}(\varphi(\vec{v}, \vec{z}) \longleftrightarrow \varphi(\vec{v}, \vec{b})))).$$

So, unless replacing $\varphi(\vec{v}, \vec{z})$ by $\varphi'(\vec{v}, \vec{z}) : \varphi(\vec{v}, \vec{z}) \wedge \eta(\vec{z})$, we see that $\vec{b}$ is the only tuple in K such that $\varphi'(\mathcal{K}^n, \vec{b}) = E(\mathcal{K}^n, \vec{a})$: in fact, if $\vec{b'}$ does not satisfy $\eta(\vec{z})$, then $\varphi(\mathcal{K}^n, \vec{b'}) = \emptyset$. This accomplishes our proof. ♣

6.4 Prime models

This section treats a fundamental feature of ω-stable theories, namely the existence and uniqueness of prime models over subsets. This property will be useful in 7.8, in proving Morley's Theorem on uncountably categorical theories; but a remarkable application will be provided also in 6.5, when dealing with differential fields; in fact we will observe that DCF_0 is ω-stable, and so the machinery of this section will apply and yield both existence and uniqueness of prime models over subsets among differentially closed fields of characteristic 0 - in algebraic terms, existence and uniqueness of differential closures among differential fields of characteristic 0 -.
The first step of our treatment is clearly to define what a prime model is. Let T be any (possibly non ω-stable) complete theory with infinite models in a countable language L. Ω still denotes a big saturated model of T.

Definition 6.4.1 *Let X be a small subset of Ω. A model $\mathcal{A}$ of T is called prime over X if and only if:*

(i) *there is an elementary function f of X into $\mathcal{A}$,*

(ii) *for every model $\mathcal{B}$ of T and elementary function g from X into $\mathcal{B}$, there is an elementary embedding h of $\mathcal{A}$ into $\mathcal{B}$ for which $h\,f = g$.*

Two problems spring quite naturally in this setting. Let $X \subseteq \Omega$ be given.

(a) (*Existence*) Is there any model $\mathcal{A}$ of T prime over X?

(b) (*Uniqueness*) Is this model unique? More precisely, let $\mathcal{A}_0$ and $\mathcal{A}_1$ be two models of T prime over X - via the elementary functions f_0, f_1 respectively -. Are $\mathcal{A}_0$ and $\mathcal{A}_1$ isomorphic by a map h satisfying $hf_0 = f_1$?

Intuitively speaking, a model of T prime over X is a "minimal" model extending X by some elementary function. Of course a sharp definition in a general setting fatally involves all the details given before. In particular uniqueness has to be required up to isomorphism, in the way stated some lines ago. On the other hand, this generality has its positive sides, and even allows some well accepted freedom. For instance, just owing to (b), there is no loss of generality in assuming that a model $\mathcal{A}$ prime over X, if any, does extend X (and so f is an inclusion).
We will see that existence and uniqueness of prime models are not guaranteed in general, for arbitrary theories T and over arbitrary sets X. However they do hold when T is ω-stable. But now let us propose some examples.

Examples 6.4.2 1. Let $T = ACF$ or, if you like to keep our completeness assumption on T, let $T = ACF_p$ for some $p = 0$ or prime. When X is an integral domain with identity in characteristic p, a model of ACF_p prime over X is the algebraic closure of the field of quotients of X (in the field theoretic sense), and is unique (as the Algebra handbooks relate).

2. Let $T = RCF$. A model of RCF prime over an ordered field X is its real closure, and is unique up to isomorphism. Again, any handbook of Algebra (such as [65]) provides the details.

3. Now let $T = DCF_0$. Take a differential field X of characteristic 0. Now Algebra does not tell us anything similar to the previous examples; in fact it clarifies neither the existence nor the uniqueness of a differential closure of X, indeed Algebra cannot either define what a differential closure is.

So the existence and uniqueness problems of prime models overlap some non trivial (unsolved?) algebraic questions.

Now let us summarize briefly how we plan to organize this section. First we introduce and discuss some notions closely related to prime models: *atomic sets* and *constructible sets*. Indeed we will see that over countable sets a model is prime if and only if it is atomic and countable, and if and only if it is constructible (warning: here *constructible* has a specific meaning, having nothing to do with what we introduced in Chapters 1, 2 and 3 in the particular framework of fields). In this perspective, a theorem of Ressayre stating the uniqueness of a model constructible over a given subset is very noteworthy, as it implies, among other things, the uniqueness of a prime model over a countable set. At this point we will discuss the existence problem of prime models. We will give a condition of topological flavour characterizing the theories T such that every small subset X of Ω admits a prime model over itself: they are exactly those where the isolated types over every X are dense in $S(X)$. This will accomplish the general analysis for arbitrary theories.
At this point we will assume T ω-stable, and we will show a theorem of Morley proving that, in this setting, the previous topological condition is satisfied and, consequently, the existence of prime models is always guaranteed. Finally we will treat a nice and deep theorem of Shelah showing the uniqueness of prime models in the ω-stable framework. Actually what Shelah proves is that, for an ω-stable T, a model prime over a set X is also constructible over X; so Ressayre's Theorem applies and gives the required uniqueness result. As already said, Shelah's Theorem is quite complicated, and needs some subtle notions and tools to be introduced later in this book. So we will postpone its proof until Chapter 7. To conclude, we will see that both existence and uniqueness of prime models fail when the ω-stability assumption is dropped.

This is the sketch of this section. Now let us begin our report by introducing atomic and constructible sets, as promised. We assume some basic acquaintance with ordinal numbers. X, Y, Z, ... denote small subsets of Ω, $\vec{a}$, $\vec{b}$, $\vec{c}$,... tuples in Ω. For simplicity we sometimes identify a tuple $\vec{a}$ and the (finite) set of the elements in $\vec{a}$.

Definition 6.4.3 *$Y \supseteq X$ is* **atomic** *over X if and only if every tuple in Y has an isolated type over X.*

Recall that every model $\mathcal{A}$ of T extending X via an elementary function realizes all the isolated types over X; on the contrary, non-isolated types can be omitted by suitably chosen models, as the Omitting Types Theorem shows. So a model prime over X, being elementarily embeddable in every

model $\mathcal{A}$ extending X as before, should not be far from being atomic over X.
Here is a list of simple and useful facts concerning atomic sets and isolated types.

Remarks 6.4.4 1. If $tp(\vec{b}, c/X)$ is isolated, then $tp(\vec{b}/X)$ is.

In fact, let $\varphi(\vec{v}, w)$ be an $L(X)$-formula isolating $tp(\vec{b}, c/X)$. An easy check shows that $\exists w \varphi(\vec{v}, w)$ isolates $tp(\vec{b}/X)$. In fact $\vec{b}$ does satisfy $\exists w \varphi(\vec{v}, w)$. Furthermore, for every $\vec{b'}$ realizing $\exists w \varphi(\vec{v}, w)$, there is some c' for which $\models \varphi(\vec{b'}, c')$. Then $(\vec{b'}, c')$ has the same type as $(\vec{b}, c)$ over X, and consequently $\vec{b'}$ has the same type as $\vec{b}$ over X.

2. If $tp(\vec{b}, \vec{c}/X)$ is isolated, then $tp(\vec{b}/X \cup \vec{c})$ is.

   In fact, if $\varphi(\vec{v}, \vec{w})$ is an $L(X)$-formula isolating $tp(\vec{b}, \vec{c}/X)$, then $\varphi(\vec{v}, \vec{c})$ isolates $tp(\vec{b}/X \cup \vec{c})$.

3. Let $X \subseteq Y$. Suppose $tp(\vec{c}/Y)$ isolated by some $L(Y)$-formula $\varphi(\vec{v}, \vec{b})$ with parameters $\vec{b}$ in Y, and $tp(\vec{b}/X)$ also isolated. Then $tp(\vec{c}/X)$ is isolated.

   In fact, take an $L(X)$-formula $\psi(\vec{w})$ isolating $tp(\vec{b}/X)$. Look at the $L(X)$-formula $\theta(\vec{v}) : \exists \vec{w}(\psi(\vec{w}) \wedge \varphi(\vec{v}, \vec{w}))$. Clearly $\Omega \models \theta(\vec{c})$. We claim that $\theta(\vec{v})$ isolates $tp(\vec{c}/X)$. For, let $\vec{c'}$ satisfy $\theta(\vec{v})$, so for some $\vec{b'}$, $\Omega \models \psi(\vec{b'}) \wedge \varphi(\vec{c'}, \vec{b'})$. In particular $\vec{b'}$ has the same type over X as $\vec{b}$ and hence there is an automorphism f of Ω fixing X pointwise and taking $\vec{b}$ to $\vec{b'}$. Then $\varphi(\vec{v}, \vec{b'})$ isolates a type over $f(Y)$; as $\Omega \models \varphi(\vec{c'}, \vec{b'})$, $tp(\vec{c'}/f(Y))$ equals this type and so contains every formula $\alpha(\vec{v}, f(\vec{d}))$ with $\alpha(\vec{v}, \vec{d}) \in tp(\vec{c}/Y)$. In particular, as f fixes X pointwise, $tp(\vec{c'}/X) = tp(\vec{c}/X)$.

   Notice that 3 implies that atomicity is transitive: for $X \subseteq Y \subseteq Z$, if Z is atomic over Y and Y is atomic over X, then Z is atomic over X.

4. Let $Y \supseteq X$ be atomic over X, $\vec{c}$ be in Y. Then Y is atomic over $X \cup \vec{c}$.

   In fact, for every $\vec{b} \in Y$, $tp(\vec{b}, \vec{c}/X)$ is isolated. So 2 implies $tp(\vec{b}/X \cup \vec{c})$ isolated, too.

Definition 6.4.5 *$Y \supseteq X$ is* **constructible** *over X if one can list the elements of Y by ordinal indexes b_ν (with $\nu < \alpha$, α a suitable ordinal) in such a way that, for every $\nu < \alpha$, b_ν has an isolated type over $X \cup \{b_\mu : \mu < \nu\}$. In this case the sequence $(b_\nu : \nu < \alpha)$ is called a* **construction** *of Y over X.*

We put for simplicity $X_\nu = X \cup \{b_\mu : \mu < \nu\}$. So $X_0 = X$, for a limit ν $X_\nu = \bigcup_{\mu<\nu} X_\mu$ and, for a successor $\nu = \mu + 1$, $X_\nu = X_\mu \cup \{b_\mu\}$. Of course $Y = \bigcup_{\nu<\alpha} X_\nu$. It is quite trivial that, if $b_\nu \in X$, then $tp(b_\nu/X)$, and consequently $tp(b_\nu/X_\nu)$, is isolated (by $v = b_\nu$).
Now we discuss the relationship between atomicity and constructibility. First we propose a simple result.

Lemma 6.4.6 *Let $Y \supseteq X$ be atomic over X. If $Y - X$ is countable, then Y is constructible over X.*

Proof. Let $b_0, b_1, \ldots, b_n, \ldots$ (n natural) enumerate $Y - X$. It is enough to show that, for every n, the type of b_n over $X_n = X \cup \{b_m : m < n\}$ is isolated. As $X_n - X$ is finite, the previous Remark 6.4.4.4 ensures that Y is isolated over X_n, and so that $tp(b_n/X_n)$ is isolated, as required. ♣

So atomicity sometimes implies constructibility. But a stronger and deeper result holds in the opposite direction: in fact, as we will see within a few lines, if $Y \supseteq X$ is constructible over X, then Y is atomic over X.
Now we relate constructibility and prime models. Fix a small $X \subseteq \Omega$ and take a model $\mathcal{A}$ of T extending X by some given elementary function. As already said, there is no loss of generality for our purposes in assuming that X is just a subset of A and that the corresponding inclusion is elementary.

Theorem 6.4.7 *If $\mathcal{A}$ is constructible over X, then $\mathcal{A}$ is prime over X.*

Proof. Let $\mathcal{B}$ be a model of T and g be an elementary function from X into $\mathcal{B}$. We are looking for an elementary embedding h of $\mathcal{A}$ into $\mathcal{B}$ extending g. Let $(b_\nu : \nu < \alpha)$ be a construction of A over X. The strategy we devise to build h is the following: we progressively extend g to larger and larger elementary functions g_ν from X_ν into $\mathcal{B}$ (for $\nu < \alpha$); at last, we check that $\bigcup_{\nu<\alpha} g_\nu$ is just the required embedding h.
We proceed by induction on ν. For $\nu = 0$, we put $g_0 = g$. When ν is limit, we set $g_\nu = \bigcup_{\mu<\nu} g_\mu$, a function from $X_\nu = \bigcup_{\mu<\nu} X_\mu$ into $\mathcal{B}$. It is easy to check that g_ν is elementary. In fact let $\vec{a}$ in X_ν satisfy $\models \varphi(\vec{a})$ for some L-formula $\varphi(\vec{v})$; for a sufficiently large $\mu < \nu$, $\vec{a}$ lies in X_μ. So the elementarity of g_μ ensures $\mathcal{B} \models \varphi(g_\mu(\vec{a}))$. As $g_\nu \supseteq g_\mu$, $\mathcal{B} \models \varphi(g_\nu(\vec{a}))$.
Finally let $\nu = \mu + 1$ be a successor ordinal. Look at the isolated type $tp(b_\mu/X_\mu)$. Call it p_μ for simplicity; $g_\mu(p_\mu)$ is in its turn an isolated type over $g_\mu(X_\mu) \subseteq B$, and so it is satisfied by some element $b'_\mu \in B$. Extend g_μ by including the new element b_μ in the domain and mapping b_μ into b'_μ. Let

g_ν be the resulting function. Then the domain of g_ν is just $X_\nu = X_\mu \cup \{b_\mu\}$; moreover, owing to the choice of b'_μ, g_ν is elementary.
At last form $f = \bigcup_{\nu<\alpha} g_\nu$. The same argument as before shows that h is an elementary embedding of $\mathcal{A}$ into $\mathcal{B}$ (recall $A = \bigcup_{\nu<\alpha} X_\nu$). ♣

So constructible models are prime. At this point, we show that constructible models satisfy the uniqueness assumption. This is a non trivial result of Ressayre. But, before stating it exactly and beginning its proof, we need some technical preliminaries. So let Y be constructible over X, fix a construction $(b_\nu : \nu < \alpha)$ of Y over X. For every $\nu < \alpha$ choose an $L(X_\nu)$-formula $\varphi_\nu(v, \vec{a_\nu})$ isolating the type of b_ν over X_ν, in particular fix the parameters $\vec{a_\nu}$ from X_ν. When $b_\nu \in X$, choose $v = b_\nu$ as an isolating formula. In this framework, a subset Z of Y is called *closed* (with respect to the given construction of Y over X) if, for every $\nu < \alpha$ with $b_\nu \in Z$, even the defining tuple $\vec{a_\nu}$ is in Z. Clearly X and Y are closed; moreover closed sets are preserved under (finite) union and intersection.

Lemma 6.4.8 *Let $X \subseteq Z \subseteq Y$, Z be closed. Then the theory of Ω_Z is axiomatized by* $\mathrm{Th}(\Omega_X) \cup \{\varphi(b_\nu, \vec{a_\nu}) : \nu < \alpha,\ b_\nu \in Z\}$.

Proof. For every $\mu \leq \alpha$ put $Z_\mu = \{b_\nu \in Z : \nu < \mu\}$. It is an easy exercise to see that Z_μ is closed for every $\mu \leq \alpha$. Moreover Z_α coincides with Z. We claim that, for every $\mu \leq \alpha$,

$$\mathrm{Th}(\Omega_{X\cup Z_\mu}) \quad \text{is axiomatized by } \mathrm{Th}(\Omega_X) \cup \{\varphi_\nu(b_\nu, \vec{a_\nu}) : b_\nu \in Z_\mu\}.$$

This is clearly enough, as the case $\mu = \alpha$ is just our thesis. Now it is obvious that

$$\mathrm{Th}(\Omega_X) \cup \{\varphi_\nu(b_\nu, \vec{a_\nu}) : b_\nu \in Z_\mu\} \subseteq \mathrm{Th}(\Omega_{X\cup Z_\mu})$$

because, for every ν and μ, when $b_\nu \in Z_\mu$, $\vec{a_\nu}$ is in Z_μ, too. So we have to prove that, conversely, every sentence in $\mathrm{Th}(\Omega_{X\cup Z_\mu})$ is a consequence of $\mathrm{Th}(\Omega_X) \cup \{\varphi_\nu(b_\nu, \vec{a_\nu}) : b_\nu \in Z_\mu\}$. We proceed by induction on μ. When $\mu = 0$, Z_0 is empty and so the claim is trivial. When μ is limit, we can observe that a sentence in $\mathrm{Th}(\Omega_{X\cup Z_\mu})$ involves only finitely many elements in Z_μ and so belongs to $Th(\Omega_{X\cup Z_\nu})$ for some suitable $\nu < \mu$. Hence the induction hypothesis applies. At last, take a successor $\mu = \nu + 1$. We can assume $b_\nu \in Z$, otherwise there is nothing to prove. As $\varphi_\nu(v, \vec{a_\nu})$ isolates $tp(b_\nu/X_\nu)$, the sentences concerning b_ν in $\mathrm{Th}(\Omega_{X_\mu})$ are proved by $\mathrm{Th}(\Omega_{X_\mu}) \cup \{\varphi_\nu(b_\nu, \vec{a_\nu})\}$. When we restrict our attention to Z, we can deduce by standard arguments of elementary logic that

$$\mathrm{Th}(\Omega_{X_\mu \cap Z}) \text{ is axiomatized by } \mathrm{Th}(\Omega_{X_\nu \cap Z}) \cup \{\varphi_\nu(b_\nu, \vec{a_\nu})\}$$

because b_ν and consequently $\vec{a_\nu}$ are in Z. Notice $X_\mu \cap Z = X \cup Z_\mu$, $X_\nu \cap Z = X \cup Z_\nu$ and deduce that

$$\mathrm{Th}(\Omega_{X \cup Z_\mu}) \text{ is axiomatized by } \mathrm{Th}(\Omega_{X \cup Z_\nu}) \cup \{\varphi_\nu(b_\nu, \vec{a_\nu})\}.$$

But owing to the induction hypothesis $\mathrm{Th}(\Omega_{X \cup Z_\nu})$ is axiomatized in its turn by $\mathrm{Th}(\Omega_X) \cup \{\varphi_\lambda(b_\lambda, \vec{a_\lambda}) : \lambda < \nu, b_\lambda \in Z\}$, whence the last set of sentences, together with $\varphi_\nu(b_\nu, \vec{a_\nu})$, axiomatizes the theory of $\Omega_{X \cup Z_\mu}$, as claimed. ♣

Still keeping the same notation, now we show:

Lemma 6.4.9 *If $Z \subseteq Y$ and Z is closed, then Y is atomic over $X \cup Z$.*

Notice that this applies also to $Z = X$ and implies

Corollary 6.4.10 *If $Y \supseteq X$ is constructible over X, then Y is atomic over X.*

Proof. (of the lemma) Let $\vec{b}$ be in Y, we have to show that the type of $\vec{b}$ over $X \cup Z$ is isolated. Extend $\vec{b}$ to a larger and closed tuple $\vec{c}$ in the following way. For every $b \in \vec{b}$, pick $\nu < \alpha$ for which $b = b_\nu$, and consider the formula $\varphi(v, \vec{a_\nu})$ isolating the type of b over X_ν. For every $a \in \vec{a_\nu}$, take $\mu \leq \nu$ such that $a = b_\mu$, and repeat the previous procedure. As there is no infinite strictly decreasing sequence of ordinals, this machinery stops after finitely many steps. Let $\vec{c}$ absorb all the elements arising in this procedure. Clearly $\vec{c}$ is closed, as well as $Z \cup \vec{c}$ and $X \cup Z$. Owing to Lemma 6.4.8, the theory of $\Omega_{X \cup Z \cup \vec{c}}$ is axiomatized by $\mathrm{Th}(\Omega_{X \cup Z}) \cup \{\varphi_\nu(b_\nu, \vec{a_\nu}) : \nu < \alpha, b_\nu \in \vec{c}\}$. Hence $\bigwedge_{b_\nu \in \vec{c}} \varphi_\nu(v_\nu, \vec{a_\nu})$ isolates the type of $\vec{c}$ over $X \cup Z$. As $\vec{c}$ enlarges $\vec{b}$, $tp(\vec{b}/X \cup Z)$ is isolated, too. ♣

Now we can state and prove Ressayre's Uniqueness Theorem.

Theorem 6.4.11 (Ressayre) *Let X be a small subset of Ω, $\mathcal{A}_0$ and $\mathcal{A}_1$ be two models of T containing X via some elementary inclusions, and constructible over X. Then there is an isomorphism h of $\mathcal{A}_0$ onto $\mathcal{A}_1$ fixing X pointwise.*

Proof. Look at the triples (Y_0, Y_1, f) where, for every $i = 0, 1$, $X \subseteq Y_i \subseteq A_i$, Y_i is closed with respect to some given construction of A_i over X and f is an elementary function of Y_0 into $\mathcal{A}_1$ acting identically on X and having Y_1 as image. The collection of these triples is not empty because it contains

(X, X, id_X) (in fact, X is closed); moreover it is partially ordered by the relation $\preceq$ such that

$$(Y_0, Y_1, f) \preceq (Y_0', Y_1', f')$$

if and only if $Y_i \subseteq Y_i'$ for every $i = 0, 1$ and $f \subseteq f'$. A straightforward check ensures that Zorn's Lemma applies to $\preceq$ and yields a maximal triple $(\overline{Y_0}, \overline{Y_1}, \overline{f})$. We claim that $\overline{Y_i} = A_i$ for every $i = 0, 1$, so $\overline{f}$ is the required isomorphism.
Suppose not. For instance, assume $\overline{Y_0} \neq A_0$. Use the same technique as in the proof of Lemma 6.4.10 and enlarge $\overline{Y_0}$ to a closed $Y_0^0 \supseteq \overline{Y_0} \cup \{b\}$ with $Y_0^0 - \overline{Y_0}$ finite. Owing to Lemma 6.4.10 the sequence $\vec{c_0}$ of the elements in $Y_0^0 - \overline{Y_0}$ has an isolated type over $\overline{Y_0}$. Call it p. Notice that even $\overline{f}(p)$ is an isolated type (over $\overline{Y_1}$) because $\overline{f}$ is elementary. In particular $\overline{f}(p)$ is realized in $\mathcal{A}_1$, say by $\vec{c_1}$. Enlarge $\overline{Y_1}$ by $\vec{c_1}$ and get a new set $Y_1^0 \supseteq \overline{Y_1}$ in A_1. Clearly $\overline{f}$ can be extended to an elementary function f^0 from Y_0^0 into $\mathcal{A}_1$ with image Y_0^1: just map $\vec{c_0}$ in $\vec{c_1}$. If Y_0^1 is closed with respect to the fixed construction of A_1 over X, then we are done; in fact, we have found a counterexample (Y_0^0, Y_1^0, f^0) to the maximality of $(\overline{Y_0}, \overline{Y_1}, \overline{f})$; this is a contradiction, and so shows $\overline{Y_0} = A_0$ as claimed. Otherwise we apply the previous procedure to Y_1^0. Accordingly we enlarge Y_1^0 to a closed $Y_1^1 \supseteq Y_1^0$ with $Y_1^1 - Y_1^0$ finite, and correspondingly we build $Y_0^0 \subseteq Y_0^1 \subseteq A_0$ and an elementary function f^1 from Y_0^1 in $\mathcal{A}_1$ with image Y_1^1. If Y_0^1 is closed, then we are done; otherwise we continue our procedure. In the worst case, we get for every natural n a triple (Y_0^n, Y_1^n, f^n) such that, for every n and $i = 0, 1$, $(\overline{Y_0}, \overline{Y_1}, \overline{f}) \preceq (Y_0^n, Y_1^n, f^n)$, $X \subseteq Y_n^i \subset Y_i^{n+1}$, f^n is an elementary map from Y_0^n into $\mathcal{A}_1$ with image Y_1^n, $Y_i^n - \overline{Y_i}$ is finite and both Y_0^{2n+1} and Y_1^{2n+2} are closed. Owing to the last condition both $\bigcup_n Y_0^n$ and $\bigcup_n Y_1^n$ are closed. So $(\bigcup_n Y_0^n, \bigcup_n Y_1^n, \bigcup_n f^n)$ again contradicts the choice of $(\overline{Y_0}, \overline{Y_1}, \overline{f})$.
In conclusion $\overline{Y_0} = A_0$, and, similarly, $\overline{Y_1} = A_1$. ♣

But now it is time to come back to the existence and uniqueness problems of prime models. As a warm-up, let us first discuss the case $X = \emptyset$. Here we show uniqueness and we characterize existence by some equivalent conditions. Notice that the assumption $X = \emptyset$ is not so restrictive as it may look. Indeed it implicitly includes the seemingly more general case when X is countable: it suffices to replace L by the (countable) language $L(X)$ and T by the theory of Ω_X, and to observe that a model of $Th(\Omega_X)$ prime over $\emptyset$ is just, as a structure of L, a model of T prime over X. Notice also that a model of T prime over $\emptyset$ is elementarily embeddable in **every** model $\mathcal{B}$ of T; in fact, as T is complete, the inclusion of $\emptyset$ in $\mathcal{B}$ is elementary. The uniqueness of a model prime over $\emptyset$ is a direct consequence of Ressayre's Theorem and the following result.

Theorem 6.4.12 *Let $\mathcal{A}$ be a model of T. Then the following propositions are equivalent:*

(1) *$\mathcal{A}$ is prime over $\emptyset$;*

(2) *$\mathcal{A}$ is countable and atomic over $\emptyset$;*

(3) *$\mathcal{A}$ is constructible over $\emptyset$.*

Proof. (1) $\Rightarrow$ (2) As already said, a model $\mathcal{A}$ of T prime over $\emptyset$ elementarily embeds itself in any model of T. This clearly implies that $\mathcal{A}$ is countable, as T does admit some countable models. Furthermore every non-isolated type over $\emptyset$ is omitted in some countable model of T (by the Omitting Types Theorem) and consequently in $\mathcal{A}$: so $\mathcal{A}$ is atomic over $\emptyset$.
(2) $\Rightarrow$ (3) Just apply Lemma 6.4.6 to $Y = A$ and $X = \emptyset$.
(3) $\Rightarrow$ (1) This is a particular case of Theorem 6.4.7. ♣

At this point Ressayre's Uniqueness Theorem applies and shows:

Theorem 6.4.13 *Any two models of T prime over $\emptyset$ are isomorphic.*

In fact these models are constructible over $\emptyset$. Now let us discuss the existence of a model of T prime over $\emptyset$. The following theorem provides a nice equivalent condition.

Theorem 6.4.14 *There exists a model of T prime over $\emptyset$ if and only if, for every positive integer n, the isolated n-types over $\emptyset$ are dense in the topological space $S_n(\emptyset)$ (this means that, for every L-formula $\varphi(\vec{v})$ in n free variables $\vec{v}$, if $\varphi(\Omega^n) \neq \emptyset$, then there is some isolated n-type over $\emptyset$ containing $\varphi(\vec{v})$).*

Proof. ($\Rightarrow$) Assume that $\mathcal{A}$ is a model of T prime over $\emptyset$. As T is complete, $\varphi(\mathcal{A})^n$ is not empty, in other words there is some tuple $\vec{a}$ in A satisfying $\varphi(\vec{v})$. As $\mathcal{A}$ is atomic over $\emptyset$, the type of $\vec{a}$ over $\emptyset$ is isolated (and includes $\varphi(\vec{v})$).
($\Leftarrow$) The strategy here is to use some classical techniques (adding constants, referring to the Tarski-Vaught Theorem and so on) and to construct a countable model $\mathcal{A}$ constructible over $\emptyset$: as said before, this is also prime over $\emptyset$. In order to form $\mathcal{A}$ we build preliminarily a sequence of isolated types $p_1 \subseteq p_2 \subseteq \ldots \subseteq p_n \subseteq \ldots$ (with n a positive integer) such that, for every n,

(a) p_n is an n-type over $\emptyset$ (in the free variables $(v_1, \ldots, v_n) = \vec{v}(n)$);

(b) if $\exists w \varphi(w, \vec{v}(n)) \in p_n$ for some formula $\varphi(w, \vec{v}(n))$, then for some integer $m > n$ $\varphi(v_m, \vec{v}(n)) \in p_m$.

These types are built by induction on n. p_1 is any isolated 1-type over $\emptyset$. Now assume that p_n is defined, we show how to form p_{n+1}. Let $\gamma_n(\vec{v}(n))$ be a formula isolating p_n, list all the formulas $\exists w \varphi(w, \vec{v}(n))$ occuring in p_n. As $p_n \supseteq p_k$ for $k \leq n$, this list includes all the formulas of the same kind occurring in p_k with $k < n$. Take the first formula $\exists w \varphi_n(w, \vec{v}(n))$ in this list not already considered in the previous steps $< n$. Observe $\exists w \varphi_n(w, \vec{v}(n)) \wedge \gamma(\vec{v}(n)) \in p_n$, so $\varphi_n(w, \vec{v}(n)) \wedge \gamma(\vec{v}(n))$ occurs in some $(n+1)$-type over $\emptyset$. Use our assumption that isolated $(n+1)$-types over $\emptyset$ are dense in $S_{n+1}(\emptyset)$ and deduce that $\varphi_n(w, \vec{v}(n)) \wedge \gamma(\vec{v}(n))$ actually occurs in some isolated $(n+1)$-type. Choose such a type as p_{n+1}.
Actually this construction should involve sooner or later any formula $\exists w \varphi(w, \vec{v})$ occurring in some p_k when k ranges over positive integers. But this can be obtained by using a suitable diagonal Cantor like procedure.
Now extend the language L of T by countably many new constants $C = \{c_h : h \text{ positive integer}\}$. For every h replace v_h by c_h in the formulas of $p = \bigcup_n p_n$. One gets a set T_p of sentences in the enlarged language $L \cup C$: T_p is a consistent theory, and indeed a complete theory, as the reader can easily check. Now take a model $\mathcal{B}_p$ of T_p. T_p is just the theory of $\mathcal{B}_p$, hence the $(L \cup C)$-sentences of the form $\exists w \varphi(w, \vec{c})$ (with $\vec{c}$ in C) true in $\mathcal{B}_p$ are exactly those occurring in T_p. Look at the set $A = \{c_h^{\mathcal{B}_p} : h \text{ positive integer}\}$ and notice that, owing to (b) and the Tarski-Vaught Theorem, A is the domain of an elementary substructure of $\mathcal{B}_p$. Restrict this substructure to L and get a model -actually a countable model- $\mathcal{A}$ of T. Moreover $\mathcal{A}$ is atomic over $\emptyset$, as required: in fact, given a sequence $\vec{c} = (c_1, \ldots, c_n)$ from C, the type of $\vec{c}^{\mathcal{B}_p}$ over $\emptyset$ contains p_n, hence equals p_n and is isolated. ♣

This concludes our discussion of the case $X = \emptyset$. As already said, when we enlarge our attention to countable sets X, we immediately obtain what follows.

Corollary 6.4.15 **(1)** *There exists a model of T prime over X if and only if, for every positive integer n, the isolated n-types over X are dense in $S_n(X)$.*

(2) *Two models of T extending X by elementary inclusions are isomorphic by a function fixing X pointwise.*

But the previous analysis discloses some useful information even in the general case, when X is any, possibly uncountable, small subset of Ω. In particular, it clarifies under which conditions on T any X admits a prime model. This is what the following theorem explains.

Theorem 6.4.16 *The following propositions are equivalent.*

(1) *For every X, there is a model of T prime over X.*

(2) *For every X and positive integer n, the isolated n-types over X are dense in $S_n(X)$.*

(3) *For every X, the isolated* 1*-types over X are dense in $S_1(X)$.*

(1)' *For every* **countable** *X, there is a model of T prime over X.*

(2)' *For every* **countable** *X and positive integer n, the isolated n-types over X are dense in $S_n(X)$.*

(3)' *For every* **countable** *X, the isolated* 1*-types over X are dense in $S_1(X)$.*

Proof. (1) $\Rightarrow$ (1)$'$, (2) $\Rightarrow$ (2)$'$ and (3) $\Rightarrow$ (3)$'$ are trivial, as well as (2) $\Rightarrow$ (3) and (2)$'$ $\Rightarrow$ (3)$'$. Moreover (1)$'$ $\Leftrightarrow$ (2)$'$ is just Corollary 6.4.15. So, in order to accomplish the proof, it suffices to show (3)$'$ $\Rightarrow$ (3), (3) $\Rightarrow$ (2) and (3) $\Rightarrow$ (1).

(3)$'$ $\Rightarrow$ (3) Suppose towards a contradiction that there are X, a tuple $\vec{a}$ in X and a formula $\varphi(v, \vec{w})$ such that $\varphi(v, \vec{a})$ occurs in no isolated 1-type over X. We want to extract a countable $X' \subseteq X$ such that X' contains $\vec{a}$ but $\varphi(v, \vec{a})$ occurs in no isolated 1-type over X': this is clearly enough because such a set X' contradicts (3)$'$. Let X_0 denote the set of the elements in $\vec{a}$. Is X_0 a reasonable candidate as X'? No, it is not, because it may happen that, in the restricted framework of X_0, some $L(X_0)$-formula $\psi(v)$ isolates a single 1-type containing $\varphi(v, \vec{a})$. However $\psi(v)$ cannot preserve this property over X, so there are at least two different 1-types over X containing $\psi(v) \wedge \varphi(v, \vec{a})$; consequently there exists some formula $\theta(v, \vec{b})$ with parameters $\vec{b}$ from X occurring in the former type but not in the latter. For every $\psi(v)$ as above, fix a corresponding $\vec{b}$ (and $\theta(v, \vec{b})$). Enlarge X_0 by adding the elements of these tuples $\vec{b}$ when $\psi(v)$ ranges over the $L(X_0)$-formulas isolating some type over X_0 containing $\varphi(v, \vec{a})$. One gets a countable $X_1 \subseteq X$ enlarging $\vec{a}$ and such that no $L(X_0)$-formula $\psi(v)$ isolates a 1-type over X_1 containing $\varphi(v, \vec{a})$. Indeed, for every $\psi(v)$, there is some formula $\theta(v, \vec{b})$ in $L(X_1)$ for which both $\psi(v) \wedge \varphi(v, \vec{a}) \wedge \theta(v, \vec{b})$ and $\psi(v) \wedge \varphi(v, \vec{a}) \wedge \neg\theta(v, \vec{b})$ enlarge to consistent 1-types over X_1. For any $L(X_0)$-formula $\psi(v)$ which cannot isolate any 1-type containing $\varphi(v, \vec{a})$ even in $S_1(X_0)$, $\theta(v, \vec{b})$ can be also found in $L(X_0)$; otherwise $\theta(v, \vec{b})$ is produced by the previous procedure. We repeat this machinery and we build, for every natural n, a countable $X_n \subseteq X$ enlarging $\vec{a}$ in a way such that, for any n, no $L(X_n)$-formula can isolate a 1-type over X_{n+1} containing $\varphi(v, \vec{a})$. At this point, consider $X' = \bigcup_n X_n$. X' in countable. Moreover, for every $L(X')$-formula $\psi(v)$, there is a suitable n for which $\psi(v)$ belongs to the language $L(X_n)$ and hence cannot isolate any 1-type containing $\varphi(v, \vec{a})$ over X_{n+1}, and consequently over X'.

(3) $\Rightarrow$ (2) We proceed by induction on n. As the case $n = 1$ is just our hypothesis, we have simply to see how to pass from any n to $n+1$. Let $\varphi(v, \vec{w})$ be an $L(X)$-formula with $n+1$ free variables $(v, \vec{w})$ ($\vec{w} = (w_1, \ldots, w_n)$) and $\varphi(\Omega^{n+1}) \neq \emptyset$. So $\exists v \varphi(v, \vec{w})$ lies in some n-type over X, and hence in some isolated n-type p over X. Let $\vec{b}$ realize p, so $\Omega \models \exists v \varphi(v, \vec{b})$, and $\varphi(v, \vec{b})$ is in some 1-type over $X \cup \vec{b}$. By hypothesis, there is some isolated 1-type q over $X \cup \vec{b}$ containing $\varphi(v, \vec{b})$. Let a realize q, so $\models \varphi(a, \vec{b})$. Look at the type of $(a, \vec{b})$ over X: it contains $\varphi(v, \vec{w})$ and is isolated because both $tp(a/X \cup \vec{b})$ and $tp(\vec{b}/X)$ are.

(3) $\Rightarrow$ (1) Fix X, we have to find a model of T prime over X. As shown in Theorem 6.4.7, it suffices to build a model $\mathcal{A}$ of T constructible over X. Let us work in a model $\mathcal{M}$ of T containing X via an elementary inclusion. To obtain $\mathcal{A}$ inside $\mathcal{M}$, we extend progressively X and form larger and larger sets $X_\mu \supseteq X$ in $\mathcal{M}$ (for μ an ordinal), all constructible over X and elementarily included in $\mathcal{M}$, until we meet in this increasing sequence the domain of an elementary substructure of $\mathcal{M}$ and, in this way, a model of T constructible over X. Start putting $X_0 = X$. For a limit μ, set $X_\mu = \bigcup_{\nu<\mu} X_\nu$, as it is right to expect: indeed this preserves constructibility and the other assumptions on X_μ. The crucial step concerns successor ordinals $\mu = \nu+1$. Suppose that X_ν is not the domain of an elementary substructure of $\mathcal{M}$ (otherwise we are done). By the Tarski-Vaught Theorem, there is an $L(X_\nu)$-formula $\varphi(v)$ such that $\exists v \varphi(v)$ is true in $\mathcal{M}$ - in other words $\varphi(\mathcal{M}) \neq \emptyset$ - but X_ν contains no realization of $\varphi(v)$. Use (3) and get some isolated 1-type p over X_ν including $\varphi(v)$. There is some element $a_\nu \in M$ satisfying p. Put $X_{\nu+1} = X_\nu \cup \{a_\nu\}$; $X_{\nu+1}$ is constructible over X because $p = tp(a_\nu/X_\nu)$ is isolated; moreover the inclusion of $X_{\nu+1}$ in $\mathcal{M}$ is elementary. Now notice that the length of this procedure cannot exceed $|M|$. Accordingly $X_\mu = X_{\mu+1}$ for some $\mu \leq |M|$; but this means that X_μ is the domain of the required model. ♣

Let us underline one more time that what we get in (3) $\Rightarrow$ (1) is actually a model of T constructible over X.
Now assume, at last, T ω-stable. We show that, under this assumption, the existence of prime models is ensured over any X. This is a result of Morley and is obtained as a direct consequence of the previous theorem.

Theorem 6.4.17 (Morley) *Let T be ω-stable. Then, for every X, there is a model of T prime over X.*

Proof. Owing to Theorem 6.4.16, it suffices to show that, for every countable X, the isolated 1-types over X are dense in $S_1(X)$. Use ω-stability and deduce that, for a countable X, $S_1(X)$ is countable, too. Now refer to topology and specifically recall a theorem of Cantor and Bendixson saying that in every countable compact Hausdorff space - like $S_1(X)$ - isolated points are dense. This is just what we need. ♣

Indeed the previous theorem says even more: in fact, for an ω-stable T, over any X we can find a constructible model.
Finally let us deal with uniqueness in the ω-stable setting.

Theorem 6.4.18 (Shelah) *Let T be ω-stable, X be a small subset of Ω, $\mathcal{A}_0$ and $\mathcal{A}_1$ be two models of T prime over X. Then there is an isomorphism h between $\mathcal{A}_0$ and $\mathcal{A}_1$ fixing X pointwise.*

Actually what Shelah proves is that, for an ω-stable T, a model $\mathcal{A}$ prime over X must be constructible over X as well. So Ressayre's Theorem applies and ensures the uniqueness of the prime model over X. However, Shelah's tools in the proof are quite sophisticated and require new progress in the general theory. We will develop these preliminaries in Chapter 7, so we have to delay the full details of the proof to that chapter.
We conclude this section observing that neither existence nor uniqueness of prime models are guaranteed when T is not ω-stable. In particular we propose a simple counterexample to existence. Uniqueness is also contradicted by some suitable theories T, but the corresponding examples are more technical and complicated, and may make this section still heavier. We omit them.

Example 6.4.19 In the language $L = \{\leq, P\}$, where P is a 1-ary relation symbol, look at the structure $(\mathbf{R}, \leq, \mathbf{Q})$. This is a dense linear order without endpoints $(\mathbf{R}, \leq)$ with a subset $\mathbf{Q}$ both dense and codense in $\mathbf{R}$. All these properties can be easily written as first order sentences in L. Let T be the L-theory axiomatized in this way. A simple arrangement of Cantor's Theorem on DLO^- shows that T is $\aleph_0$-categorical and so, by Vaught's Test, complete. Hence $T = Th(\mathbf{R}, \leq, \mathbf{Q})$. Of course T is not ω-stable (just as DLO^-). Now suppose that T has a prime model over $\mathbf{Q}$ inside $(\mathbf{R}, \leq, \mathbf{Q})$. Up to isomorphism this model will be of the form $(A, \leq, \mathbf{Q})$ for some countable subset A of $\mathbf{R}$ properly including $\mathbf{Q}$. Let $a \in A - \mathbf{Q}$. If we take a away from A, then we get a new model $(A - \{a\}, \leq, \mathbf{Q})$ of T; this contains the rationals, but cannot embed $(A, \leq, \mathbf{Q})$. In fact the cut a fills in A over $\mathbf{Q}$ is not realized in $(A - \{a\}, \leq, \mathbf{Q})$.

6.5 DCF_0 revisited

We pursue here the analysis of existentially closed differential fields of characteristic 0 begun in Chapter 2. We said at that time that their class is elementary and we provided (without proof) their nice first order characterization DCF_0 due to Lenore Blum, according to which they are the differential fields $\mathcal{K}$ of characteristic 0 such that, for every non zero $f(x)$ and $g(x)$ in $K\{x\}$ such that the order of $g(x)$ is smaller than the order of $f(x)$, there is some $\alpha \in K$ such that $f(\alpha) = 0$ and $g(\alpha) \neq 0$ (let us momentarily call these fields differentially closed fields; actually we are going to show that they are just the existentially closed differential fields). Now it is the right moment to prove at last that result, as well as the related fact, mentioned several times, that every differential field of characteristic 0 has its own differential closure. In fact, we will show that DCF_0 is ω-stable, and so the machinery developed in the last section will apply and produce a prime model of DCF_0 (so a differential closure) over any differential field of characteristic 0. The ω-stability of DCF_0 will be also used to prove, in a way closely resembling that pursued for ACF in 6.3, that DCF_0 uniformly eliminates the imaginaries.

This is the plan of this section. In order to begin our treatment and, in particular, to focus existentially closed differential fields, let us preliminarily examine how a differential field $\mathcal{K}$ can be enlarged and which is the structure of a differential extension $\mathcal{H}$ of $\mathcal{K}$. In particular, for $\alpha \in H - K$, we want to single out in $\mathcal{H}$ the smallest differential subfield $\mathcal{K}\langle\alpha\rangle$ containing K and α (i. e. the intersection of all the differential subfields of $\mathcal{H}$ containing both K and α); this requires to clarify and understand how α is related to $\mathcal{K}$, in other words which differential polynomials over K α satisfies. Of course this is basic (Differential) Algebra rather than Model Theory; and in fact there is

a straightforward and general algebraic technique to approach this question, that is to consider the function F_α from the differential domain $\mathcal{K}\{x\}$ into $\mathcal{H}$ taking any differential polynomial $f(x) \in K\{x\}$ to its value $f(\alpha)$ in α. This is a differential ring homomorphism, and its kernel $I(\alpha/K) = \{f(x) \in K\{x\} : f(\alpha) = 0\}$ is a proper prime ideal of $\mathcal{K}\{x\}$, and even a *differential ideal* (i. e. an ideal closed under derivation D, as it is easy to check).
Incidentally define α *differentially algebraic* over $\mathcal{K}$ if $I(\alpha/K) \neq 0$ (so if $f(\alpha) = 0$ for some nonzero differential polynomial $f(x) \in K\{x\}$), and *differentially transcendental* otherwise.
When α is differentially transcendental over $\mathcal{K}$, there is nothing to add. For, F_α is a differential ring isomorphism between $\mathcal{K}\{x\}$ and a suitable differential subring of $\mathcal{K}$ containing K and α. It is straightforward to deduce that the differential field $\mathcal{K}\langle\alpha\rangle$ we are looking for is isomorphic to the field of quotients of $\mathcal{K}\{x\}$ (which inherits a natural structure of differential field regulated by the usual derivation rules for quotients).
But the most critical point of our analysis concerns a differentially algebraic element α over $\mathcal{K}$. In fact, on the one hand it is easy to realize that even in this case the quotient ring $\mathcal{K}\{x\}/I(\alpha/K)$ gets a natural structure of differential domain, just because $I(\alpha/K)$ is a prime differential ring and so it makes sense to put, for $f(x) \in K\{x\}$,

$$D(f(x) + I(\alpha/K)) = Df(x) + I(\alpha/\mathcal{K});$$

the field of quotients of $\mathcal{K}\{x\}/I(\alpha/K)$ inherits in its turn an obvious structure of differential field, and $\mathcal{K}\langle\alpha\rangle$ is isomorphic to this field. This is quite general, and formally satisfactory. But we aim at understanding the internal structure of $\mathcal{K}\langle\alpha\rangle$, and this clearly requires to study the prime differential ideal $I(\alpha/K)$ and to provide an intrinsic description of its polynomials.
First let us observe that every prime differential ideal $I \neq \{0\}$, $K\{x\}$ of $\mathcal{K}\{x\}$ can be represented as $I = I(\alpha/K)$ for some suitable $\mathcal{H}$ and $\alpha \in H$ (differentially algebraic over $\mathcal{K}$). This is straightforward and quite general. In fact, look at the differential domain $\mathcal{K}\{x\}/I$, form its differential field of quotients $\mathcal{Q}$, notice that $\mathcal{Q}$ extends $\mathcal{K}$ provided one identifies each $k \in K$ and its class $I + k$ in Q, and that $\alpha = I + x$ just annihilates the differential polynomials in I over K.
So take any nontrivial prime differential ideal I of $\mathcal{K}\{x\}$. Choose a nonzero polynomial $f(x) \in I$ having

$\star$ a minimal order n,

then

⋆ a minimal degree in $D^n(x)$,

and finally

⋆ a minimal total degree.

As I is prime, $f(x)$ is irreducible as an algebraic polynomial in $\mathcal{K}[D^m(x) : m$ natural]. We wonder whether I equals by chance the differential ideal $\langle f(x) \rangle$ generated by $f(x)$, in other words the algebraic ideal of $\mathcal{K}[D^m(x) : m$ natural] generated by $f(x)$ and its derivatives. This is not true in general, and I is somewhat more complicated. But clarifying this point requires a further notion.

Definition 6.5.1 *Let $f(x) \in K\{x\}$ have order n. The formal partial derivative of $f(x)$ with respect to $D^n(x)$ is called the* **separant** *of $f(x)$ and denoted $s_f(x)$.*

For example, the separant of $f(x) = (D^3x)^2 + x \cdot D^3x + Dx \cdot D^2x$ is $2D^3x + x$. Now, given a differential polynomial $f(x) \in K\{x\}$, look at the set $I(f(x))$ of the polynomials $t(x) \in K\{x\}$ such that for some natural m

$$s_f^m(x) \cdot t(x) \in \langle f(x) \rangle;$$

one sees that $I(f(x))$ is a differential ideal and, when $f(x)$ is irreducible, it is also prime. Of course the membership to $I(f(x))$ is not easy and immediate to check; but one can show that, for an irreducible $f(x)$ of order n, a polynomial $g(x) \in I(f(x))$ must have order $\geq n$ and, if its order is just n, then $f(x)$ directly divides $g(x)$. This analysis implies that, if α is a root of $f(x)$, then $I(f(x)) = I(\alpha/K)$.
More generally, let us come back to our nontrivial prime differential ideal I and to the polynomial $f(x) \in I$ chosen before. One proves

Theorem 6.5.2 $I = I(f(x))$.

This is a noteworthy and deep algebraic fact; its proof can be found in the references quoted at the end of this chapter. This also concludes our preliminary outline of basic Differential Algebra. Now let us deal at last with Model Theory. In fact, owing to what we said about $I(f(x))$ we are in a position to show:

Theorem 6.5.3 *Existentially closed differential fields $\mathcal{K}$ of characteristic 0 are differentially closed.*

Proof. We have to check that $\mathcal{K}$ satisfies Blum's axioms: for every nonzero polynomials $f(x)$, $g(x) \in K\{x\}$ such that the order of $f(x)$ is larger than the order of $g(x)$, there is α is K satisfying $f(\alpha) = 0$ and $g(\alpha) \neq 0$. As the existence of α can be expressed by an existential sentence with parameters in K and $\mathcal{K}$ is existentially closed, it suffices to find such an element α in some differential field $\mathcal{H}$ extending $\mathcal{K}$. Pick an irreducible factor $f'(x)$ of $f(x)$ in $K\{x\}$ of the same order as $f(x)$, and form $I = I(f'(x))$. Clearly $f(x) \in I$, while $g(x)$ cannot belong to I because its order is less than the order of $f'(x)$. Now enlarge $\mathcal{K}$ to the differential field of quotients $\mathcal{H}$ of $\mathcal{K}\{x\}/I$ and notice that in H $I + x$ satisfies $f(v) = 0$ and cannot realize $g(v) = 0$, because $f(x) \in I$ and $g(x) \notin I$. ♣

In order to conclude that DCF_0 is just the theory of existentially closed differential fields of characteristic 0, we have to show the inverse implication (saying that any differentially closed field of characteristic 0 is existentially closed) and definitively to check two more points:

(i) every differential field of characteristic 0 has a differentially closed extension;

(ii) DCF_0 is model complete.

Here are their proofs.

Theorem 6.5.4 *Every differential field $\mathcal{K}$ of characteristic 0 has a differentially closed extension.*

Proof. This can be shown by using some familiar chain arguments. In detail, list in some (possible transfinite) way the pairs $(f(x), g(x))$ of nonzero polynomials in $K\{x\}$ such that the order of $f(x)$ is larger than the order of $g(x)$. Apply the same technique as in Theorem 6.5.3 and, for every pair $(f(x), g(x))$, enlarge the ground field in order to include a root α of $f(x)$ such that $g(\alpha) \neq 0$. Repeat this procedure and eventually obtain a differential extension $\mathcal{K}'$ of $\mathcal{K}$ with the following property: for every choice of $f(x)$, $g(x)$ in $K\{x\} - \{0\}$ such that the order of $f(x)$ is larger than the order of $g(x)$, there is some $\alpha \in K'$ such that $f(\alpha) = 0$ and $g(\alpha) \neq 0$. However this does not mean that $\mathcal{K}'$ is differentially closed (in fact, what happens for $f(x)$ and $g(x)$ in $K'\{x\}$?). But now we can form a new sequence of differential extensions $\mathcal{K}_n$ of $\mathcal{K}$, by putting $\mathcal{K}_0 = \mathcal{K}$ and, for every n, $\mathcal{K}_{n+1} = \mathcal{K}_n'$ (a differential field enlarging $\mathcal{K}_n$ and containing, for every choice of nonzero polynomials $f(x)$, $g(x)$ in $K_n\{x\}$ such that the order of $f(x)$ is larger than the order of $g(x)$, a root α of $f(x)$ that does not annihilate $g(x)$). $\tilde{K} = \bigcup K_n$

is the domain of a differential field extending $\mathcal{K}$. A straightforward check shows that $\tilde{\mathcal{K}}$ is differentially closed. ♣

Theorem 6.5.5 *DCF_0 is complete and eliminate the quantifiers in the language $L = \{+, \cdot, -, 0, 1, D\}$. In particular DCF_0 is model complete.*

Proof. First we show that DCF_0 is complete, and hence that any two models $\mathcal{K}$, $\mathcal{K}'$ of DCF_0 are elementarily equivalent. As every structure has some $\aleph_0$-saturated elementary extension, we can assume that both $\mathcal{K}$ and $\mathcal{K}'$ are $\aleph_0$-saturated. As partial isomorphism implies elementary equivalence, it suffices to show $\mathcal{K} \simeq_p \mathcal{K}'$. Accordingly let J be the set of all the isomorphisms between finitely generated differential subfields of $\mathcal{K}$ and $\mathcal{K}'$ respectively. J is not empty because both $\mathcal{K}$ and $\mathcal{K}'$ include the rational field (with a zero derivative) as the minimal differential subfield generated by $\emptyset$. Now let $\mathcal{K}_0$ and $\mathcal{K}_0'$ be two finitely generated differential subfields of $\mathcal{K}$, $\mathcal{K}'$ respectively, corresponding to each other by some isomorphism h. Of course, if $\vec{a}$ is a tuple generating $\mathcal{K}_0$, then $h(\vec{a}) = \vec{a'}$ generates $\mathcal{K}_0'$. So, algebraically speaking, any element of K_0 can be obtained from $\vec{a}$ by the usual elementary operations and D, and the same can be said about K_0' and $\vec{a'}$. Incidentally notice that $K_0 \subseteq dcl(\vec{a})$ and, parallely, $K_0' \subseteq dcl(\vec{a'})$. We have to control that h satisfies the back-and-forth properties (i) and (ii). Clearly it suffices to deal with (i), as (ii) can be handled in a similar way. So take $\alpha \in K - K_0$. We look for two finitely generated extensions $\mathcal{K}_1$, $\mathcal{K}_1'$ of $\mathcal{K}_0$, $\mathcal{K}_0'$ respectively such that $\alpha \in K_1$ and h can be enlarged to an isomorphism between $\mathcal{K}_1$ and $\mathcal{K}_1'$. We distinguish two cases, according to whether α is differentially algebraic, or differentially transcendental over $\mathcal{K}_0$.
When α is differentially algebraic, consider the nonzero prime differential ideal $I(\alpha/K_0)$ and the (irreducible) polynomial $f(x)$ in $I(\alpha/K_0)$ having minimal order n, then minimal degree in $D^n x$ and, finally, minimal total degree. Then $f'(x) = h(f(x))$ is an irreducible polynomial of order n in $\mathcal{K}_0\{x\}$. Look at the type p given by the formulas

$$f'(v) = 0, \quad \neg(g'(v) = 0)$$

where $g'(x)$ ranges over the nonzero polynomials of order $< n$ in $K_0'\{x\}$. As $K_0' \subseteq dcl(\vec{a'})$, p can be viewed as a type over $\vec{a'}$. As $\mathcal{K}'$ is differentially closed, any finite portion of p is satisfied in $\mathcal{K}'$. As $\mathcal{K}'$ is $\aleph_0$-saturated, p itself is realized in $K' - K_0'$ by a suitable element α'. Moreover $I(f'(x)) = I(\alpha'/K_0')$. So one can easily enlarge h to an isomorphism between $\mathcal{K}_0\langle\alpha\rangle$ and $\mathcal{K}_0'\langle\alpha'\rangle$

mapping α in α' (for, $I(\alpha/K_0)$ and $I(\alpha'/K_0')$ correspond to each other by h). Accordingly $\mathcal{K}_1 = \mathcal{K}_0\langle\alpha\rangle$ and $\mathcal{K}_1' = \mathcal{K}_0'\langle\alpha'\rangle$ yield (i).

The case when α is differentially transcendental over $\mathcal{K}_0$ is simpler. In fact, using the $\aleph_0$-saturation of $\mathcal{K}'$, one easily finds an element $\alpha' \in K'$ algebraically transcendental over $K_0' \subseteq dcl(\vec{a'})$. At this point, one observes that $\mathcal{K}_0\langle\alpha\rangle$ and $\mathcal{K}_0'\langle\alpha'\rangle$ are isomorphic by a function extending h and taking α to α'. This ensures (i) even in this case, and eventually accomplishes the proof that DCF_0 is complete.

Now we prove that DCF_0 eliminates the quantifiers in L. Owing to completeness, every L-sentence is DCF_0-equivalent to $0 = 0$ when belonging to DCF_0, and to $0 = 1$ otherwise. So we have to eliminate the quantifiers in the L-formulas $\varphi(\vec{v})$ when the sequence $\vec{v}$ of free variables is not empty. For this purpose we arrange the previous completeness proof in the following way. Let $\vec{a}$, $\vec{a'}$ be two tuples in the universe Ω of DCF_0 satisfying the same quantifier free L-formulas; we claim that they have the same type over $\emptyset$. There is no loss of generality in assuming $\vec{a}$ in K and $\vec{a'}$ in K' for some suitable $\aleph_0$-saturated differentially closed fields $\mathcal{K}$ and $\mathcal{K}'$. The hypothesis that $\vec{a}$ and $\vec{a'}$ realize the same quantifier free formulas easily implies that $\vec{a}$, $\vec{a'}$ generate isomorphic differential subrings, and consequently isomorphic differential subfields in $\mathcal{K}$, $\mathcal{K}'$ respectively, by $\vec{a} \mapsto \vec{a'}$. We look now at the set of the isomorphisms between finitely generated differential subfields of $\mathcal{K}$ containing $\vec{a}$ and finitely generated differential subfields of $\mathcal{K}'$ containing $\vec{a'}$ which extend $\vec{a} \mapsto \vec{a'}$. By proceeding as before, one obtains this time $(\mathcal{K}, \vec{a}) \equiv (\mathcal{K}', \vec{a'})$ in the language enlarging L by suitably many new constants to be interpreted in $\vec{a}$, $\vec{a'}$ respectively. In other words, $\vec{a}$, $\vec{a'}$ have the same type over $\emptyset$, as claimed. Hence, two sequences satisfying the same quantifier free formulas do admit the same type; in order to conclude, we have simply to realize that this is just the elimination of quantifiers for L-formulas $\varphi(\vec{v})$ with a non-empty $\vec{v}$. Let us see why.

If $\varphi(\vec{v})$ does not occur in any type over $\emptyset$, then it is clearly equivalent to $\neg(v_1 = v_1)$. Otherwise we have just seen that, for every type p containing $\varphi(\vec{v})$, $\varphi(\vec{v})$ is a consequence of the quantifier free formulas in p. By compactness, only finitely many formulas of this kind, and even a single quantifier free formulas $\varphi_p(\vec{v})$ -their conjunction-, are enough to imply $\varphi(\vec{v})$ within DCF_0. Topologically speaking, when p ranges over the types over $\emptyset$ containing $\varphi(\vec{v})$, the neighbourhoods $U_{\varphi_p(\vec{v})}$ form an open covering of the closed, hence compact set $U_{\varphi(\vec{v})}$. Again using compactness, one finds a finite set P_0 of types p containing $\varphi(\vec{v})$ such that $U_{\varphi(\vec{v})}$ equals $\bigcup_{p \in P_0} U_{\varphi_p(\vec{v})}$. In other words

$$\forall \vec{v}(\varphi(\vec{v}) \longleftrightarrow \bigvee_{p \in P_0} \varphi_p(\vec{v})) \in DCF_0.$$

As the latter formula is quantifier free, we are done. ♣

Hence DCF_0 is the model companion of the theory of differential fields of characteristic 0. But the previous analysis says more and in particular ensures:

Theorem 6.5.6 *DCF_0 is ω-stable.*

Proof. What we have to check is that, for every differentially closed field $\mathcal{K}$ in characteristic 0, $|S_1(K)| = |K|$. Actually we will show something more, resembling what we did for algebraically closed fields. In fact, we will see that, for every $\mathcal{K}$, there is a natural bijection between 1-types over K in DCF_0 and (proper) prime differential ideals in $\mathcal{K}\{x\}$: owing to what we saw about these ideals and their structure, this implies $|S_1(K)| \leq |K\{x\}| = |K|$, whence $|S_1(K)| = |K|$.
First take a 1-type p over K and form

$$I(p) = \{f(x) \in K\{x\} \, : \, f(v) = 0 \in p\}.$$

Checking that $I(p)$ is a prime differential ideal in $\mathcal{K}\{x\}$ is a straightforward exercise. Indeed every (proper) prime differential ideal I of $\mathcal{K}\{x\}$ can be obtained in this way: in fact, build the (differential) field of quotients $\mathcal{Q}$ of $\mathcal{K}\{x\}/I$, enlarge $\mathcal{Q}$ to a differentially closed $\tilde{\mathcal{K}}$ and, at last, take the type of $I + x$ over (the isomorphic copy of) $\mathcal{K}$ (in $\tilde{\mathcal{K}}$): clearly a polynomial $f(x) \in K\{x\}$ satisfies $f(v) = 0 \in p$ if and only if $f(x) \in I$. Furthermore two different 1-types $p \neq q$ over $\mathcal{K}$ define different ideals. In fact there is some $L(K)$-formula $\varphi(v)$ in p and not in q. Now use elimination of quantifiers: $\varphi(v)$ can be chosen as a disjunction of conjunctions -and even as a unique conjunction- of equations $f(v) = 0$ or negations $\neg(g(v) = 0)$ with $f(x)$ and $g(x)$ in $K\{x\}$. But then any single equation and disequation is in p, and at least one of them is not in q. Consequently $I(p) \neq I(q)$. ♣

As a corollary, let us point that the constant subfield $C(\mathcal{K}) = \{a \in K : Da = 0\}$ of a differentially closed $\mathcal{K}$, as a structure definable in $\mathcal{K}$, is ω-stable, and so algebraically closed (due to Macintyre's Theorem).
Incidentally, we should also say that DCF_0 is not strongly minimal; indeed its Morley rank is ω and its degree is 1. However, owing to the theorems of Morley and Shelah on the existence and uniqueness of prime models over subsets in ω-stable theories, we can deduce:

Corollary 6.5.7 *Over any differential field $\mathcal{K}$ of characteristic 0 there is a prime model of DCF_0 (hence a differential closure), and this is unique up to isomorphism fixing K pointwise.*

We already emphasized several times that this approach -using Model Theory, in particular ω-stability, Morley's Theorem and Shelah's Theorem- is the very first proof, and virtually the only one known till now, of the existence and uniqueness of differential closures.
Again using ω-stability we show now the last result of this section.

Theorem 6.5.8 *DCF_0 uniformly eliminates the imaginaries.*

Proof. We repeat almost wholly the approach followed in 6.3 for algebraically closed fields. Indeed the first two algebraic preliminary steps in that proof concern arbitrary fields and ideals, so apply to our present setting as well. The same can be said about the use of ω-stability, owing to Theorem 6.5.6, although now types correspond to (proper) prime differential ideals, and not directly to prime ideals. In conclusion what we have to check is that the third preliminary step in the proof of Theorem 6.3.2 still works in the new framework. Accordingly take a differential field $\mathcal{K}$ of characteristic 0, and $I_0, \ldots, I_m$ prime differential ideals of $\mathcal{K}\{x\}$ in the same conjugacy class (with respect to differential field automorphisms of $\mathcal{K}$). We look for a(n even algebraic) subfield $\mathcal{K}_0$ of $\mathcal{K}$ such that, for every automorphism σ of $\mathcal{K}$ as a differential field,

$$\sigma \text{ permutes } I_0, \ldots, I_m \text{ if and only if } \sigma \text{ fixes } K_0 \text{ pointwise.}$$

This can be obtained by rearranging the approach in the field case. Form $I = \bigcap_{j \leq m} I_j$. This is a differential ideal, and even a radical ideal in the usual sense: for every $f(\vec{x}) \in K\{\vec{x}\}$, if $f^t(\vec{x}) \in I$ for some positive integer t, then $f(\vec{x})$ itself is in I. A differential version of the *Basis Theorem*, due to Ritt and Raudenbush and working over differential fields of characteristic 0, says that every differential ideal of $\mathcal{K}\{\vec{x}\}$ -like I, but also $I_0, \ldots, I_m$- is finitely generated (as a differential ideal). Fix a finite set of differential polynomials including generators of $I, I_0, \ldots, I_m$. Let H be their maximal order, so all these polynomials can be viewed as algebraic polynomials over K in the unknowns $D^h x_i$ for $h \leq H$, $1 \leq i \leq m$. Put $\mathcal{R} = \mathcal{K}[D^h x_i : h \leq H, 1 \leq i \leq m]$, and look at $I \cap R$, $I_0 \cap R, \ldots, I_m \cap R$. Then $I \cap R = \bigcap_{j \leq m}(I_j \cap R)$, and $I_0 \cap R, \ldots, I_m \cap R$ are prime ideals in $\mathcal{R}$ in the same conjugacy class with respect to field automorphisms of $\mathcal{K}$: in fact, for $j, j' \leq m$, there is some automorphism of $\mathcal{K}$ as a differential field, hence as a field, taking I_j to

I'_j, and so $I_j \cap R$ to $I'_j \cap R$. Use Step 3 in Theorem 6.3.2 and deduce that there is an algebraic subfield $\mathcal{K}_0$ of $\mathcal{K}$ such that, for every automorphism σ of $\mathcal{K}$, viewed as a field and in particular as a differential field, σ permutes the $I_j \cap R$'s if and only if it fixes K_0 pointwise. But each $I_j \cap R$ generates I_j as a differential ideal in $\mathcal{K}\{\vec{x}\}$, so, for j, $j' \leq m$, $\sigma(I_j \cap R) = I'_j \cap R$ implies $\sigma(I_j) = I'_j$. Consequently, for every automorphism σ of the differential field $\mathcal{K}$, σ permutes $I_0, \ldots, I_m$ if and only if σ acts identically on K_0. ♣

6.6 Ryll-Nardzewski's Theorem, and other things

We conclude our outline of Morley's ideas in these two chapters with a perhaps oblique and superficially related argument. In fact we want to treat $\aleph_0$-categorical theories. As already underlined, they behave quite autonomously and include examples which are not categorical in any uncountable cardinal, and neither are ω-stable: for instance, think of DLO^-. On the other hand, there do exist uncountably categorical structures which are not $\aleph_0$-categorical (such as ACF_p for any fixed p). However $\aleph_0$-categorical theories have their peculiar and specific properties. In particular they enjoy the following nice characterization, due to Ryll-Nardzewki and others.

Theorem 6.6.1 (Ryll-Nardzewski) *Let T be a complete theory in a countable language L. Then the following propositions are equivalent:*

(i) *T is $\aleph_0$-categorical;*

(ii) *for every positive integer n, there are only finitely many n-types over $\emptyset$ in T;*

(iii) *for every positive integer n, there are only finitely many formulas in n free variables pairwise inequivalent in T.*

Proof. The equivalence between (ii) and (iii) is clear: it can be deduced directly, or using the classical duality between Boolean spaces and Boolean algebras (recall that, for every n, the n-types over $\emptyset$ form the dual space $S_n(\emptyset)$ of the Boolean algebra of $\emptyset$-definable subsets of Ω^n).
So let us compare (i) and (ii). First suppose $S_n(\emptyset)$ infinite for some n. As $S_n(\emptyset)$ is compact, it contains some non-isolated type p. The Omitting Types Theorem ensures that p is omitted in some countable model of T; on the other hand, there does exist some countable model of T realizing p. The latter model cannot be isomorphic to the former; hence T is not

$\aleph_0$-categorical. Now assume $S_n(\emptyset)$ finite for every n. Consequently $S_n(\emptyset)$ is a discrete space, and so every n-type over $\emptyset$ is isolated. As this is true for every n, every model of T is atomic over $\emptyset$, hence every countable model of T is prime over $\emptyset$. Then all the countable models of T are pairwise isomorphic. ♣

Let us underline a simple consequence showing in the countable case the more general fact that, for a λ-categorical T, the only model of power λ is saturated (in the uncountable case, this result is much more complicated and refers to Morley's Categoricity Theorem).

Corollary 6.6.2 *Let T be an $\aleph_0$-categorical theory. Then the only countable model $\mathcal{M}$ of T is $\aleph_0$-saturated.*

Proof. Let $\vec{a}$ be a tuple (of length m) in M. For every positive integer n, there are only finitely many n-types over $\vec{a}$, indeed only finitely many $(n+m)$-types over $\emptyset$. So every n-type over $\vec{a}$ is isolated, and consequently is realized in $\mathcal{M}$. ♣

Now let us propose some examples illustrating Ryll-Nardzewski's Theorem.

Examples 6.6.3 1. We know that dense linear orders (with or) without endpoints have an $\aleph_0$-categorical theory (see Chapter 1). Let us confirm this result via the Ryll-Nardzewski Theorem. First take a positive integer m and $a_0, \ldots, a_m, b_0, \ldots, b_m$ in the universe of DLO^-, with $a_0 < \ldots < a_m$, $b_0 < \ldots < b_m$. Then there is an automorphism of Ω mapping a_j in b_j for every $j \leq m$. In fact the intervals in Ω

$$]-\infty, a_0[,\ ]-\infty, b_0[,$$

$$]a_j, a_{j+1}[,\ ]b_j, b_{j+1}[\quad \forall j < m,$$

$$]a_m, +\infty[,\ ]b_m, +\infty[$$

are models of DLO^- and are saturated in the same power as Ω. Consequently they are isomorphic to each other. In particular $(a_0, \ldots, a_m)$, $(b_0, \ldots, b_m)$ have the same type over $\emptyset$. This implies that, for every positive integer n and sequence $(c_1, \ldots, c_n) \in \Omega^n$, the type of $(c_1, \ldots, c_n)$ over the empty set is fully determined by the isomorphism type of the ordered set $(\{c_1, \ldots, c_n\}, \leq)$. So we get only finitely many n-types over $\emptyset$ for every n, and the Ryll-Nardzewski Theorem applies to confirm that DLO^- is $\aleph_0$-categorical.

2. On the contrary, ACF_p is not $\aleph_0$-categorical for any $p = 0$ or prime. Let us see why by using the Ryll-Nardzewski Theorem. Is suffices to notice that there are infinitely many pairwise distinct $\emptyset$-definable subsets in the universe Ω of ACF_p, and even that there are infinitely many pairwise distinct *finite* $\emptyset$-definable subsets of Ω, in other words that $acl(\emptyset)$ is infinite. But this is well known because $acl(\emptyset)$ equals the (field theoretic) algebraic closure of the prime subfield of Ω.

Of course, no (expansion of a) field of characteristic 0 can admit an $\aleph_0$-categorical theory, for the same reason as in Example 6.6.3, 2. This applies in particular to real closed fields. Checking what happens for differentially closed fields, or separably closed fields, or $ACFA$, could be a useful exercise.

We conclude this section and the whole chapter by discussing a related matter, again concerning definable sets and types. In particular we wonder which information we can obtain about a complete theory T by looking at the isomorphism types of the Boolean algebras of definable sets of its models. Can the knowledge of these algebras say anything essential about the model theoretic complexity of T? Incidentally, recall that, at least in the countable case, there does exist a satisfactory classification of Boolean algebras up to isomorphism, provided by Ketonen.
In this framework, λ-categoricity can be replaced, for every infinite cardinal λ, by a seemingly weaker notion, called Boolean λ-categoricity. A complete theory T is said to be Booleanly λ-categorical if and only if all its models of power λ have isomorphic Boolean algebras of definable 1-ary sets. Clearly categoricity implies Boolean categoricity in every λ. Notably

Theorem 6.6.4 (Mangani-Marcja) *For an uncountable λ, a complete theory T is Boolean λ-categorical if and only if is λ-categorical.*

But this is false when $\lambda = \aleph_0$. Indeed

Theorem 6.6.5 (Marcja-Toffalori) *Every $\aleph_1$-categorical theory is Boolean $\aleph_0$-categorical.*

So algebraically closed fields have Boolean $\aleph_0$-categorical theories. One can see that the same is true also for real closed and differentially closed fields, as well as for every module.
Other connections between isomorphism types of Boolean algebras of definable sets and structural properties of theories are investigated in the papers quoted at the end of the chapter.

6.7 References

ω-stable theories and structures are dealt with in [8]. The details about Morley rank and degree in strongly minimal theories can be found in [108], or in the Ziegler contribution to [16]. [132] (and its recent english translation) provide a nice and clear treatment of ω-stable groups; also [15] is a rich source on this subject. Macintyre's Theorem on ω-stable fields is in [90]; by the way, let us mention its generalization to superstable fields in [25] saying that any superstable field is algebraically closed (superstability is a notion enlarging ω-stability, and will be introduced in the next chapter).
Prime models over the empty set were studied by Vaught [174]; the analysis over arbitrary sets was developed in [118]. Ressayre's contributions to constructible sets, and in particular the Uniqueness Theorem, were never published, while Shelah's Uniqueness Theorem is in [148].
As already said, the theory of differentially closed fields of characteristic 0 was developed in Blum's Ph.D. Thesis [12] (see also [146]). Good references are also [110] or [131].
Ryll-Nardzewski's Theorem is in [145]; see also [43] and [156].
The classification of countable Boolean algebras is in [69]. Boolean λ-categoricity was introduced in [104] and intensively studied in [105] and [165]; actually, Boolean categoricity was called pseudo categoricity in those papers. The study of the structural properties of theories by the isomorphism types of Boolean algebras of definable sets of their models is also pursued in [106] and [107].

Chapter 7

Classifying

7.1 Shelah's Classification Theory

A central subject in Mathematics is classifying, that is characterizing the objects of a given class up to equivalence relations, for instance the structures of a given language, or the models of a given theory up to isomorphism. Let us mention some classical examples.

Examples 7.1.1 1. An infinite set (without additional structure) is fully characterized up to isomorphism by its power, so by a(n infinite) cardinal number.

2. Let $\mathcal{K}$ be a fixed (countable) field. A non-zero vectorspace over $\mathcal{K}$ is completely determined up to isomorphism by its dimension over $\mathcal{K}$, so again by a cardinal number.

3. An algebraically closed field of a given characteristic is fully determined up to isomorphism by its transcendence degree over its prime subfield, so, once again, by a cardinal number.

   Notice that the previous examples concern elementary classes and even strongly minimal theories (provided we restrict our attention to infinite vectorspaces in 7.1.1, 2). But, of course, the classification purposes touch larger horizons.

4. It is well known that every finitely generated abelian group $\mathcal{A}$ decomposes as a finite direct sum of cyclic groups, and even of copies of $\mathbf{Z}$, $\mathbf{Z}/p^n$ where p ranges over primes and n over positive integers. Moreover the latter decomposition is unique up to isomorphism (and up

to permuting the summands); so the isomorphism type of $\mathcal{A}$ is determined by saying how many copies of $\mathbf{Z}$, $\mathbf{Z}/p^n$ (p prime, n positive integer) are involved in this decomposition of $\mathcal{A}$. Notice that finitely generated abelian groups are not an elementary class (why?).

5. Consider torsionfree abelian groups of rank at most n, where n is a given positive integer: they are the subgroups of the additive group $(\mathbf{Q}^n, +)$. Baer found in the thirties a nice classification of these groups up to isomorphism when $n = 1$. However it has been a longstanding open question to accomplish a general satisfactory classification for every $n \geq 2$, and actually no classification is presently known. We will go deep in this question later.

6. Let $\mathcal{K}$ be a given field, and look at the ring $\mathcal{K}\langle x, y\rangle$ of polynomials with coefficients in K and two non-commutings unknowns x and y. Consider $\mathcal{K}\langle x, y\rangle$-modules finite dimensional over $\mathcal{K}$ towards a possible characterization of their isomorphism classes. Well, this problem is generally believed intractable. In fact, finite dimensional $\mathcal{K}\langle x, y\rangle$-modules interpret the word problem of groups, and classifying them as said means solving this problem.

Of course, classification problems do not restrain themselves to the isomorphism relation; for instance one can try to classify square matrices by similarity, and so on. Also, it should be underlined that a classification is sometimes (often?) very unlikely to be obtained (as Examples 7.1.1, 5 and 7.1.1, 6 witness). However looking for a classification, and even realizing its unfeasibility, can disclose new connections between different areas, generate new techniques and, definitively, enlight new horizons within Mathematics. From an abstract point of view, a natural problem arises in this framework, that is developing a theoretical treatment to recognize which classes and relations admit a classification, and to propose general tools to accomplish this classification when possible, or to measure its difficulty in the hard cases. A closely related and preliminary question is just: what does classifying mean?
These issues are intensively studied within Mathematical Logic. For instance, Descriptive Set Theory (and consequently, through it, both Set Theory and Recursion Theory) are concerned with the classification matter. In fact Descriptive Set Theory deals with definable sets of $\mathbf{R}$, viewed as a separable complete metric space, and of similar topological spaces (*Polish spaces*). This does overlap classification. Indeed the objects that are to classified often belong to a Polish space X, or at least to a Borel subset of

X, and the classifying equivalence relation E is a Borel, or analytic, etc., subset of X^2: in fact, this is the case of the similarity relation for square matrices, as well as the isomorphism relation for denumerable structures (such as groups and fields) in a given language. In this setting it is sometimes possible to associate in a Borelian way to any object of X a point (called *"invariant"*) in another Polish space such that two different objects have the same associated point if and only if they are equivalent in E. The relation E is called *smooth* when it satisfies these requirements. For example, Jordan's canonical form is an invariant of the similarity relation for matrices, which is therefore smooth. In Ergodic Theory, smooth relations are just those which are considered actually classifiable. But not all equivalence relations one meets are smooth. In the latest years a general theory of classification of equivalence relations which live in Polish spaces has been developed within Descriptive Set Theory: it allows to establish, for E and F equivalence relations, when classifying modulo E is more complex than classifying modulo F. In particular it includes some recent and beautiful results of Simon Thomas (and others) on isomorphism for torsionfree abelian groups of rank n; these works show that this relation is of increasing complexity when n grows, and is not smooth (and so is likely to be intractable) when $n \geq 2$.
But now let us come back to Model Theory. Of course, also Model Theory is interested in classifying (structures, definable sets, and so on). But its approach is peculiar, and differs from the perspectives and the aims of Descriptive Set Theory: and indeed, very roughly speaking, one could say that Model Theory and Descriptive Set Theory are complementary in this framework. In fact, one could agree that the latter is mainly concerned in measuring how hard and difficult classifying is in the worst *"wild"* cases, while Model Theory aims at determining the *"tame"* settings in Mathematics, and accordingly at clarifying where and why a classification can be done.
Classification in Model Theory promptly recalls Shelah. In fact, it was Shelah who approached, treated and essentially solved the classification problem in Model Theory since the end of the sixties until the early eighties. So Shelah's formidable work dates back to almost twenty years ago, but is far from having exhausted its stimulus, as we will explain later in more detail. The Shelah strategy was to determine a series of successive key properties (simplicity, stability, superstability and so on) concerning complete theories and having a twofold role: in fact, each of them allows a new significant step towards classifiability, while its negation excludes any hope of classification. At the end of these successive *dichotomies* Shelah exactly determines which abstract conditions ensure classifiability.

The aim ot this chapter is to outline Shelah's classification program, its concepts, tools and techniques (forking, stability, superstability and so on), as well as its recent developments regarding simplicity and spectrum problem. We will lay emphasis on ideas and motivations rather than on proofs. So we will sacrifice full details in general. However we will provide a complete report in the ω-stable case.

The first point to be clarified is: what do we want to classify? Here the answer is quick and ready: we will deal with classes of structures. Just to avoid too many complications, we will limit our analysis here to elementary classes, and even to classes $Mod(T)$ of models of countable complete first order theories T. We illustrated this choice in Chapter 1. As usual, we will work inside a big saturated model Ω of T.

The second preliminary question is: with respect to which relation will we classify? Also in this case the reply is fast and easy: we will classify up to isomorphism.

But now we have to answer a more delicate and fundamental question, that is: what does it mean to classify the models of a complete T up to isomorphism? By which *"invariants"* will we accomplish our classification program?

When T is strongly minimal - just as in the Examples 7.1.1, 1-3 quoted before - classifying means assigning to every model a cardinal number - its dimension - in such a way that two models are isomorphic if and only if they are given the same cardinal. But we already saw that, for some theories, a classification cannot be accomplished by assigning a single cardinal and often requires something more complicated, like pairs of cardinals, or even (possibly infinite) sequences of cardinals, and so on. This is what happens in Example 7.1.1, 4, where actually the involved class is not elementary; other elementary cases were discussed in Chapter 5. Here is another example.

Example 7.1.2 Consider the theory of the structures (A, E_1, E_2) where E_1, E_2 are equivalence relations, $E_2 \subseteq E_1$ and

- every E_2-class has infinitely many elements,
- every E_1-class contains infinitely many E_2-classes;
- there are infinitely many E_1-classes.

T is complete (for instance because it is categorical in $\aleph_0$). In a model of T any E_1-class of power $\aleph_\alpha$ is determined up to isomorphism by the function mapping any infinite cardinal $\aleph_\beta \leq \aleph_\alpha$ to the number of the E_2-subclasses of power $\aleph_\beta$, and hence by an ordered sequence of α cardinals

$\leq \aleph_\alpha$. Consequently, if $\mathcal{A} = (A, E_1, E_2)$ is a model of T of power λ, then the isomorphism type of $\mathcal{A}$ is completely given by the function associating to any ordered sequence of α cardinals $\leq \aleph_\alpha$ ($\aleph_\alpha \leq \lambda$) the number of the E_1-classes in $\mathcal{A}$ corresponding to this sequence.

This construction can be iterated to produce more and more complicated examples, needing more and more sophisticated isomorphism invariants. This suggests the following definition.

Definition 7.1.3 *Let C denote the class of cardinal numbers. For every ordinal α, we define a class C^α by induction on α in the following way.*

- *C^0 is C,*
- *if $\alpha = \beta + 1$ is a successor ordinal, then we put $C^\alpha = C^\beta \cup \mathcal{P}(C^\beta)$;*
- *if α is limit, then we put $C^\alpha = \bigcup_{\beta<\alpha} C^\beta$.*

Definition 7.1.4 *Let T be a complete theory, α be an ordinal. T is said to have an* **invariant system** *of rank α if and only if there is a function f associating every model of T with an element of C^α such that two models $\mathcal{A}$ and $\mathcal{B}$ of T are isomorphic if and only if $f(\mathcal{A}) = f(\mathcal{B})$.*

Examples 7.1.5

1. A strongly minimal theory has an invariant system of rank 0.
2. An ordered sequence of cardinals is an element of C^2. So the theories in Example 5.8.2 have an invariant system of rank 2.
3. The theory in Example 7.1.2 has an invariant system of rank 4.

Shelah's proposal is to assume that a complete theory T is *classifiable* if and only if T has an invariant system of rank α for some ordinal α.
We will discuss Shelah's point of view later in this chapter, after describing Shelah's classification theory. In particular, we will see that there are some good reasons to agree with it. Once this is done, we can deduce:

Theorem 7.1.6 *Let T be a complete theory such that, for every uncountable power λ, T has 2^λ pairwise non isomorphic models of cardinality λ. Then T is not classifiable.*

In fact some cardinal and ordinal computations exclude that such a T has an invariant system of any possible rank. This is quite remarkable, because there are some very familiar theories satisfying the assumptions of Theorem

7.1.6. For instance, this is the case of the theory DLO^- of dense linear orders without endpoints: in fact DLO^- is $\aleph_0$-categorical and so has only one countable model (up to isomorphism) but it gets 2^λ pairwise non isomorphic models in any uncountable power λ. Accordingly DLO^- is not classifiable. More generally Theorem 7.1.6 applies, as we will see in the next section, to any complete theory of linearly ordered infinite structures. None of these theories is classifiable. In particular, no o-minimal theory is classifiable, although any such theory satisfies the conditions (D1)-(D4) of the dependence relation

$$a \prec A \qquad \Leftrightarrow \qquad a \in acl(A)$$

for $a \in \Omega$, A a small subset of Ω. This is a little surprising and will be discussed in more detail later in this chapter, and then in Chapter 9.
Anyhow, if we (momentarily) agree with Shelah's definition of classifiable theory, then we have to take note that too many models (in other words 2^λ non-isomorphic models in every uncountable λ) exclude classifiability. On the other side, we would like also to determine the key criteria ensuring the classifiability of an arbitrary complete theory: this will be the matter of the forthcoming sections. To conclude the present one, let us spend some more words about the close relationship between classifying theories T (in the Shelah sense) and counting the models of T. This is already explicit in Theorem 7.1.6. More generally, for every countable complete T, one can define a function $I(T, \ldots)$ associating to every infinite cardinal λ

$$I(T, \lambda) = \text{number of the isomorphim types of models of } T \text{ of power } \lambda.$$

$\lambda \mapsto I(T, \lambda)$ is called the *spectrum function* of T. Recall that $1 \leq I(T, \lambda) \leq 2^\lambda$ for every λ, owing to the Löwenheim-Skolem Theorem and cardinal computations. The content of Theorem 7.1.6 is just that $I(T, \lambda) = 2^\lambda$ for every uncountable λ excludes the classifiability of T. Indeed there was a conjecture of Morley, preceding Shelah's work and, in some sense, originating it, saying:

Conjecture 7.1.7 (Morley) *Let T be a countable complete first order theory. Then the spectrum function $\lambda \mapsto I(T, \lambda)$ is increasing among uncountable cardinals: for $\aleph_0 < \lambda \leq \mu$, $I(T, \lambda) \leq I(T, \mu)$.*

Shelah positively answered this conjecture, as a non-minor consequence of his classification analysis. The problem of determining all the possible spectrum functions $\lambda \mapsto I(T, \lambda)$ when T ranges over countable complete theories (and λ is uncountable) was solved only in 2000 by Hart, Hrushovki

and Laskowski, who explicitly listed all these functions. Notably their proof involves some arguments from Descriptive Set Theory.
Finally, what can we say when $\lambda = \aleph_0$? As we observed in the last chapter when talking about $\aleph_0$-categoricity, this case is a little oblique with respect to the general analysis and requires peculiar approaches and techniques. A classical conjecture in this framework was raised by Vaught.

Conjecture 7.1.8 (Vaught) *For a countable complete first order theory T, either $I(T, \aleph_0) \leq \aleph_0$ or $I(T, \aleph_0) = 2^{\aleph_0}$ (apart from the Continuum Hypothesis, of course).*

As in the case of Morley's Conjecture, this question is not only a mere cardinal investigation; what is more relevant is to understand the structure of the countable models of T. Shelah (together with Harrington and Makkai) positively answered Vaught's Conjecture for ω-stable theories T. Other partial positive answers are known. Indeed in the latest months the news of a counterexample (a theory with exactly $\aleph_1$ countable models) due to R. Knight [75] has been spreading, but this negative solution still seems (october 2002) under examination.

7.2 Simple theories

All throughout this section T is a complete first order theory in a countable language L, T has no finite models and Ω denotes the universe of T.
We aim at determining which key properties make T classifiable. A classification is very easy in the strongly minimal case. In fact, when T is strongly minimal, every model of T is labelled by a cardinal number - its dimension - classifying it up to isomorphism; what rules this dimension and its assignment is a notion of dependence, based on the model theoretic algebraic closure *acl*. Unfortunately the *acl* dependence does not work any more when we enlarge our setting and we leave the strongly minimal framework.
So we need a more general notion of (in)dependence, still including the classical cases of linear independence in vectorspaces, algebraic independence in algebraically closed fields and, definitively, *acl* independence in strongly minimal theories, but applying to a wider context. In other words we aim at defining for any T

$\vec{a}$ is independent from B over A

where $\vec{a}$ is a tuple in Ω and $A \subseteq B$ are small subsets of Ω. As notions require abstract symbols to be presented let us denote by I (I for independence) the set of all the triples $(\vec{a}, B, A)$ in Ω such that

$\vec{a}$ is independent from B over A

in a sense to be made more precise. It is reasonable to expect that I satisfies the following properties:

(I1) (*invariance*) for every $(\vec{a}, B, A) \in I$ and automorphism f of Ω, $(f(\vec{a}), f(B), f(A))$ is still in I;

(I2) (*local character*) for every $\vec{a}$ and B, there is a countable $A \subseteq B$ such that $(\vec{a}, B, A) \in I$;

(I3) (*finite character*) for every $\vec{a}$, A and B, $(\vec{a}, B, A) \in I$ if and only if, for all finite tuples $\vec{b}$ in B, $(\vec{a}, A \cup \vec{b}, A) \in I$;

(I4) (*extension*) for every $\vec{a}$, A and B, there is a tuple $\vec{a}'$ having the same length and the same type over A as $\vec{a}$ such that $(\vec{a}', B, A) \in I$;

(I5) (*symmetry*) for every $\vec{a}$, $\vec{b}$ and A, $(\vec{a}, A \cup \vec{b}, A) \in I$ if and only if $(\vec{b}, A \cup \vec{a}, A) \in I$;

(I6) (*transitivity*) for every $\vec{a}$ and $A \subseteq B \subseteq C$, $(\vec{a}, C, A) \in I$ if and only if $(\vec{a}, B, A) \in I$ and $(\vec{a}, C, B) \in I$.

Definition 7.2.1 *A set I of triples $(\vec{a}, B, A)$ with $\vec{a}$ in Ω and $A \subseteq B$ small subset of Ω is called an* **independence system** *of T if and only if I satisfies (I1) - (I6).*

An easy application of (I3) and (I5) shows that, if B, $B' \supseteq A$ are small subsets of Ω, then

$$(\vec{b}, B', A) \in I \quad \forall \vec{b} \in B$$

if and only if

$$(\vec{b'}, B, A) \in I \quad \forall \vec{b'} \in B'.$$

We will say that B and B' are independent over A when this happens.
Notice also that, if $\vec{a}'$ is a subsequence of $\vec{a}$ and $(\vec{a}, B, A) \in I$, then $(\vec{a}', B, A) \in I$ as well. By (I3), it suffices to check $(\vec{a}', A \cup \vec{b}, A) \in I$ for all $\vec{b}$ in B. We know $(\vec{a}, A \cup \vec{b}, A) \in I$. By symmetry (I5), $(\vec{b}, A \cup \vec{a}, A) \in I$, whence $(\vec{b}, A \cup \vec{a}', A) \in I$ by (I3), and $(\vec{a}', A \cup \vec{b}, A) \in I$ by (I5) once again.

Let us propose some examples of independence systems, both to illustrate the meaning of (I1) - (I6) and to confirm that they provide a reasonable axiomatic ground to introduce an abstract notion of independence.

Examples 7.2.2 1. First let us check that the old independence notion in strongly minimal theories T corresponds naturally to an indendence system in the new sense. In fact, let T be any theory. For $a \in \Omega$ and $A \subseteq B$ small subsets of Ω put

$$(a, B, A) \in I \text{ if and only if } a \in acl(B) \text{ implies } a \in acl(A)$$

or, equivalently but more transparently,

$$(a, B, A) \notin I \text{ if and only if } a \in acl(B) - acl(A).$$

We claim that, if the acl dependence relation in T satisfies $(D1)$, $(D2)$ and $(D4)$ (in particular, if T is strongly minimal, or also o-minimal), then the triples (a, B, A) in I do satisfy (I1) - (I6). We underline that we are momentarily dealing with elements a rather than with tuples $\vec{a}$ in Ω.

(I1) is clear.

(I2) follows from (D2), which says even more; in fact, it ensures that, for every a and B, if $a \in acl(B)$, then there is a **finite** $A \subseteq B$ such that $a \in acl(A)$.

(I3) says in our particular setting that, for every a and $A \subseteq B$ with $a \notin acl(A)$, $a \in acl(B)$ if and only if there exists a tuple $\vec{b}$ in B such that $a \in acl(A \cup \vec{b})$. So the direction from left to right just follows from the definition of acl, and the converse is a direct consequence of (D2).

(I4) Let $(a, B, A) \notin I$, so $a \in acl(B) - acl(A)$. $tp(a/A)$ is not algebraic, and so is realized even out of $acl(B)$. Take any $a' \notin acl(B)$ such that $tp(a'/A) = tp(a/A)$ and notice that $(a', B, A) \in I$.

(I5) when restricted to elements a and b in Ω is just the Exchange Principle (D4).

(I6) is clear.

It is straightforward to see that the previous analysis extends to arbitrary tuples $\vec{a}$ in Ω, provided we put, for $\vec{a} = (a_1, \ldots, a_n)$,

$$(\vec{a}, B, A) \in I$$

if and only if, for every $i \leq n$,

$$(a_i, B \cup \{a_0, \ldots, a_{i-1}\}, A \cup \{a_0, \ldots, a_{i-1}\}) \in I.$$

The resulting I is an independence system of T. As already said, this example includes both strongly minimal and o-minimal theories. In particular it applies to the following classes of structures.

(a) (Pure) infinite sets: here, for a, A, B as above, $(a, B, A) \notin I$ (so a dependent from B over A) means $a \in B - A$.

(b) Vectorspaces over a given (countable) field $\mathcal{K}$: in this case, $(a, B, A) \notin I$ means that a is in the subspace spanned by B but is not in the subspace spanned by A; hence I recovers linear independence.

(c) Algebraically closed fields: this time, for A and B subfields, $(a, B, A) \notin I$ means that a is algebraic over B but is transcendental over A; accordingly I recovers algebraic independence in this case.

(d) Real closed fields (such as the ordered field of reals): recall that these fields are o-minimal; moreover, for ordered subfields A and B, the model theoretic algebraic closure acl equals the real closure in the usual algebraic sense; accordingly, for a, A and B as before, $(a, B, A) \notin I$ means that a is in the real closure of B but is not in the real closure of A.

But independence systems do go beyond the strongly minimal and o-minimal settings, and apply to larger sceneries, for instance to ω-stable theories, and other theories as well.

2. Let T be ω-stable. Define I by putting, for $\vec{a}$, A and B as before,

$$(\vec{a}, B, A) \in I \iff RM(tp(\vec{a}, B)) = RM(tp(\vec{a}, A)).$$

Then I is an independence system of T. The details will be provided in Section 7.5 below. We give here just a few comments. Firstly recall that $RM(tp(\vec{a}, B)) \leq RM(tp(\vec{a}, A))$ for every $\vec{a}$ and $A \subseteq B$. Secondly notice that, for a strongly minimal T and for $a \in \Omega$, $RM(tp(a, B)) < RM(tp(a, A))$ just means $a \in acl(B) - acl(A)$. In other words, when we restrict our attention to strongly minimal theories, we recover the independence system in 1. But, of course, independence concerns now a much larger framework, including, for instance, differentially closed fields of characteristic 0 (the differential case will be discussed in more detail in Section 7.10).

3. Now consider *random graphs*. They are (infinite) graphs (G, R) (with R a symmetric irreflexive binary relation on G) satisfying the following additional condition:

   $(\star)$ for every $m, n \in \mathbf{N}$ and $a_0, \ldots, a_n, b_0, \ldots, b_m \in G$, there is some $g \in G$ satisfying $R(g, a_i)$ for all $i \leq n$ and $\neg R(g, b_j)$ for all $j \leq m$.

   Clearly $(\star)$ can be expressed by infinitely many first order sentences in the language $L = \{R\}$, so random graphs are an elementary class. Their theory T is ω-categorical and consequently complete; furthermore it eliminates the quantifiers in L. Notice that T is not ω-stable. In fact, given a random graph (G, R), for any subset S of G, the formulas

   $$R(v, s)\ \forall s \in S, \quad \neg R(v, g)\ \forall g \in G - S$$

   defines a 1-type over G, and different subsets produce different types; so $S_1(G)$ contains at least $2^{|G|}$ elements. This implies that T is not ω-stable.

   Now define, as in the case of infinite sets, for a in Ω and $A \subseteq B$,

   $$(a, B, A) \in I \text{ if and only if } a \in B - A.$$

   It is easy to check that one determines in this way an independence system I.

4. Any completion of the theory $ACFA$ of existentially closed fields with an automorphism has an independence system: this will be described in more detail in the final section of this chapter, among other algebraic examples.

However theories having an independence system (such as those considered in the previous examples) do not behave in the same way. For instance, one realizes that the following amalgamation property is a crucial dividing line:

(I7) (amalgamation) let $\mathcal{A}$ be a model of T, $B, B' \supseteq A$ be independent over A, $\vec{b}, \vec{b'}$ be tuples in Ω having the same type p over A and satisfying $(\vec{b}, B, A) \in I$, $(\vec{b'}, B', A) \in I$ respectively; then there is some $\vec{c}$ in Ω realizing the same type as $\vec{b}$ over B and as $\vec{b'}$ over B', and satisfying $(\vec{c}, B \cup B', A) \in I$.

Definition 7.2.3 *An independence system I of T is* **good** *if and only if it satisfies (I7).*

Let us run through the previous examples to illustrate (I7) and its relevance.

Examples 7.2.4 1. Let T be strongly minimal, I be defined as before. Then (I7) holds. Let us check this when dealing with tuples of length 1, so for elements of Ω. If b or b' is in A, then $b = b'$ and $c = b = b'$ works. So assume $b, b' \notin A$; independence implies $b \notin acl(B)$ and $b' \notin acl(B')$, in other words b, b' realize the only non-algebraic 1-type over B, B' respectively; so any realization of the unique non-algebraic 1-type over $B \cup B'$ works as c. Notice that, in this case, (I7) is a direct consequence of the fact that, for every small subset X of Ω, there is a unique non-algebraic 1-type p_X over X and consequently over every small $Y \supseteq X$ there is a unique non-algebraic 1-type extending p_X.

However (I7) fails within o-minimal theories: I is not good in this setting. Let us see why in the particular case when $T = RCF$ is the theory of real closed fields. Consider the ordered field $\mathbf{R}$ of reals and, in a suitably saturated extension Ω of $\mathbf{R}$, 4 positive elements

$$b < a < a' < b'$$

each infinitely larger than the previous ones, and b infinitely larger than $\mathbf{R}$. Then a and a' are independent over $\mathbf{R}$ $((a, \mathbf{R} \cup \{a'\}, \mathbf{R}) \in I)$, b, b' realize the same type over $\mathbf{R}$ and

$$(b, \mathbf{R} \cup \{a\}, \mathbf{R}) \in I, \quad (b', \mathbf{R} \cup \{a'\}, \mathbf{R}) \in I.$$

However no $c \in \Omega$ can satisfy the same type as b over $\mathbf{R} \cup \{a\}$ and as b' over $\mathbf{R} \cup \{a'\}$ because $b < a < a' < b'$ and so $c < a$ excludes $a' < c$.

2. The independence system associated with an ω-stable T as before is good. This will be checked in detail in 7.5.

3. Also the independence system of random graphs enjoys amalgamation: the reader may easily check this.

4. The same is true for $ACFA$ and its completions.

Definition 7.2.5 *T is called* **simple** *when T has a good independence system. A structure $\mathcal{A}$ is* **simple** *when its theory is.*

Therefore ω-stable theories (and in particular strongly minimal theories) are simple; random graphs are simple, as well as any existentially closed field with an automorphism: in all these cases, an independence notion satisfying not only (I1) - (I6) but also the additional condition (I7) can be introduced. But we can say even more. Indeed, for a simple T, this independence notion is unique; in other words, there is only one good independence system I of T making T simple. I can be characterized as follows.

Definition 7.2.6 (Shelah) *Let $A \subseteq B$ be small subsets of Ω, p be a type over B. p is said to* **fork** *over A if and only if there are a formula $\varphi(\vec{v}, \vec{w})$ and tuples $\vec{b}_\nu$ (where $\nu < \lambda$ and λ is a suitable infinite ordinal), all having the same length as $\vec{w}$, such that*

(i) *the $\vec{b}_\nu$'s have the same type over A;*

(ii) $\varphi(\vec{v}, \vec{b}_0) \in p$;

(iii) $\bigcap_{\nu<\lambda} \varphi(\Omega^n, \vec{b}_\nu) = \emptyset$ *(where n is, obviously, the length of $\vec{v}$).*

In any simple theory T, a good independence system necessarily arises from forking: this is the content of the following theorem of Kim and Pillay.

Theorem 7.2.7 (Kim - Pillay) *Let T be simple, I be a good independence system of T. Then, for every tuple $\vec{a}$ in Ω and $A \subseteq B$ small subsets of Ω, $(\vec{a}, B, A) \in I$ if and only if $tp(\vec{a}/B)$ does not fork over A.*

So the only possible good independence system of a simple theory T is the set I of the triples $(\vec{a}, B, A)$ as before such that $tp(\vec{a}/B)$ does not fork over A: this is what is necessarily meant when we say that $\vec{a}$ independent from B over A in a simple T; we will write $\vec{a} \downarrow_A B$ when this happens. When A and B are any small sets (so possibly $A \not\subseteq B$), $\vec{a} \downarrow_A B$ abbreviates $\vec{a} \downarrow_A A \cup B$; one usually omit the subscript A when A is empty; so $\downarrow$ just means $\downarrow_\emptyset$; for B and B' sets, $B \downarrow_A B'$ abridges $\vec{b} \downarrow_A B'$ for every $\vec{b}$ in B, or equivalently $\vec{b}' \downarrow_A B$ for every $\vec{b}'$ in B'.
For technical convenience, let us restate (I1) - (I7) in this new notation for a simple T.

(I1) (*Invariance*) for every $\vec{a}$, B and A as before and every automorphism f of Ω, if $\vec{a} \downarrow_A B$, then $f(\vec{a}) \downarrow_{f(A)} f(B)$;

(I2) (*local character*) for every $\vec{a}$ and B, there is a countable $A \subseteq B$ such that $\vec{a} \downarrow_A B$;

(I3) (*finite character*) for every $\vec{a}$, A and B, $\vec{a} \downarrow_A B$ if and only if, for all finite tuples $\vec{b}$ in B, $\vec{a} \downarrow_A \vec{b}$.

(I4) (*extension*) for every $\vec{a}$, A and B, there is a tuple $\vec{a}'$ having the same length and the same type over A as $\vec{a}$ and satisfying $\vec{a}' \downarrow_A B$;

(I5) (*symmetry*) for every $\vec{a}$, $\vec{b}$ and A, $\vec{a} \downarrow_A \vec{b}$ if and only if $\vec{b} \downarrow_A \vec{a}$;

(I6) (*transitivity*) for every $\vec{a}$ and $A \subseteq B \subseteq C$, $\vec{a} \downarrow_A C$ if and only if $\vec{a} \downarrow_A B$ and $\vec{a} \downarrow_B C$.

(I7) (*amalgamation*) let $\mathcal{A}$ be a model of T, B, $B' \supseteq A$ satisfying $B \downarrow_A B'$, $\vec{b}$, $\vec{b'}$ be tuples in Ω having the same type p over A and satisfying $\vec{b} \downarrow_A B$, $\vec{b'} \downarrow_A B'$ respectively; then there is some $\vec{c}$ in Ω realizing the same type as $\vec{b}$ over B and as $\vec{b'}$ over B', and satisfying $\vec{c} \downarrow_A B \cup B'$.

The reader may directly realize what these properties imply when tuples $\vec{a}$ are replaced by arbitrary small subsets, so when one deals with $B \downarrow_A B'$, and rewrite them in this enlarged setting.
Remarkably symmetry - namely (I5) - is a key property of the forking independence within simple theories: indeed, simplicity is equivalent to the symmetry of forking. Also transitivity (I6) and local character (I2) have the same crucial role. This was observed by Kim and Pillay as to local character, and has been recently shown by Kim in the other two cases.

Theorem 7.2.8 *A theory T is simple if and only if, one of the following propositions holds:*

(i) *the forking independence satisfies symmetry* **(I5)***;*

(ii) *the forking independence satisfies transitivity* **(I6)***;*

(iii) *the forking independence satisfies local character* **(I2)**.

In conclusion, simple theories are those where independence can be reasonably introduced and developed in the axiomatic way we said before. Of course, this is a positive property towards classifiability. But one can wonder which is the reverse of the medal, in other words what happens in non-simple theories. Well, Shelah proved a quite negative result about them: in fact, they are not classifiable, and so must be rejected in our classification program.

Theorem 7.2.9 (Shelah) *If T is not simple, then $I(T, \lambda) = 2^\lambda$ for every uncountable cardinal λ.*

Simple theories exclude o-minimal theories, as observed before. More generally one can show that no complete theory of linear orders can be simple.

Theorem 7.2.10 *No linearly ordered infinite structure has a simple theory. In particular, no complete theory of linearly ordered infinite structures is classifiable.*

It is perhaps worth repeating that, in spite of this result, infinite linearly ordered structures are not so bad. On the contrary, some of them, like dense or discrete linear orders, real closed fields and, more generally, o-minimal models satisfy some nice and powerful model theoretic properties; in all these frameworks a suitable independence notion, satisfying the axioms (I1)- (I6) can be introduced. So things are not so sick as they look. We will come back to this point at the end of Chapter 9.
To conclude this section, let us sketch a brief history of simple theories. Simplicity was first introduced by Shelah in 1980, but its relevance was not completely realized at that time; indeed Shelah laid a major emphasis on a stronger notion - stability, the matter of the next section - and regarded simplicity as a generalization of stability, and not directly as a key dichotomy. The role of simplicity in introducing independence and so towards classifiability was neglected until the nineties, when one observed that several interesting algebraic structures - most notably, existentially closed fields with an automorphism - have a simple unstable theory. Then Kim in 1996 showed that the forking independence satisfies symmetry (I5) within simple theories and, together with Pillay, proved the close connection between independence and simplicity introduced before (in Theorem 7.2.7). Kim again realized in 2001 the key role of symmetry for the forking independence in the simple framework.
Now simplicity theory has deserved its own room in Model Theory, as an autonomous and relevant part of the classification program.

7.3 Stable theories

As said in the last section, Shelah's original analysis put its emphasis on stability rather than simplicity. Stability looked the key property towards independence. But, after Kim and Pillay, one realized that it is simplicity that plays the central role here, while stability strengths simplicity in the sense we are going to explain.
There are several ways to introduce stability, and their equivalence is not so immediate as one would like. We adopt the following definition, which underlines the connection with simplicity: T still denotes a complete first order theory with no finite models in a countable language L, and Ω is a big saturated model of T.

Definition 7.3.1 *T is* **stable** *if and only if T has an independence system I satisfying the following additional assumption:*

(I8) *(**stationarity over models**) for every model $\mathcal{A}$ of T, tuples $\vec{a}$, $\vec{a'}$ in Ω and small subset $B \supseteq A$ in Ω, if $\vec{a}$, $\vec{a'}$ have the same type over A and are independent from B over A (with respect to I), then they have the same type also over B.*

T is called **unstable** *when it is not stable. A structure is said to be* **stable** *or* **unstable** *when its theory is.*

The content of (I8) is clear: there is a unique way to extend a type p over a model $\mathcal{A}$ of T to a type over $B \supseteq A$ of tuples independent from B over A. Let us propose an easy example to illustrate this condition (I8), and so stability.

Example 7.3.2 Just for a change(?!), we deal with a strongly minimal T and 1-types over a model $\mathcal{A}$ of T. The algebraic 1-types are realized in $\mathcal{A}$ and so extend uniquely over any $B \supseteq A$. But the only non-algebraic $p \in S_1(A)$ may have several extensions over B; however only one of them does not fork over A, and this is the unique non-algebraic 1-type over B, in other words the type of any element out of $acl(B)$. What happens if we enlarge our analysis from models of T to arbitrary small sets A? Now even an algebraic 1-type p over A may have several extensions over a $B \supseteq A$, and all of them do not fork over A; however their number is finite and cannot exceed $|\varphi(\Omega)|$ where $\varphi(v)$ is an $L(A)$-formula isolating p; on the contrary the non-algebraic 1-type over A has again a unique nonforking extension over p, that is the non-algebraic 1-type over B.

Hence strongly minimal theories are stable. But this property may fail elsewhere; indeed we are going to see that it does not hold any longer in non-simple theories, and even in certain simple structures, like random graphs (and others).

Theorem 7.3.3 *If T is stable, then T is simple.*

Proof. Let I be an independence system of T satisfying (I8). We have to check amalgamation (I7). So take a model $\mathcal{A}$ of T, two sets B and B' independent over A (with respect to I), and two tuples $\vec{b}$, $\vec{b'}$ having the same type over A and satisfying $(\vec{b}, B, A) \in I$, $(\vec{b'}, B', A) \in I$ respectively. Using extension (I3), one finds two tuples $\vec{d}$, $\vec{d'}$ having the same type as $\vec{b}$ over B and as $\vec{b'}$ over B' respectively, and both independent over $B \cup B'$ over A.

In particular, $\vec{d}$, $\vec{d}'$ have the same type over A. By (I8), they have the same type also over $B \cup B'$. Hence $\vec{d}$ has the same type as $\vec{d}'$, and consequently as $\vec{b}'$, over B'; furthermore $\vec{d}$ has the same type as $\vec{b}$ over B. So $\vec{d}$ satisfies (I7) as $\vec{c}$. ♣

Then any independence system I satisfying (I8) in a stable T is good, and must equal the forking independence: for $\vec{a}$, A and B as usual,

$$(\vec{a}, B, A) \in I \iff \vec{a} \downarrow_A B.$$

Accordingly (I8) can be restated as follows for a simple T.

(I8) Let $\mathcal{A}$ be a model of T, B be a small subset of Ω including A. Then every type over A has a unique non-forking extension over B.

Once again recall that the existence of a non-forking extension is just stated in (I4). Uniqueness may fail for a stable T over arbitrary subsets A of Ω; but also in this extended framework stability bounds the number of the possible non-forking extensions over B of a type over A, in the following sense.

Theorem 7.3.4 *T is stable if and only if T is simple and, for every small $A \subseteq \Omega$ and $p \in S(A)$, there is a cardinal κ (less than $|\Omega|$) such that, for any small $B \supseteq A$, p has at most κ non-forking extensions in $S(B)$.*

Another equivalent way of defining stability relies upon a counting type characterization (resembling ω-stability).

Theorem 7.3.5 *T is stable if and only if there is some infinite cardinal λ such that, for every small $A \subseteq \Omega$ of power λ, $|S(A)| = \lambda$.*

We wish to underline another equivalent characterization, saying that stability just excludes definable infinite linear orders, in the following sense.

Theorem 7.3.6 *T is unstable if and only if it satisfies the* **order property***: there is a formula $\varphi(\vec{v}, \vec{w})$ - where $\vec{v}$ and $\vec{w}$ have the same length n - such that $\varphi(\Omega^{2n})$ linearly orders an infinite (possibly non definable) subset of Ω^n.*

So the order property is the key obstruction to stability. On the other side we saw in the last section that no infinite linearly ordered structure has a classifiable theory; this applies more generally to theories with the order property, hence to unstable theories.

Theorem 7.3.7 (Shelah) *If T is not stable, then for every uncountable power λ $I(T, \lambda) = 2^\lambda$. In particular, T is not classifiable.*

Hence instability is a negative condition in the classification perspective. On the other hand, which are the benefits of the stability assumption? Some of them are already described by the definition and the equivalent (positive) characterizations listed before. In particular we emphasize once again that, for a stable T and for a model $\mathcal{A}$ of T, every type p over A enlarges uniquely to a non-forking extension over any set $B \supseteq A$. Let $p|B$ denote this extension in $S(B)$; $p|B$ can be characterized as follows

$\star$ for every L-formula $\varphi(\vec{v}, \vec{w})$, if there is some $\vec{b}$ in B for which $\varphi(\vec{v}, \vec{b})$ is in $p|B$, then there is $\vec{a}$ in A such that $\varphi(\vec{v}, \vec{a}) \in p$;

$\star\star$ for every formula $\varphi(\vec{v}, \vec{b}) \in p|B$, there is some $\vec{a}$ in A for which $\models \varphi(\vec{a}, \vec{b})$.

For an arbitrary T, a type q over B satisfying $\star$ (in the place of $p|B$) is called an *heir* of p, and a type q over B satisfying $\star\star$ is called a *coheir* of p. So, within stable theories and for types over models,

heir = *coheir* = *non-forking extension*.

We will check these equalities in the restricted ω-stable framework in 7.5. Another benefit of the stability assumption concerns definability. To describe this, we need the following definition.

Definition 7.3.8 *Let A be any small subset of Ω. A type p over A is said to be* **definable** *over A if and only if, for every L-formula $\varphi(\vec{v}, \vec{w})$, the set of the tuples $\vec{a}$ for which $\varphi(\vec{v}, \vec{a})$ occurs in p is A-definable; in other words, there is an $L(A)$-formula $d\varphi(\vec{w})$ (possibly with parameters) such that, for every $\vec{a}$ in A, $\varphi(\vec{v}, \vec{a}) \in p$ if and only if $\models d\varphi(\vec{a})$.*

Theorem 7.3.9 *T is stable if and only if every type over a model $\mathcal{A}$ of T is definable; moreover, for every small $B \supseteq A$, even $p|B$ is definable over B, and indeed, for every formula $\varphi(\vec{v}, \vec{w})$ the same $L(A)$-formula $d\varphi(\vec{w})$ working for p and A goes right also over B.*

The last theorem implies in its turn

Corollary 7.3.10 *Let $\mathcal{A} \prec \mathcal{B}$ be models of a stable theory T. If $\varphi(\vec{v})$ is a formula* **with parameters from** *B (and n is the length of $\vec{v}$), then $\varphi(A^n)$ is A-definable.*

We will check these results in detail in 7.5 in the restricted framework of ω-stable theories. Now, to conclude this section, let us propose some examples of stable or unstable theories.

Examples 7.3.11 1. Every ω-stable theory T is stable. This will be deduced by the definition of stability in 7.5 below; it follows even more directly from other characterizations (for instance, from Theorem 7.3.5). Indeed, the particular case of strongly minimal theories was already dealt with a few lines ago. So any infinite vectorspace, as well as any algebraically closed field, or any differentially closed field of characteristic 0, has a stable theory.

2. Any module (over a countable ring R) has a stable theory: this extends what we have just observed about vectorspaces. The proof requires very basic preliminaries about the model theory of modules, and can be found in the references quoted at the end of the chapter.

3. Real closed fields do not have a stable theory. Indeed RCF is not even simple, and anyhow it houses infinite linear orders.

4. The theory of random graphs is unstable, although simple. In fact, let $\mathcal{A} = (A, R)$ be a random graph, $B \supseteq A$ be a small subset of Ω. Use compactness and find a and a' in Ω such that $\Omega \models R(a, x)$ for every $x \in B$ and $\Omega \models R(a, b)$ for a unique $b \in B$, with $b \notin A$. Then a and a' have the same type over A and each of them is independent from B over A. But they do not have the same type over B.

5. Similarly, existentially closed fields with an automorphism are simple and unstable.

6. On the contrary, separably closed fields and differentially closed fields in prime characteristics are stable.

7.4 Superstable theories

Also in this section T is a complete first order theory with infinite models in a countable language L, and Ω denotes a big saturated model of T.
Let us underline once again a particular feature of the strongly minimal case: this assumption implies, among other things, that T is stable and, over any small subset A of Ω, there is a unique non-algebraic 1-type p, which is realized by all the elements outside $acl(A)$; moreover, for a and B in $p(\Omega)$, so out of $acl(A)$,

$$a \not\downarrow_A B \text{ means } a \in acl(B).$$

In particular $\not\downarrow_A$ satisfies within $p(\Omega)$ the properties (D1) - (D4) stated in Chapter 5.
Now consider an arbitrary stable T. In this enlarged setting there may be several non-algebraic 1-types over a small A; fix one of them, call it p and look at the set $p(\Omega)$ of its realizations in Ω. We wonder whether $\not\downarrow_A$ still satisfies (D1) - (D4). Assume momentarily the following condition.

($\star$) For every small $B \supseteq A$, p has a unique non-forking extension over B.

Recall that p has at least one non-forking extension over any B: this is just what is stated in (I4). ($\star$) requires that all the elements $a \in p(\Omega)$ satisfying $a \downarrow_A B$ realize the same type also over B. Notice that, as T is stable, every p over a model of T satisfies ($\star$). Call p *stationary* when ($\star$) holds.
So take a non-algebraic stationary type p over a small A. It is easy to check that $\not\downarrow_A$ satisfies (D1), (D2) and (D4) in $p(\Omega)$. In fact, let a, $b \in p(\Omega)$ and $B \subseteq p(\Omega)$.

(D1) If $a \in B$, then $a \not\downarrow_A B$.

As p in not algebraic we can find infinitely many elements $b_0 = a$, $b_1, \ldots,$ $b_n, \ldots$ realizing p in Ω. The formula $v = b_0$ is in $tp(a/B)$ but no other b_n can satisfy it. Hence $tp(a/B)$ forks over A according to Shelah's definition, and so $a \not\downarrow_A B$.

(D2) If $a \not\downarrow_A B$, then there is some finite $B_0 \subseteq B$ for which $a \not\downarrow_A B_0$.

This follows directly from (I3).

(D4) If $a \not\downarrow_A B \cup \{b\}$ but $a \downarrow_A B$, then $b \not\downarrow_A B \cup \{a\}$.

Otherwise transitivity (I6) implies $b \downarrow_B a$, whence $a \downarrow_B b$ by symmetry (I5); as $a \downarrow_A B$, (I6) applies again and gives $a \downarrow_A B \cup \{b\}$.

On the other side, there do exist non algebraic stationary p's for which $\not\downarrow_A$ does not obey the last condition:

(D3) if $a \not\downarrow_A B \cup \{b\}$ and $b \not\downarrow_A B$, then $a \not\downarrow_A B$.

Here is a counterexample.

Example 7.4.1 Consider the theory of the usual Cartesian plane $\mathbf{R}^2$ with the two projections

$$(x, y) \mapsto (x, 0), \quad (x, y) \mapsto (0, y), \quad \forall x, y \in \mathbf{R}$$

onto the x-axis and the y-axis respectively. There are three non-algebraic 1-types over the empty set; they correspond in $\mathbf{R}^2$ to:

(i) any element $(x, 0)$ with $x \neq 0$,

(ii) any element $(0, y)$ with $y \neq 0$,

(iii) any element (x, y) with $x, y \neq 0$

respectively (the only algebraic 1-type concerns $(0, 0)$, that is the unique point lying in both the projection sets). Let p denote the third type in the previous list. Now let us consider the algebraic 1-types over $\mathbf{R}^2$; they can be partitioned into three classes:

(a) a new element (∞, y) with an old ordinate $y \in \mathbf{R}$ and $\infty \notin \mathbf{R}$;

(b) a new element (x, ∞) with an old absciss $x \in \mathbf{R}$ and $\infty \notin \mathbf{R}$;

(c) a entirely new (∞, ∞) with $\infty \notin \mathbf{R}$.

Actually the same analysis works over every model of T. The types in (a) and (b) have Morley rank 1, while the third type has Morley rank 2. Accordingly T is ω-stable with Morley rank 2 and Morley degree 1. The third type is the only non-forking extension of p; actually, even the types in (a), (b) with $x, y \neq 0$ enlarge p, but they fork over $\emptyset$ (when $y = 0$ and $x = 0$ the types in (a) and (b) extend those in (i), (ii) respectively). It is also straightforward to check that p is stationary. Now take

$$A = \emptyset,\ B = \mathbf{R}^2,\ a = (\infty, \infty),\ b = (x, \infty)$$

with $x \neq 0$ in $\mathbf{R}$. Clearly a and b realize p, in particular $a \downarrow \mathbf{R}^2$. But $a \not\downarrow \mathbf{R}^2 \cup \{b\}$ and $b \not\downarrow \mathbf{R}^2$. This contradicts (D3).

Definition 7.4.2 *A (non-algebraic stationary) 1-type p over a small A is called* **regular** *when (D3) holds in $p(\Omega)$.*

Our classification purposes suggest to handle stable theories T admitting suitably many regular types. In fact, roughly speaking, the more they are, the better one can expect to approach T. Now put:

Definition 7.4.3 *A stable theory T is called* **superstable** *if and only if, for every small B and $p \in S(B)$, there is a finite subset A of B over which p does not fork. A structure $\mathcal{A}$ is said to be* **superstable** *when its theory is.*

Notice that superstability generalizes what happens in the strongly minimal case. But what is more remarkable for our purposes is the following result.

Theorem 7.4.4 *Let T be a superstable theory, $\mathcal{A}$ and $\mathcal{B}$ be two models of T such that $\mathcal{A}$ is an elementary substructure of $\mathcal{B}$ and $A \neq B$. Then there is some $b \in B - A$ such that the type of b over A is regular.*

So regular types are not rare within superstable theories. On the other hand, if superstability fails, then no classification can be expected.

Theorem 7.4.5 *Let T be stable unsuperstable. Then, for every uncountable power λ, $I(T, \lambda) = 2^\lambda$. In particular T is not classifiable.*

Accordingly superstability provides a new dichotomy in Shelah's approach to classification. For, it ensures sufficiently many regular types, while its negation excludes any classification. Superstability can be introduced in other equivalent ways. Let us quote a characterization based on a counting types criterion (just as we did in the stable setting).

Theorem 7.4.6 *T is superstable if and only if, for every $\lambda \geq 2^{\aleph_0}$ and every set A of power λ, there are at most λ 1-types over A.*

Let us conclude this section by listing some examples and counterexamples of superstable theories.

Examples 7.4.7

1. Separably closed fields, or even differentially closed fields in a prime characteristic are not superstable (although they are stable).
2. On the other hand, all ω-stable theories are superstable. This can be easily deduced from the previous definition, but we will postpone the details to 7.5 below.
3. There do exist superstable non ω-stable theory. For instance the theory of the additive group of integers is so.

7.5 ω-stable theories

Just to have a short break in our outline of Shelah's classification, let us come back in this section to a more familiar setting, the one of ω-stable theories. What we plan to do is to examine closerly within this particular framework the main notions introduced so far in our report. Accordingly we will check that ω-stable theories are simple, stable and superstable; we will characterize forking, independence and so on in the ω-stable setting; we

will provide proofs and details of the main theorems stated till now. So, all throughout this section, T will denote an ω-stable theory in a countable language L.
The first fact we want to check is that ω-stable theories are stable as well. Indeed within Shelah's dichotomies, and in our outline, simplicity preceded stability, and consequently one might reasonably expect to meet simplicity before stability at this point; we will explain later why we are reversing this order and postponing simplicity in this section. We provided several equivalent characterizations of stability in Section 7.3; in particular we said that stable theories are just those failing to have the order property. And actually it is easy enough to show that ω-stability implies stability if we refer to this characterization.

Theorem 7.5.1 *No ω-stable theory T satisfies the order property. Equivalently, any ω-stable T is stable.*

Proof. Let T be ω-stable. Suppose towards a contradiction that there are an L-formula $\varphi(\vec{v}, \vec{w})$ and tuples $\vec{a_i}$ $(i \in \mathbf{N})$ in Ω (say of length n) such that, for every choice of i and j in $\mathbf{N}$,

$$\Omega \models \varphi(\vec{a_i}, \vec{a_j}) \quad \Leftrightarrow \quad i \leq j.$$

A straightforward application of the Compactness Theorem shows that we can replace $\mathbf{N}$ by $\mathbf{Q}$ above; in other words, there are tuples $\vec{a_i}$ $(i \in \mathbf{Q})$ in Ω such that, for every choice of i and j in $\mathbf{Q}$,

$$\Omega \models \varphi(\vec{a_i}, \vec{a_j}) \quad \Leftrightarrow \quad i \leq j.$$

Notice that there do exist definable subsets X in Ω such that the indexes $i \in \mathbf{Q}$ such that $\vec{a_i} \in X$ form an interval of positive length $I(X)$ in $\mathbf{Q}$; for instance, when $X = \Omega^n$, $I(\Omega^n) = \mathbf{Q}$ is so. Choose X as before with minimal Morley rank and degree. Fix a point j in the interior of the corresponding interval $I(X)$ and look at

$$X_0 = X \cap \varphi(\Omega^n, \vec{a_j}), \quad X_1 = X - X_0.$$

Both X_0 and X_1 are definable subsets of X. Furthemore $I(X_0) = \{i \in I(X) : i \leq j\}$ and $I(X_1) = \{i \in I(X) : i > j\}$ are intervals of positive length. So X_0 and X_1 have the same Morley rank and degree as X; but this contradicts the fact that they partition X. ♣

Another equivalent way of introducing stable theories concerns definability of types: let us check that ω-stability implies stability by referring to this characterization. Recall that a type p over a small subset A of the universe Ω of T is said to be *definable* over A if and only if, for every L-formula $\varphi(\vec{v}, \vec{w})$, the set of the tuples $\vec{a}$ in Ω for which $\varphi(\vec{v}, \vec{a}) \in p$ is A-definable. T is stable exactly when any type over any model $\mathcal{M}$ of T is definable over M: this is what we want to check for an ω-stable T. We begin by two technical lemmas.

Lemma 7.5.2 *Let $X \subseteq \Omega^n$ be a definable set, A be a small subset of Ω such that every automorphism f of Ω fixing A pointwise fixes X setwise. Then X is A-definable.*

Proof. Let $\varphi(\vec{v})$ be a formula defining X, and let P denote the set of the types p over A of elements $\vec{a}$ in X. Notice that, in this case, every realization $\vec{b}$ of p is in X; in fact there is an A-automorphism f of Ω sending $\vec{a}$ to $\vec{b}$; f fixes X setwise, and so, as $\vec{a}$ is in X, $\vec{b}$ is in X, too. In conclusion, for every $p \in P$ and $\vec{a} \models p$, $\Omega \models \varphi(\vec{a})$. Now repeat the same argument as in Theorem 6.5.5: by compactness, for every $p \in P$ there is a finite conjunction of formulas in p, and hence a single formula $\varphi_p(\vec{v})$ of p such that

$$\Omega \models \forall \vec{v}(\varphi_p(\vec{v}) \to \varphi(\vec{v})).$$

Furthermore, for every $\vec{a} \in X$, $p = tp(\vec{a}/A)$ is in P, and $\vec{a} \in \varphi_p(\Omega^n)$. In conclusion $X = \bigcup_{p\in P} \varphi_p(\Omega^n)$, in other words $\{\mathcal{U}_{\varphi_p} : p \in P\}$ is an open covering of $\mathcal{U}_\varphi$. By compactness again, there is a finite subset P_0 of P such that $\{\mathcal{U}_{\varphi_p} : p \in P_0\}$ covers $\mathcal{U}_\varphi$, hence $X = \bigcup_{p\in P_0} \varphi_p(\Omega^n)$. Accordingly X is defined by the $L(A)$-formula $\bigvee_{p\in P_0} \varphi_p(\vec{v})$. ♣

Lemma 7.5.3 *Let T be an ω-stable theory, $\mathcal{M}$ be an $\aleph_0$-saturated model of T, $X \subseteq \Omega^n$ be a non-empty M-definable set, Y be any definable (possibly non M-definable) subset of Ω such that $Y \subseteq X$ and Y has the same Morley rank as X. Then $Y \cap M^n \neq \emptyset$.*

Proof. Put $\alpha = RM(X)$, $d = GM(X)$. Recall $0 \leq \alpha < \infty$. Then proceed by induction on the pair (α, d) (with respect to the lexicographic order). If $\alpha = 0$, then X is finite, and so $X \subseteq M^n$. This implies $Y \subseteq M^n$. Furthermore $Y \neq \emptyset$ because has the same Morley rank as X.
Now suppose $\alpha > 0$, $d = 1$. Then $RM(X - Y) < \alpha$. Put $\beta = RM(X - Y)$. Owing to Proposition 5.7.9, there are infinitely many pairwise disjoint M-definable subsets X_i ($i \in \mathbf{N}$) of Morley rank β inside X. In particular there exists some natural i for which $RM((X - Y) \cap X_i) < \beta$. Accordingly $RM(Y \cap X_i) = \beta$, and by induction $Y \cap X_i \cap M^n \neq \emptyset$. Hence $Y \cap M^n \neq \emptyset$.

At last assume $\alpha > 0$ and $d > 1$. By Proposition 5.7.9 once again, X can be decomposed as the disjoint union of d M-definable sets $X_0, \ldots, X_{d-1}$ having Morley rank α and degree 1. For some $i < d$, $Y \cap X_i$ has Morley rank α and hence Morley degree 1. By induction $Y \cap X_i \cap M^n \neq \emptyset$, whence $Y \cap M^n \neq \emptyset$. ♣

The previous lemmas have a prevalent technical flavour. The next lemma is more substantial and will be used several times later in this section; it ensures that, in an ω-stable theory T, for every definable $X \subset \Omega^n$ and formula $\varphi(\vec{v}, \vec{w})$, the set of the tuples $\vec{a}$ for which $\varphi(\Omega^n, \vec{a}) \cap X$ has the same Morley rank as X is definable.

Lemma 7.5.4 *Let T be an ω-stable theory, A be a small subset of Ω, n be a positive integer, X be an A-definable subset of Ω^n, α be the Morley rank of X. Let $\varphi(\vec{v}, \vec{w})$ be a formula of the language L of T (and n be the length of $\vec{v}$, m denote the length of $\vec{w}$). Then both the sets of the tuples $\vec{a} \in \Omega^m$ satisfying*

$$RM(\varphi(\Omega^n, \vec{a}) \cap X) < \alpha, \quad RM(\varphi(\Omega^n, \vec{a}) \cap X) = \alpha$$

respectively are A-definable.

Proof. For every $\vec{a} \in \Omega^m$, $RM(\varphi(\Omega^n, \vec{a}) \cap X) \leq \alpha$; accordingly it is sufficient to show that the former set - that of the tuples $\vec{a} \in \Omega^m$ such that $RM(\varphi(\Omega^n, \vec{a}) \cap X) < \alpha$ is A-definable; for, the latter set is just its complement. Moreover notice that any A-automorphism di Ω fixes the set of the tuples $\vec{a} \in \Omega^m$ for which $RM(\varphi(\Omega^n, \vec{a}) \cap X) < \alpha$; consequently, if this set is definable, then it is even A-definable, owing to Lemma 7.5.2. In conclusion, it is enough to prove that the former set is definable. Now look at the Morley degree d of X: X decomposes as the disjoint union of d definable subsets, all having Morley rank α (and degree 1) $X = \bigcup_{i<d} X_i$. Clearly, $\forall \vec{a} \in \Omega^m$,

$$RM(\varphi(\Omega^n, \vec{a}) \cap X) < \alpha \Leftrightarrow \forall i < d, \ RM(\varphi(\Omega^n, \vec{a}) \cap X_i) < \alpha.$$

Hence it suffices to show that, $\forall i < d$, the set of the tuples $\vec{a} \in \Omega^m$ satisfying $RM(\varphi(\Omega^n, \vec{a}) \cap X_i) < \alpha$ is definable. In other words, we can also assume $d = 1$.

Now let us sketch the plan of the proof.

The first step will produce a finite subset D of X such that, for every $\vec{a} \in \Omega^m$, if $D \cap \varphi(\Omega^n, \vec{a}) = \emptyset$, then $RM(\varphi(\Omega^n, \vec{a}) \cap X) < \alpha$.

The second step will even prove that, for every $\vec{a} \in \Omega^m$ satisfying $RM(\varphi(\Omega^n, \vec{a}) \cap X) < \alpha$, there exists a finite subset $D(\vec{a})$ such that $D(\vec{a}) \cap \varphi(\Omega^n, \vec{a}) = \emptyset$ and, for $\vec{a'} \in \Omega^m$ satisfying $D(\vec{a}) \cap \varphi(\Omega^n, \vec{a'}) = \emptyset$, $RM(\varphi(\Omega, \vec{a'}) \cap X) < \alpha$.
Assume momentarily these preliminary steps shown, and consider all the finite sets $D \subseteq X$ such that, $\forall \vec{a} \in \Omega^m$,

$$D \cap \varphi(\Omega^n, \vec{a}) = \emptyset \ \Rightarrow \ RM(\varphi(\Omega^n, \vec{a}) \cap X) < \alpha.$$

Let Δ denote the class of these sets (Δ may have the same power as Ω). Notice that, for every $D \in \Delta$, one can write a suitable formula $\alpha_D(\vec{w})$ (possibly with parameters) just stating "$D \cap \varphi(\Omega^n, \vec{w}) = \emptyset$". Observe that $\bigcup_{D \in \Delta} \alpha_D(\Omega^m)$ is the set of the tuples $\vec{a} \in \Omega^m$ such that

$$RM(\varphi(\Omega^n, \vec{a}) \cap X) < \alpha.$$

In fact, if $\vec{a} \in \Omega^m$ satisfies $\alpha_D(\vec{w})$ for some $D \in \Delta$, then $D \cap \varphi(\Omega^n, \vec{a}) = \emptyset$ and hence $RM(\varphi(\Omega^n, \vec{a}) \cap X) < \alpha$. Conversely, suppose $RM(\varphi(\Omega^n, \vec{a}) \cap X) < \alpha$ for $\vec{a} \in \Omega^m$ and look at the corresponding set $D(\vec{a})$ (determined in the second step); $D(\vec{a})$ is in Δ and satisfies $D(\vec{a}) \cap \varphi(\Omega^n, \vec{a}) = \emptyset$, whence $\vec{a} \in \alpha_D(\Omega^m)$. So the third (and final) step of the proof will aim at showing that $\bigcup_{D \in \Delta} \alpha_D(\Omega^m)$ is definable. Actually, there is a subtle objection one might make with respect to this claim. In fact, one might notice that $\bigcup_{D \in \Delta} \alpha_D(\Omega^m)$ joins $|\Delta|$ many sets, and Δ may admit the same power as Ω, and so have an uncountable inaccessible cardinality; accordingly our proof might require here some hypotheses on inaccessible cardinals. However we can avoid any reference to these deep set theoretic conditions by proceeding in the way we are going to describe. Let $\vartheta(\vec{v})$ be a formula defining X, fix an ω-saturated model $\mathcal{M}$ of T containing the parameters in $\vartheta(\vec{v})$, and rearrange the first two steps of the proof as follows.
First step: there exists a finite subset $D \subseteq \vartheta(\mathcal{M}^n)$ such that, for every $\vec{a} \in \Omega^n$,

$$D \cap \varphi(\Omega^n, \vec{a}) = \emptyset \ \Rightarrow \ RM(\varphi(\Omega^n, \vec{a}) \cap X) < \alpha.$$

Otherwise, for every natural i, one can find two tuples $\vec{a_i} \in \Omega^m$, $\vec{b_i} \in \Omega^n$ such that, for any $i, j \in \mathbf{N}$,

$\vec{b_i} \in \vartheta(\mathcal{M}^n)$ (in particular, $b_i \in M^n$),

$RM(\varphi(\Omega^n, \vec{a_i}) \cap X) = \alpha$,

$\Omega \models \neg\varphi(\vec{b_j}, \vec{a_i}) \iff j < i.$

Let us see how to introduce $\vec{a_i}$, $\vec{b_i}$ for a given natural i. Suppose $\vec{a_j}$, $\vec{b_j}$ given for all $j < i$. Then $D = \{\vec{b_j} : j < i\}$ is a finite subset of $\vartheta(\mathcal{M}^n)$, and consequently there is $\vec{a_i} \in \Omega^m$ for which

$$\forall j < i, \quad \vec{b_j} \notin \varphi(\Omega^n, \vec{a_i})$$

but

$$RM(\varphi(\Omega^n, \vec{a_i}) \cap X) = \alpha.$$

As $RM(X) = \alpha$ and $GM(X) = 1$, $RM(X \cap \bigcap_{j\leq i} \varphi(\Omega^n, \vec{a_j})) = \alpha$. By Lemma 7.5.3, as X is M-definable, there is some element $\vec{b_i} \in M^n$ in $X \cap \bigcap_{j\leq i} \varphi(\Omega^n, \vec{a_j})$ such that $\forall j \leq i \ \Omega \models \varphi(\vec{b_i}, \vec{a_j})$. This yields the tuples $\vec{a_i}$, $\vec{b_i}$ as required. But at this point it is easy to contradict the ω-stability (indeed the stability) of T. In fact the formula $\varphi'(\vec{v}, \vec{w}; \vec{v'}, \vec{w'}) : \neg\varphi(\vec{v}, \vec{w'})$ satisfies

$$\Omega \models \varphi'(\vec{b_j}, \vec{a_j}; \vec{b_i}, \vec{a_i}) \Leftrightarrow \Omega \models \varphi(\vec{b_i}, \vec{a_j}) \Leftrightarrow j < i$$

for every i and j in $\mathbf{N}$, and so obeys the order property.

In the same way one shows the next step.

Second step: for every $\vec{a} \in \Omega^m$ for which $RM(\varphi(\Omega^n, \vec{a}) \cap X) < \alpha$, there is a finite set $D(\vec{a})$ contained in $\vartheta(\mathcal{M}^n)$ such that $D(\vec{a}) \cap \varphi(\Omega^n, \vec{a}) = \emptyset$ and, for $\vec{a'} \in \Omega^m$,

$$D(\vec{a}) \cap \varphi(\Omega^n, \vec{a'}) = \emptyset \quad \Rightarrow \quad RM(\varphi(\Omega^n, \vec{a'}) \cap X) < \alpha.$$

It suffices to apply the same procedure as in the first step to $\vartheta(\mathcal{M}^n) - \varphi(\Omega^n, \vec{a})$ (and use the fact that $X - \varphi(\Omega^n, \vec{a})$ has the same Morley rank and degree as X).
At this point, just as said before, we can consider the **set** Δ of all the finite subsets $D \subseteq \vartheta(\mathcal{M}^n)$ such that, $\forall \vec{a} \in \Omega^m$,

$$D \cap \varphi(\Omega^n, \vec{a}) = \emptyset \quad \Rightarrow \quad RM(\varphi(\Omega^n, \vec{a}) \cap X) < \alpha.$$

Δ has power $< |\Omega|$, and $\bigcup_{D\in\Delta} \alpha_D(\Omega^m)$ (where, for every $D \in \Delta$, α_D is introduced as before, and hence is M-definable because $D \subseteq M^n$) equals the set of the tuples $\vec{a} \in \Omega^m$ for which $RM(\varphi(\Omega^n, \vec{a}) \cap X) < \alpha$. It remains to show

Third step: $\bigcup_{D\in\Delta} \alpha_D(\Omega^m)$ is definable.
To get this, first notice that what we have proved with respect to $\varphi(\vec{v}, \vec{w})$ applies to $\neg\varphi(\vec{v}, \vec{w})$ as well. So the set of the tuples $\vec{a} \in \Omega^m$ satisfying

$RM(X - \varphi(\Omega^n, \vec{a})) < \alpha$, that is $RM(\varphi(\Omega^n, \vec{a}) \cap X) = \alpha$, decomposes as a union of M-definable sets $\beta_G(\Omega^m)$ (where G ranges over a suitable set Γ). Accordingly the formulas $\alpha_D(\vec{w})$ with $D \in \Delta$ and $\beta_G(\vec{w})$ with $G \in \Gamma$ form an open covering of $S_m(M)$. By compactness we can extract a finite subcovering, and in particular we can find a finite subset $\Delta_0 \subseteq \Delta$ such that, $\forall \vec{a} \in \Omega^m$,

$$RM(\varphi(\Omega^n, \vec{a}) \cap X) < \alpha \Leftrightarrow \vec{a} \in \bigcup_{D \in \Delta_0} \alpha_D(\Omega^m).$$

As $\bigcup_{D\in\Delta_0} \alpha_D(\Omega^m)$ is definable, we are done. ♣

At last we are in a position to show:

Theorem 7.5.5 *Let T be an ω-stable theory. Then every type p over a small $A \subseteq \Omega$ is A-definable.*

Proof. Let $\vartheta(\vec{v})$ be a formula of p having the same Morley rank and degree as p. In particular, let α denote the rank of p. For every formula $\varphi(\vec{v}, \vec{w})$ in the language of T and $\vec{a} \in A^m$ (where m is the length of $\vec{w}$),

$$\varphi(\vec{v}, \vec{a}) \in p \Leftrightarrow \vartheta(\vec{v}) \wedge \varphi(\vec{v}, \vec{a}) \in p.$$

Moreover, if this is the case, then $\vartheta(\vec{v}) \wedge \varphi(\vec{v}, \vec{a})$ has the same Morley rank and degree as p, and so $RM(\vartheta(\Omega^n) - \varphi(\Omega^n, \vec{a})) < \alpha$. Conversely, $RM(\vartheta(\Omega^n) - \varphi(\Omega^n, \vec{a})) < \alpha$ implies that $\vartheta(\vec{v}) \wedge \neg\varphi(\vec{v}, \vec{a}) \notin p$, that is $\vartheta(\vec{v}) \wedge \varphi(\vec{v}, \vec{a}) \in p$. In conclusion, for every formula $\varphi(\vec{v}, \vec{w})$ and $\vec{a} \in A^m$,

$$\varphi(\vec{v}, \vec{a}) \in p \Leftrightarrow RM(\vartheta(\Omega^n) - \varphi(\Omega^n, \vec{a})) < \alpha.$$

By Lemma 7.5.4, the tuples $\vec{a} \in \Omega^m$ satisfying

$$RM(\vartheta(\Omega^n) - \varphi(\Omega^n, \vec{a})) < \alpha$$

(that is $RM(\vartheta(\Omega^n) \cap \varphi(\Omega^n, \vec{a})) = \alpha$) form an A-definable class. The $L(A)$-formula $d\varphi(\vec{w})$ defining this class is just what we are looking for. ♣

Another remarkable consequence of Lemma 7.5.4 is the following

Theorem 7.5.6 *Let p be a type over any model $\mathcal{M}$ of an ω-stable theory T. Then p has Morley degree* 1.

Proof. Let $\vartheta(\vec{v})$ be a formula of p having the same Morley rank α and the same Morley degree d as p. Accordingly there are d formulas $\vartheta_i(\vec{v}, \vec{x}_i)$ $(i < d)$ with parameters $\vec{x}_i$ in Ω such that

(i) $\vartheta(\Omega^n) \subseteq \bigcup_{i<d} \vartheta_i(\Omega^n, \vec{x_i})$,

(ii) $\vartheta_i(\Omega^n, \vec{x_j}) \cap \vartheta_j(\Omega^n, \vec{x_i}) = \emptyset$ for $j < i < d$,

(iii) $RM(\vartheta(\Omega^n) \cap \vartheta_i(\Omega^n, \vec{x_i})) = \alpha$ for every $i < d$.

The class of the sequences $\vec{x_i}$ in Ω satisfying (iii) for a given $i < d$ is M-definable because $\vartheta(\vec{v})$ has got its own parameters in M. Accordingly the existence of some sequences $\vec{x}_0, \ldots, \vec{x}_{d-1}$ satisfying (i), (ii) and (iii) can be expressed by a suitable first order sentence with parameters in M. Consequently these tuples $\vec{x}_0, \ldots, \vec{x}_{d-1}$ can be chosen in M. Correspondingly we can assume that, for every $i < d$, $\vartheta(\vec{v}, \vec{x_i})$ is a formula with parameters in M; so there is $i < d$ such that $\vartheta(\vec{v}, \vec{x_i}) \in p$, as $\bigvee_{i<d} \vartheta(\vec{v}, \vec{x_i}) \in p$. If $d > 1$, then (iii) implies $RM(\vartheta(\vec{v}) \wedge \vartheta_j(\vec{v}, \vec{x_j})) = \alpha$ for every $j < d$, $j \neq i$, and hence $GM(\vartheta(\Omega^n) \cap \vartheta_i(\Omega^n, \vec{x_i})) < d$. But this contradicts $\vartheta(\vec{v}) \wedge \vartheta_i(\vec{v}, \vec{x_i}) \in p$. So $d = 1$, as claimed. ♣

Now we turn our attention to simplicity. Our aim is to prove that ω-stable theories are simple. More specifically, we will show that, for an ω-stable theory T, the triples $(\vec{a}, B, A)$ such that $\vec{a}$ is a tuple in Ω, $B \supseteq A$ are small subsets of Ω and $RM(tp(\vec{a}/A)) = RM(tp(\vec{a}/B))$ form a good independence system of T. This confirms that T is simple, and proves also that, for $\vec{a}$, A and B as before, $\vec{a}$ is independent of B over A $\vec{a} \downarrow_A B$ if and only if $tp(\vec{a}/A)$ has the same Morley rank as its extension $tp(\vec{a}/B)$. Incidentally recall that, for any $\vec{a}$, A and B, $tp(\vec{a}/B)$ always includes $tp(\vec{a}/A)$, and consequently $RM(tp(\vec{a}/B)) \leq RM(tp(\vec{a}/A))$.

Theorem 7.5.7 *Let T be an ω-stable, and let I be the set of the triples $(\vec{a}, B, A)$ where $\vec{a}$ is a tuple in Ω, $A \subseteq B$ are small subsets of Ω and $RM(tp(\vec{a}/B)) = RM(tp(\vec{a}/A))$. Then I is a good independence system of T.*

Proof. We have to check that I satisfies the conditions (I1)-(I7) listed in Section 7.2.

(I1) is trivially true, as automorphisms preserve both Morley rank and Morley degree.

(I2) Fix $\vec{a}$, B, and pick a formula $\varphi(\vec{v})$ of $tp(\vec{a}/B)$ having the same Morley rank (and degree) as $tp(\vec{a}/B)$. Let A be the set of the parameters in $\varphi(\vec{v})$. Then A is finite (hence countable) and $\varphi(\vec{v})$ belongs to $tp(\vec{a}/A)$, whence the Morley rank of $tp(\vec{a}/A)$ equals that of $\varphi(\vec{v})$, and so that of $tp(\vec{a}/B)$.

Before (I3), let us treat (I6).

(I6) Fix $A \subseteq B \subseteq C$ and $\vec{a}$. Recall $RM(tp(\vec{a}/A)) \geq RM(tp(\vec{a}/B)) \geq RM(tp(\vec{a}/C))$. Consequently $RM(tp(\vec{a}/A))$ and $RM(tp(\vec{a}/C))$ coincide if and only if both these ranks equal $RM(tp(\vec{a}/B))$.

(I3) Take $A \subseteq B$ e $\vec{a}$. If $RM(tp(\vec{a}/A)) = RM(tp(\vec{a}/B))$, then we can apply what we have just checked in (I6) and deduce $RM(tp(\vec{a}/A \cup \vec{b})) = RM(tp(\vec{a}/A))$ for every $\vec{b} \in B$.
Conversely suppose $RM(tp(\vec{a}/A)) > RM(tp(\vec{a}/B))$. Choose a formula $\varphi(\vec{v}, \vec{b})$ in $tp(\vec{a}/B)$ having the same Morley rank as $tp(\vec{a}/B)$. Look at $tp(\vec{a}/A \cup \vec{b})$: this type includes $\varphi(\vec{v}, \vec{b})$ and consequently has the same Morley rank as $tp(\vec{a}/B)$, and anyhow a Morley rank smaller than $tp(\vec{a}/A)$.

(I4) Now we have to show that, for every choice of $A \subseteq B$ and $\vec{a}$, there is a tuple $\vec{a'}$ having the same type over A as $\vec{a}$ and satisfying $RM(tp(\vec{a'}/A)) = RM(tp(\vec{a'}/B))$. In other words, we claim that every type p over A has some extension q in $S(B)$ having the same Morley rank (to realize this, put $p = tp(\vec{a'}/A)$, $q = tp(\vec{a'}/B)$). Let α be the Morley rank of p. Choose a formula $\varphi(\vec{v})$ of p having the same Morley rank α and the same Morley degree as p. We know that there do exist (finitely many) types q in $S(B)$ containing $\varphi(\vec{v})$ and having rank α; indeed the sum of their Morley degrees equals the degree of $\varphi(\vec{v})$, and so the degree of p. It suffices to show that the extensions of p in $S(B)$ with rank α coincide with these types. Now it is obvious that any extension of p contains $\varphi(\vec{v})$. On the other side, take any type q over B containing $\varphi(\vec{v})$ and having rank α. Pick $\vartheta(\vec{v}) \in p$, then $\varphi(\vec{v}) \wedge \vartheta(\vec{v})$ is also in p, and consequently has the same Morley rank and degree as p. Accordingly the Morley rank of $\varphi(\vec{v}) \wedge \neg\vartheta(\vec{v})$ is less than α and $\varphi(\vec{v}) \wedge \neg\vartheta(\vec{v})$ cannot belong to p. So neither $\neg\vartheta(\vec{v})$ is in p. As q is complete, q includes $\vartheta(\vec{v})$. In conclusion $p \subseteq q$.
In particular observe that, if $\mathcal{M}$ is a model of T and B is a small subset of Ω including M, then every type p over M has a unique extension q over B with the same Morley rank. In fact, as we saw before in this section, the Morley degree of p is 1. This remark is useful to show (I7).

(I7) Let $\mathcal{M}$ be a model of T, B and B' be two small sets including M. Owing to the last observation, any type p over M has a unique extension with the same Morley rank over any set extending M, in particular over B, over B' and over $B \cup B'$. A tuple realizing the last type (over $B \cup B'$) must satisfy the only extension of p of the same rank both over B and over B'.

It remains to show (I5).

(I5) Let A be a small subset of Ω, $\vec{a}$, $\vec{b}$ be two sequences in Ω. We have to prove that $RM(tp(\vec{a}/A \cup \vec{b})) = RM(tp(\vec{a}/A))$ implies $RM(tp(\vec{b}/A \cup \vec{a})) =$

$RM(tp(\vec{b}/A))$. Let n, m denote the length of $\vec{a}$, $\vec{b}$ respectively.
First assume that $A = M$ is the domain of an $\aleph_0$-saturated model of T. Let $\varphi(\vec{v})$ be a formula of $tp(\vec{a}/M)$ with the same Morley rank α as $tp(\vec{a}/M)$ (and degree 1), and let $\psi(\vec{v})$ be a formula of $tp(\vec{b}/M)$ with the same Morley rank β as $tp(\vec{b}/M)$ (and degree 1). Suppose towards a contradiction $RM(tp(\vec{b}/M \cup \vec{a})) < \beta$. Then there is a formula $\chi(\vec{v}, \vec{w}) \in tp(\vec{a}, \vec{b}/M)$ such that $\chi(\vec{a}, \Omega^m)$ has Morley rank $< \beta$. Without loss of generality we can assume

$$\Omega \models \forall \vec{v} \forall \vec{w} (\chi(\vec{v}, \vec{w}) \to \varphi(\vec{v}) \wedge \psi(\vec{w}));$$

otherwise we replace $\chi(\vec{v}, \vec{w})$ by its conjunction with $\varphi(\vec{v}) \wedge \psi(\vec{w})$. The set of the tuples $\vec{c}$ in Ω for which $RM(\chi(\vec{c}, \Omega^m)) < \beta$ (equivalently $RM(\psi(\Omega^m) \cap \chi(\vec{c}, \Omega^m)) < \beta$) is M-definable because $\psi(\Omega^m)$ is. Furthermore $RM(\chi(\vec{a}, \Omega^m)) < \beta$. Accordingly we can assume $RM(\chi(\vec{c}, \Omega^m)) < \beta$ for every tuple $\vec{c}$ in Ω such that $\chi(\vec{c}, \Omega^m) \neq \emptyset$ (just replace $\varphi(\vec{v})$ by $\varphi(\vec{v}) \wedge$ "$RM(\chi(\vec{v}, \Omega^m)) < \beta$"). It follows $\chi(\Omega^n, \vec{b}) \cap M^n = \emptyset$, otherwise, if $\vec{a'}$ is any element in this intersection, then $RM(\chi(\vec{a'}, \Omega^m) \geq \beta$ as $\chi(\vec{a'}, \vec{w}) \in tp(\vec{b}/M)$. By Lemma 7.5.3, $RM(\chi(\Omega^n, \vec{b})) < RM(\varphi(\Omega^n)) = \alpha$ because $\chi(\Omega^n, \vec{b}) \subseteq \varphi(\Omega^n)$, but $\chi(\Omega^n, \vec{b}) \cap M^n = \emptyset$. On the other side $\chi(\vec{v}, \vec{b}) \in tp(\vec{a}/M \cup \vec{b})$, so $RM(tp(\vec{a}/M \cup \vec{b})) < \alpha = RM(tp(\vec{a}/M))$ (a contradiction).
Now let A be any small subset of Ω. Let $\mathcal{M}$ be an $\aleph_0$-saturated model of T extending A. We can assume $RM(tp(\vec{b}/M)) = RM(tp(\vec{b}/A))$ and $RM(tp(\vec{a}/M \cup \vec{b})) = RM(tp(\vec{a}/A \cup \vec{b}))$. In fact, if this is not the case, then use (I4) and get a tuple $\vec{b'}$ realizing $tp(\vec{b}/A)$ and satisfying $RM(tp(\vec{b'}/A)) = RM(tp(\vec{b}/A))$. As $\vec{b}$, $\vec{b'}$ have the same type over A, there is an A-automorphism f of Ω mapping $\vec{b}$ into $\vec{b'}$. Of course $f(\vec{a})$ has the same type over A as $\vec{a}$. Now use again (I4) and replace $f(\vec{a})$ by a tuple $\vec{a'}$ having the same type over $A \cup \vec{b'}$ as $f(\vec{a})$ and satisfying $RM(tp(\vec{a'}/M \cup \vec{b'})) = RM(tp(\vec{a}/A \cup \vec{b'}))$. As before, there is an automorphism g of Ω fixing $A \cup \vec{b'}$ pointwise and mapping $f(\vec{a})$ into $\vec{a'}$. So gf is an A-automorphism sending $\vec{a}$ into $\vec{a'}$ and $\vec{b}$ into $\vec{b'}$. Owing to (I1), if $\vec{a'}$ and $\vec{b'}$ satisfy our claim, then the same holds for $\vec{a}$ and $\vec{b}$. In conclusion, we can assume $RM(tp(\vec{b}/M)) = RM(tp(\vec{b}/A))$ and $RM(tp(\vec{a}/M \cup \vec{b})) = RM(tp(\vec{a}/A \cup \vec{b}))$, as said before. Now recall what our hypothesis says, that is $RM(tp(\vec{a}/A \cup \vec{b})) = RM(tp(\vec{a}/A))$. Furthermore $RM(tp(\vec{a}/M \cup \vec{b})) = RM(tp(\vec{a}/A \cup \vec{b}))$. Hence $RM(tp(\vec{a}/M \cup \vec{b})) = RM(tp(\vec{a}/A))$ and consequently $RM(tp(\vec{a}/M \cup \vec{b})) = RM(tp(\vec{a}/M))$. Owing to what we showed before when dealing with an $\aleph_0$-saturated model, $RM(tp(\vec{b}/M \cup \vec{a})) = RM(tp(\vec{b}/M))$. As $RM(tp(\vec{b}/M)) = RM(tp(\vec{b}/A))$, it follows $RM(tp(\vec{b}/M \cup \vec{a})) = RM(tp(\vec{b}/A))$, whence $RM(tp(\vec{b}/A \cup \vec{a})) = RM(tp(\vec{b}/A))$. This accomplishes the proof of (I5) and the whole theorem.
♣

At this point we can also check (I8) and confirm once again in this way that an ω-stable theory T is stable. Indeed, given any type p over a small $A \subseteq \Omega$ and $B \supseteq A$, the number of non-forking extensions of p in $S(B)$ cannot exceed the Morley degree of p, and so is anyhow finite. So, when A is the domain of some model of T, (I8) is just a rephrasing of Theorem 7.5.6: in fact, the Morley degree of p is 1 in this case.
At this point, it is immediate to deduce also that ω-stability implies superstability.

Corollary 7.5.8 *An ω-stable theory T is superstable.*

Proof. T is stable, and the set A in (I2) is finite. ♣

Now we want to discuss in our ω-stable setting the concepts of heir and coheir introduced in Section 7.3, and to show their equivalence with the notion of non-forking extension. Let us recall briefly how heir and coheir are defined. First we fix our framework: T is a(n ω-stable) theory, $\mathcal{M}$ is a model of T, B is a small subset of Ω including M, n is a positive integer, p is an n-type over M and q is an extension of p in $S_n(B)$. In this setting

- q is an *heir* of p if and only if, for every formula $\vartheta(\vec{v}, \vec{b}) \in q$, there exists a tuple $\vec{a}$ in M such that $\vartheta(\vec{v}, \vec{a}) \in p$,
- q is a *coheir* of p if and only if, for every formula $\vartheta(\vec{v}, \vec{b}) \in q$, $\vartheta(\mathcal{M}^n, \vec{b}) \neq \emptyset$.

We are going to show the equalities *heir* = *coheir* = *non-forking extension* among the types over a small set B extending a given type p over a model $\mathcal{M}$ of T. First let us recall what we showed before in this section, namely that a type p over any small set A is A-definable and so, for every L-formula $\vartheta(\vec{v}, \vec{w})$, finds an $L(A)$-formula $d\vartheta(\vec{w})$ such that a tuple $\vec{a}$ in A satisfies $d\vartheta(\vec{w})$ if and only if $\vartheta(\vec{v}, \vec{a}) \in p$; indeed, if $\varphi(\vec{v})$ is a formula of p with the same Morley rank and degree as p, then $d\vartheta(\vec{w})$ defines the set of all tuples $\vec{a}$ in A for which $RM(\varphi(\vec{v}) \wedge \neg\vartheta(\vec{v}, \vec{a})) < RM(p)$. But now, after seeing how independence in T is characterized (by the Morley rank), we can say even more. In fact, suppose that $A = M$ is the domain of some model $\mathcal{M}$ of T and let B be a small set including M. Then p (and $\varphi(\vec{v})$) have Morley degree 1, and so there is a unique non-forking extension q of p over B, and $\varphi(\vec{v})$ is a formula of q having its Morley rank and degree. Accordingly, for every L-formula $\vartheta(\vec{v}, \vec{w})$, the $L(M)$-formula $d\vartheta(\vec{w})$ still satisfies $\Omega \models d\vartheta(\vec{b}) \Leftrightarrow \vartheta(\vec{v}, \vec{b}) \in q$ for every $\vec{b}$ in B.
Now we can state our theorem (we are still keeping here the notation introduced a few lines ago).

Theorem 7.5.9 *The following propositions are equivalent:*

(i) *q does not fork over A;*

(ii) *q is an heir of p;*

(iii) *q is a coheir of p.*

In particular the heir (and coheir) of p over B is unique.

Proof. Let $q \in S(B)$ be a non-forking extension of p. As already underlined, q is unique as p has Morley degree 1, and, for every L-formula $\vartheta(\vec{v}, \vec{w})$, there is an $L(M)$-formula $d\vartheta(\vec{w})$ such that, for every $\vec{b}$ in B,

$$\vartheta(\vec{v}, \vec{b}) \in q \Leftrightarrow \Omega \models d\vartheta(\vec{b})$$

(and, for every $\vec{a}$ in M, $\vartheta(\vec{v}, \vec{a}) \in p$ if and only if $\mathcal{M} \models d\vartheta(\vec{a})$). So suppose that, for some $\vec{b}$ in Ω, $\vartheta(\vec{v}, \vec{b}) \in q$. Then $\Omega \models \exists \vec{w}\, d\theta(\vec{w})$, and consequently $\mathcal{M} \models \exists \vec{w}\, d\theta(\vec{w})$ because $d\theta(\vec{w})$ is an $L(M)$-formula. Let $\vec{a} \in M$ satisfy $\mathcal{M} \models d\theta(\vec{a})$, then $\vartheta(\vec{v}, \vec{a})$ is in p and in q. So q is an heir of p. Now suppose that $q' \in S(B)$ is another heir of p. There are a formula $\vartheta(\vec{v}, \vec{w})$ in L and a tuple $\vec{b}$ in B such that $\vartheta(\vec{v}, \vec{b}) \in q - q'$; in particular $\Omega \models d\vartheta(\vec{b})$, and hence

$$\neg\vartheta(\vec{v}, \vec{a'}) \wedge d\vartheta(\vec{a'}) \in p.$$

So $\mathcal{M} \models d\vartheta(\vec{a'})$, while $\vartheta(\vec{v}, \vec{a'}) \notin p$: this is a contradiction. It follows that q is the only heir of p in $S(B)$. This implies that (i) and (ii) are equivalent. Now we check (ii) $\Leftrightarrow$ (iii). First suppose $B - M$ finite; let $\vec{b}$ list the elements of $B - M$, and let $\vec{a}$ realize q. Observe that

$tp(\vec{a}/M \cup \vec{b})$ is an heir of $tp(\vec{a}/M)$

if and only if

for every $L(M)$-formula $\varphi(\vec{v}, \vec{w})$ satisfying $\Omega \models \varphi(\vec{a}, \vec{b})$, there exists $\vec{m} \in M$ such that $\Omega \models \varphi(\vec{a}, \vec{m})$

and hence if and only if

$tp(\vec{b}/M \cup \vec{a})$ is a coheir of $tp(\vec{b}/M)$.

So the equivalence (i) $\Leftrightarrow$ (ii) and (I5) imply that $tp(\vec{a}/M \cup \vec{b})$ is an heir of $tp(\vec{a}/M)$ if and only if $tp(\vec{b}/M \cup \vec{a})$ is a coheir of $tp(\vec{b}/M)$. This establishes the equivalence of (ii) and (iii) when $B - M$ is finite. To extend this conclusion to the general setting, it suffices to observe that q is an heir (a coheir) of p

if and only if, for every finite subset B_0 of B, the restriction of p to $M \cup B_0$ is an heir (a coheir respectively) of p. ♣

At this point we can complete what we observed in 6.2 about the connected component of an ω-stable group.

Theorem 7.5.10 *Let $\mathcal{G}$ be an ω-stable group. Then the connected component G^0 of $\mathcal{G}$ has Morley degree 1.*

Proof. Take two types p, q over G having the same Morley rank as G and containing the formula $\varphi^0(v)$ defining G^0. Let $x \models p$, $y \models q$, $x \downarrow_G y$ (namely $RM(tp(x/G)) = RM(tp(x/G \cup \{y\}))$). For every formula $\varphi(v, \vec{w})$ and $\vec{a}$ in G, $\varphi(v, \vec{a}) \in tp(yx/G \cup \{y\})$ if and only if $\varphi(yv, \vec{a}) \in tp(x/G \cup \{y\})$ (or also $\varphi^0(y) \wedge \varphi(yv, \vec{a}) \in tp(x/G \cup \{y\})$). As $tp(x/G \cup \{y\})$ is the heir of $tp(x/G)$, there is some $b \in G^0 = \varphi^0(G)$ such that $\varphi(bv, \vec{a}) \in tp(x/G) = p$. As G^0 is the left stabilizer of p, $\varphi(v, \vec{a}) \in p$. Hence the type of yx over G coincides with p. But just inverting the roles of x and y one deduces in the same way that yx realizes q. So $p = q$. ♣

As said before, ω-stability implies superstability. Recall that, roughly speaking, the superstable theories T are those ensuring that every proper elementary extension $\mathcal{M} < \mathcal{N}$, $M \neq N$ between models of T realizes some regular type over M; and the regular types are just the non-algebraic types p over M such that $\not\downarrow_A$ satisfies (D3), and consequently (D1) - (D4), in $p(\Omega)$.
Our aim is to examine closerly these matters in the ω-stable framework, where an elegant approach proposed by Lascar allows a simpler treatment. Indeed stronger tools are available in the restricted ω-stable setting; for instance, we will use over and over again (existence and uniqueness of) prime models over subsets below: this is a feature of ω-stable theories which may fail in arbitrary superstable theories. Accordingly fix an ω-stable T and a model $\mathcal{M}$ of T; we wish to examine types over M. By the way, let us introduce the following notation: for X a small subset of Ω, let $\mathcal{M}(X)$ denote the model of T prime over $M \cup X$; when X is the domain of some finite tuple $\vec{a}$, we write $\mathcal{M}(\vec{a})$ instead of $\mathcal{M}(X)$; notice that, if $\vec{a}$ and $\vec{a'}$ have the same type over $\mathcal{M}$, then $\mathcal{M}(\vec{a})$ and $\mathcal{M}(\vec{a'})$ are isomorphic by a function fixing M pointwise and mapping $\vec{a}$ into $\vec{a'}$; in other words, the isomorphism type of $\mathcal{M}(\vec{a})$ over $\mathcal{M}$ does not depend directly on $\vec{a}$, but on the type of $\vec{a}$ over M.
What we first need in our analysis is a criterion comparing types p and q over M, and measuring whether realizing p is easier or harder than realizing q. This is provided by the so called *Rudin-Keisler relation* $\geq_{RK}$ (although it

was Lascar who introduced this notion, faintly inspired by a Rudin-Keisler order relation concerning ultrafilters). Here is its definition.

Definition 7.5.11 *Let p and q in $S(M)$. We put $p \geq_{RK} q$ if and only if, for every tuple $\vec{a}$ satisfying p, the model $\mathcal{M}(\vec{a})$ prime over $M \cup \vec{a}$ contains some realization $\vec{b}$ of q, and we put $p \sim_{RK} q$ if and only if $p \geq_{RK} q$ and $q \geq_{RK} p$.*

Notice that, owing to what we said about $\mathcal{M}(\vec{a})$ before, we could equivalently write "for some $\vec{a}$" instead of "for every $\vec{a}$" in this definition.
$\geq_{RK}$ is a preorder relation: it is reflexive and transitive, but it is not antisymmetric. So $\sim_{RK}$ is an equivalence relation and $\geq_{RK}$ determines an order relation in the quotient set $S(M)/\sim_{RK}$ among $\sim_{RK}$-classes of types. Roughly speaking, $p \geq_{RK} q$ means that, wherever p is realized, q is realized as well; and consequently $p \sim_{RK} q$ says that p and q are realized by the same models extending $\mathcal{M}$.
It is also clear that the types over M minimal with respect to $\geq_{RK}$ are the algebraic ones, as they are already satisfied in M. But they do not interest us, so let us exclude them and call a type p over M *RK-minimal* when it is minimal with respect to $\geq_{RK}$ among non-algebraic types.

Definition 7.5.12 *A non-algebraic type $p \in S(M)$ is RK-**minimal** if it is $\sim_{RK}$-equivalent to every non-algebraic type q over M such that $p \geq_{RK} p$.*

Notice that every RK-minimal type p is $\sim_{RK}$-equivalent to some 1-type. In fact take $\vec{a} \models p$; as p is not algebraic, there exists some a in $\vec{a}$ out of M; then $tp(a/M)$ is not algebraic and $tp(a/M) \leq_{RK} p$, whence $tp(a/M) \sim_{RK} p$ by RK-minimality.
There is another relevant relation between types arising in this setting, and closely connected with $\sim_{RK}$; it is called *orthogonality* and is defined as follows.

Definition 7.5.13 *Let p and q in $S(M)$. We say that p is* **orthogonal** *to q and we write $p \perp q$ if and only if, for every $\vec{a} \models p$ and $\vec{b} \models q$, $\vec{a} \downarrow_M \vec{b}$.*

Notice that $\perp$ is symmetric. Roughly speaking, $p \perp q$ means that realizing p and realizing q are "independent" facts. Here are some simple examples concerning $\geq_{RK}$, $\sim_{RK}$ and $\perp$.

Examples 7.5.14 1. Let $\mathcal{K}$ be an algebraically closed field. Recall that there is a unique non-algebraic 1-type over K, that of the transcendental elements over K. More generally look at arbitrary types p and

q over K. Let $\vec{a} \models p$ and $\vec{b} \models q$ (where $\vec{a}$, $\vec{b}$ may have different lengths). Then $p \leq_{RK} q$ if and only if the transcendence degree of $\mathcal{K}(\vec{a})$ over $\mathcal{K}$ is not smaller than the transcendence degree of $\mathcal{K}(\vec{b})$ over $\mathcal{K}$, and $p \sim_{RK} q$ if and only if these transcendence degrees coincide. In particular the Rudin-Keisler relation determines a total order in the quotient set $S(K)/\sim_{RK}$, and the unique non-algebraic 1-type over K is the only RK-minimal type (up to $\sim_{RK}$).

Notice that the same analysis can be repeated in any strongly minimal structure $\mathcal{M}$ (with respect to the *acl* dependence relation): for p, $q \in S(M)$, $\vec{a} \models p$ and $\vec{b} \models q$, $p \leq_{RK} q$ if and only if the dimension of $\vec{a}$ over M is not smaller than the dimension of $\vec{b}$ over M, and $p \sim_{RK} q$ if and only if these dimensions coincide. So the Rudin-Keisler order in $S(M)/\sim_{RK}$ is still linear, and the only RK-minimal class if that of the unique non-algebraic 1-type over M.

2. Take the theory T of an equivalence relation E with infinitely many infinite classes (and no finite class). T is complete (why?). Let us describe the non-algebraic 1-types over a model $\mathcal{M} = (M, E)$ of T. First, for every $a \in M$, we have to consider the 1-type p_a of a new element in the E-class of a; moreover there is another 1-type q, concerning the elements in a new E-class; no further 1-type arises. In particular $|S_1(M)| = |M|$, and T is ω-stable. Notice also that, for every type p listed before and $x \models p$, the domain of $\mathcal{M}(x)$ is just $M \cup \{x\}$ ($M \cup x/E$ with a countable x/E in the case of q). Accordingly these types are pairwise $\sim_{RK}$-inequivalent and RK-minimal. Notice also that they are pairwise orthogonal.

Now let us introduce *strongly regular* types. As we will see below, they can reasonably replace regular types in the ω-stable framework. $\mathcal{M}$ still denotes a model of a countable ω-stable theory T.

Definition 7.5.15 *A 1-type p over M is called* **strongly regular** *if p is not algebraic and there is some formula $\varphi(v) \in p$ such that, for every $a \models p$ and $b \in M(a) - M$, $b \models p$ if and only if $\models \varphi(b)$.*

Notice that, owing to what we observed before about $\mathcal{M}(a)$, we could say "for some $a \models p$" instead of "for every $a \models p$" in this definition. Now let us propose some examples.

Examples 7.5.16 1. The only non-algebraic 1-type over an algebraically closed field, or also over an arbitrary strongly minimal structure, is strongly regular: $v = v$ works as $\varphi(v)$.

2. Take a structure (M, E) where E is an equivalence relation with infinitely many infinite classes and no finite class, as in Example 7.5.13, 2. Each type listed in 7.5.13, 2 is strongly regular, p_a because of $E(v, a)$ and q because of $v = v$.

The basic fact about strongly regular types is:

Theorem 7.5.17 *Let T be ω-stable, $\mathcal{M}$ and $\mathcal{N}$ be models of T such that $\mathcal{M}$ is an elementary substructure of $\mathcal{N}$ and $M \neq N$. Then there is some $a \in N - M$ for which $tp(a/M)$ is strongly regular.*

Proof. Let $a \in N - M$ be such that $p = tp(a/M)$ has a minimal Morley rank, and let the formula $\varphi(v)$ isolate p among the types of the same Morley rank in $S_1(M)$; in particular $\varphi(v)$ has the same Morley rank as p. We claim that $\varphi(v)$ makes p strongly regular. In fact $\varphi(v) \in p$. Furthermore, let $b \in \varphi(M(a)) - M$; owing to the choice of a, $RM(tp(b/M)) \geq RM(tp(a/M))$; as $\varphi(v) \in tp(b/M)$, $RM(tp(b/M)) = RM(tp(a/M))$ and even $tp(b/M) = tp(a/M)$. ♣

Corollary 7.5.18 *For every non-algebraic $q \in S(M)$ there is some strongly regular type $p \in S_1(M)$ for which $p \leq_{RK} q$. In particular every RK-minimal q is $\sim_{RK}$-equivalent to some strongly regular p.*

Proof. Let $\vec{b} \models q$. As q is not algebraic, $\mathcal{M}(\vec{b})$ properly extends $\mathcal{M}$. So apply the previous theorem and gets some strongly regular p realized in $M(\vec{b})$: $p \leq_{RK} q$. ♣

On the other side

Theorem 7.5.19 *Every strongly regular $p \in S_1(M)$ is RK-minimal.*

Proof. This needs a more laborious approach, and some preliminary steps which perhaps have their own intrinsic interest. First let us fix our setting. There is some $L(M)$-formula $\varphi(v)$ witnessing that p is strongly regular. We take a non-algebraic $q \in S(M)$ satisfying $q \leq_{RK} p$ and we have to show that $q \sim_{RK} p$. There is no loss of generality in assuming $q \in S_1(M)$. Given $a \models p$, there is some $b \in M(a) - M$ realizing q. It is an easy exercise to realize that $b \not\downarrow_M M(a)$ (just compare the Morley rank of the types of b over $M(a)$ and over M). The first preliminary step states that actually $b \not\downarrow_M a$. In fact we have:

Step 1. Let $a \notin M$, $b \downarrow_M a$. Then $b \downarrow_M M(a)$.

Notice that we do not use here the hypothesis that p is strongly regular. What we want to show is that $tp(b/M(a))$ does not fork over M, more precisely is the heir of $tp(b/M)$. Accordingly take a formula $\chi(v, \vec{c})$ in $tp(b/M(a))$ with parameters $\vec{c}$ in $M(a)$. The type of $\vec{c}$ over $M \cup \{a\}$ is isolated, say by $\eta(\vec{w}, a)$. We claim that $\eta(\vec{w}, a)$ isolates $tp(\vec{c}/M \cup \{a, b\})$ as well. Otherwise there is some tuple $\vec{c'}$ such that $\models \eta(\vec{c'}, a)$ but $tp(\vec{c}/M \cup \{a, b\}) \neq tp(\vec{c'}/M \cup \{a, b\})$, and so there exists some formula $\psi(\vec{w}, a, b)$ (possibly with further parameters from M) for which $\models \psi(\vec{c}, a, b) \wedge \neg\psi(\vec{c'}, a, b)$. Consequently $\exists \vec{w} \exists \vec{w'}(\eta(\vec{w}, v) \wedge \eta(\vec{w'}, v) \wedge \psi(\vec{w}, v, b) \wedge \neg\psi(\vec{w'}, v, b))$ is in $tp(a/M \cup \{b\})$. As $a \downarrow_M b$ and so $tp(a/M \cup \{b\})$ is the heir of $tp(a/M)$, there is some $m \in M$ satisfying

$$\exists \vec{w} \exists \vec{w'}(\eta(\vec{w}, v) \wedge \eta(\vec{w'}, v) \wedge \psi(\vec{w}, v, m) \wedge \neg\psi(\vec{w'}, v, m)) \in tp(a/M);$$

but this gives a contradiction because $\eta(\vec{w}, a)$ isolates a single type over $M \cup \{a\}$. Accordingly $\eta(\vec{w}, a)$ isolates the type of $\vec{c}$ also over $M \cup \{a, b\}$. This type contains $\chi(b, \vec{w})$, so

$$\forall \vec{w}(\eta(\vec{w}, a) \to \chi(v, \vec{w})) \wedge \exists \vec{w} \eta(\vec{w}, a) \in tp(b/M \cup \{a\}),$$

whence, as before, $\chi(v, \vec{m}) \in tp(b/M)$ for some $\vec{m}$ in M. In conclusion, $tp(b/M(a))$ is the heir of $tp(b/M)$, as claimed.

The second preliminary step proves:
Step 2. $\varphi(\mathcal{M}(b)) \neq \varphi(\mathcal{M})$.
Otherwise $a \downarrow_M b$, which contradicts the previous conclusion. In fact, let $\theta(v, a)$ be a formula in $tp(b/M \cup \{a\})$. As $b \in M(a)$, $\mathcal{M}(a) \models \theta(b, a)$, and consequently $\mathcal{M}(a) \models \exists w(\theta(b, w) \wedge \varphi(w))$ (a sentence with parameters in $M \cup \{b\}$). As $\mathcal{M}(b)$ is an elementary substructure of $\mathcal{M}(a)$, even $\mathcal{M}(b)$ satisfies this sentence. So $\models \theta(b, c)$ for some $c \in \varphi(\mathcal{M}(b)) \subseteq M$. Hence $\theta(v, c) \in tp(b/M)$. In conclusion, $tp(b/M \cup \{a\})$ is the heir of $tp(b/M)$.

Now we can conclude our proof. Let $d \in \varphi(\mathcal{M}(b)) - M$; so $d \in \varphi(\mathcal{M}(a))$ as well, and consequently $tp(d/M)$ equals p. As $d \in M(b)$, $p \leq_{RK} q$, whence $p \sim_{RK} q$. ♣

Using more or less the same technical premises one shows also the following notable characterization of $\sim_{RK}$ within RK-minimal types.

Theorem 7.5.20 *Two RK-minimal types over M are $\sim_{RK}$-equivalent if and only if they are orthogonal.*

We omit the details. There are other relevant things to underline about strongly regular types. First of all, strongly regular types are regular as well: checking this requires the same ingredients as in the last theorem and some equivalent characterizations of both the involved notions. Again, we omit a full treatment of this point (this section is going to be quite long, and actually we already proved about strongly regular types what we will need in the rest of the chapter). However, it is worthy pointing out that there do exist even in the ω-stable setting some regular types which are not strongly regular (and indeed are not $\sim_{RK}$-equivalent to any strongly regular type). There is another point in this framework deserving a few words. In fact, $\leq_{RK}$ (and consequently $\sim_{RK}$) as well $\perp$ are preserved under passing from types p, q over a model $\mathcal{M}$ of an ω-stable theory T to their non-forking extensions over some elementary extension $\mathcal{M}'$: $p \leq_{RK} q$, $p \sim_{RK} q$, $p \perp q$ if and only if $p|M' \leq_{RK} q|M'$, $p|M' \sim_{RK} q|M'$, $p|M' \perp q|M'$ respectively. The same can be said about RK-minimality. With respect to strongly regular types, one sees that, if $\mathcal{M} < \mathcal{M}'$, p is a 1-type over M and $\varphi(v)$ is an $L(M)$-formula making p strongly regular, then the same formula makes $p|M'$ strongly regular as well.

The last matter we want to deal with in this section still concerns non-forking extensions and ω-stable theories. It is a theorem (actually valid even for stable theories) which will turn out to be a useful tool in the proof of Shelah's Uniqueness Theorem in Section 7.7. It is generally called the *Open Mapping Theorem*, a name clearly evoking topology; of course, the topological spaces we are concerned with are the Boolean spaces of types over small subsets of Ω. In particular, we consider two small subsets $A \subseteq B$ in an ω-stable theory T, and we look at the spaces $S(A)$ and $S(B)$. It is easy to realize that the restriction map r from $S(B)$ in $S(A)$, sending every type q over B into its restriction to A (so into the set of the $L(A)$-formulas of q) is continuous and surjective: continuity follows from the trivial remark that $L(A) \subseteq L(B)$, hence $L(A)$-formulas are $L(B)$-formulas as well, and any open set in $S(A)$ can be regarded as the image under r of some open set in $S(B)$; surjectivity is obvious.
Now consider the set of the types q over B that do not fork over A. Let $N(B, A)$ denote this set. First notice

Proposition 7.5.21 *$N(B, A)$ is a closed subset of $S(B)$.*

Proof. The claim is obvious when A is the domain of some model $\mathcal{A}$ of T. In fact we know that, for every $q \in S(B)$, $q \in N(B, A)$ if and only if q is the coheir of its restriction to A, and it is easy to realize that the last

condition just ensures that q is in the closure of the set of the types over B of the tuples in A. Accordingly $N(B, A)$ equals this closed set.
Now let A be any subset of B. Use extension (I3) and compactness, and build a model $\mathcal{M}$ of T independent of B over A. Notice that every type in $N(B, A)$ has a non-forking extension over $B \cup M$, and so by transitivity enlarges to some type in $N(B \cup M, A)$, which, again by transitivity, lies also in $N(B \cup M, M)$. On the other hand, every type $q \in N(B \cup M, M)$ restricts to a type in $N(B, A)$. In fact, let $\vec{c}$ realize q, hence let $\vec{c} \downarrow_M B$. Use symmetry and deduce $B \downarrow_M M \cup \vec{c}$. As $B \downarrow_A M$, transitivity yields $B \downarrow_A M \cup \vec{c}$ and consequently $B \downarrow_A A \cup \vec{c}$. Use symmetry and deduce $\vec{c} \downarrow_A B$, in other words that the type of $\vec{c}$ does not fork over A, as claimed. Then the restriction map from $S(B \cup M)$ onto $S(B)$ - a continuos function between compact spaces - sends $N(B \cup M, M)$ just onto $N(B, A)$. As the former set is closed, its image is, too. ♣

Now we are in a position to prove the Open Mapping Theorem.

Theorem 7.5.22 *Let $A \subseteq B$ be small sets, r denote the restriction map from $N(B, A)$ onto $S(A)$. Then r is open. In other words, for every $L(B)$-formula $\theta(\vec{v})$ there is an $L(A)$-formula $\theta'(\vec{v})$ such that a type q over B non-forking over A contains $\theta(\vec{v})$ if and only its restriction to A includes $\theta'(\vec{v})$.*

Proof. There is no loss of generality in replacing B by a model $\mathcal{M}$ of T elementarily including B and saturated in some power $> |A|$. In fact, let r', r'' denote the restriction maps from $N(M, A)$ onto $S(A)$ and from $N(M, A)$ in $S(B)$ respectively; it is an easy exercise to check that the latter function maps $N(M, A)$ onto $N(B, A)$ and that r' is just the composition of r'' and r. Consequently, if r' is open, then r is, as r'' is continuous. Hence we can replace B by $\mathcal{M}$, as claimed; accordingly, let r denote from now on the restriction map from $N(M, A)$ onto $S(A)$. Let U be an open set of $N(M, A)$. We can assume that U is the set of the types over M which do not fork over A and include a given formula $\varphi(\vec{v}, \vec{b})$ with parameters $\vec{b}$ from M. We have to show that $r(U)$ is open. Let $U_\star$ denote $r^{-1}r(U)$ (so the set of the types $q \in N(M, A)$ having the same restriction to A as some type in U); elementary topology ensures that it suffices to prove that $U_\star$ is open: in fact, in this case, the complement $U_\star'$ of U is closed, as well as its image $r(U_\star') = (r(U))'$ because r is closed; whence $r(U)$ is open, as claimed. To show that $U_\star$ is open, first observe that $U_\star$ equals $\bigcup_g \{q \in N(M, A) : \varphi(\vec{v}, g(\vec{b})) \in q\}$ where g ranges over the automorphisms of $\mathcal{M}$ fixing A pointwise. $\supseteq$ is clear. In fact, for $\varphi(\vec{v}, g(\vec{b})) \in q$, $\varphi(\vec{v}, \vec{b}) \in g^{-1}(q)$, and the last type $g^{-1}(q)$ does not fork over A as non-forking is preserved under automorphism; hence

$g^{-1}(q) \in U$, and so, as q and $g^{-1}(q)$ have the same restriction to A, $q \in U_\star$. Now let us deal with the other implication. Let $q \in U_\star$, hence there is some type $p \in U$ such that p and q have the same restriction to A. For every formula $\chi(\vec{w})$ of $tp(\vec{b}/A)$, $\varphi(v, \vec{b}) \wedge \chi(\vec{b})$ is in p; as p does not fork over A, there is some tuple $\vec{a}$ in A for which $\varphi(\vec{v}, \vec{a}) \wedge \chi(\vec{a})$ occurs in the restriction of p to A and consequently in q. Recall that $q \in S(M)$ is definable, whence there is some $L(M)$-formula "$\varphi(\vec{v}, \vec{w}) \in q$" defining the tuples $\vec{c}$ such that $\varphi(\vec{v}, \vec{c}) \in q$. Accordingly the set of $L(M)$-formulas

$$\{\text{“}\varphi(\vec{v}, \vec{w}) \in q\text{”}\} \cup tp(\vec{b}/A)$$

determines a (possibly incomplete) type over M. As $\mathcal{M}$ is saturated in some power $> |A|$, we can even find a tuple $\vec{c}$ in M realizing these formulas; in particular $\varphi(v, \vec{c}) \in q$. Enlarge the identity map of A by $\vec{b} \mapsto \vec{c}$, and get an automorphism g of $\mathcal{M}$ such that $\varphi(\vec{v}, g(\vec{b})) \in q$, as required. ♣

7.6 Classifiable theories

We continue and conclude in this section our outline of Shelah's classification program. T still denotes a complete first order theory with infinite models in a countable language L, Ω a big saturated model of T. As unsuperstable theories have too many models and are not classifiable, we should assume T superstable. However, to simplify our exposition, we will even assume T ω-stable, and we will treat in detail this restricted framework. In fact ω-stable theories are superstable as well (but not conversely) and enjoy some stronger properties (such as existence and uniqueness of prime models) making them more tractable.

To introduce our next steps, let us refer once again to strongly minimal theories, and even to the more particular case when $T = ACF_p$ for a given $p = 0$ or prime. Recall Steinitz's analysis of an algebraically closed field $\mathcal{K}$ of characteristic p: given

- the prime subfield $\mathcal{K}_{<>}$ of $\mathcal{K}$,
- a transcendence basis B of $\mathcal{K}$ (over $\mathcal{K}_{<>}$),

$\mathcal{K}$ is fully determined up to isomorphism as the algebraic closure of the extension of $\mathcal{K}_{<>}$ by B, which in its turn depends only on the cardinality of B (the transcendence degree of $\mathcal{K}$). This provides a quite satisfactory answer to our classification purposes. But, in view of a possible generalization, we have to be very careful in weighing the specific advantages of the particular algebraically closed (or also strongly minimal) setting and in checking which of them can be extended to a broader framework. Above all, we have to recall that over $K_{<>}$ (as well as over any subfield of $\mathcal{K}$) there is a unique non-algebraic 1-type, that of transcendental elements; by the way, this type is strongly regular.

Now take any ω-stable T and a model $\mathcal{M}$ of T. Keeping the previous example in mind, we examine the structure of $\mathcal{M}$. $\mathcal{M}$ (elementarily) includes the model $\mathcal{M}_{<>}$ of T prime over $\emptyset$. To construct $\mathcal{M}$ upon $\mathcal{M}_{<>}$, look at the non-algebraic 1-types over $\mathcal{M}_{<>}$ realized in M. Unlike the example of fields, in this general setting we should expect to meet quite a lot of types; however we can restrict our attention to the RK-minimal ones, partition them according to the equivalence relation $\sim_{RK}$ and choose a strongly regular representative in each $\sim_{RK}$-class. Incidentally recall that the resulting types are pairwise orthogonal. At this point take a maximal independent set of realizations for each of these types, glue these sets, form their union X and the model $\mathcal{M}_{<>}(X)$ prime over $M_{<>} \cup X$ in $\mathcal{M}$. If this model equals $\mathcal{M}$ (just as in algebraically closed fields), then we are done: the isomorphism type of $\mathcal{M}$ should be given by the sizes of the independent sets of realizations of the involved strongly regular types. But it may happen that, given a strongly regular type over $M_{<>}$ realized in M, say by a, after forming the model $\mathcal{M}_{<>}(a)$ of T prime over $M_{<>} \cup \{a\}$, one meets some new non-algebraic types over $M_{<>}(a)$ which are orthogonal to any type over $M_{<>}$ and are realized in M (as the next Example 7.6.1 will show). One sees that this excludes $\mathcal{M} = \mathcal{M}_{<>}(X)$. Accordingly we have to repeat the previous machinery over and over again until $\mathcal{M}$ is reached (provided that $\mathcal{M}$ can be eventually reached).

Example 7.6.1 Let T be the theory of a 1-ary function f such that each element has infinitely many preimages via f, there is an element 0 for which $f(0) = 0$ and no $a \neq 0$ satisfies $f^n(a) = a$ for any positive integer n. The reader may check that T is complete: we do not wish to linger over this point and we prefer to describe in detail the non-algebraic 1-types over a model $\mathcal{M}$ of T. In this setting, first we have to consider, for any positive integer n and $a \in M$, the type $p_{\mathcal{M}}(n, a)$ of an element satisfying $f^n(v) = a, \quad \neg(f^{n-1}(v) = b)$ for every $b \in M$. In addition there is a type $q_{\mathcal{M}}$ determined by the formulas $\neg(f^n(v) = a)$ for any $a \in M$ and positive integer n. No further 1-type arises. In particular $S_1(M)$ is countable when M is, and so T is ω-stable. Incidentally notice that the set theoretic union of two models of T is again a model of T.

Now observe that all the 1-types mentioned above are strongly regular. Let us check this. First consider $p_{\mathcal{M}}(n,a)$ for $a \in M$, n a positive integer. Let $x \models p_{\mathcal{M}}(n,a)$. To form $\mathcal{M}(x)$, one takes M, and one adds first $x, f(x), \ldots, f^{n-1}(x)$, and then $\aleph_0$ preimages for each of them, and for every preimage, and so on. Consequently the only elements of $M(x) - M$ realizing $p_{\mathcal{M}}(n,a)$ are those satisfying the formula $f^n(v) = a$. This confirms that $p_{\mathcal{M}}(n,a)$ is strongly regular. Now look at $x \models q_{\mathcal{M}}$. This time, to form $\mathcal{M}(x)$ we take M and we add $f^n(x)$ for every positive integer n, and then $\aleph_0$ preimages for each of them, and for every preimage, and so on. All these elements satisfy $q_{\mathcal{M}}$, and so the points of $M(x) - M$ realizing $q_{\mathcal{M}}$ are exactly those satisfying, for instance, $v = v$. So also $q_{\mathcal{M}}$ is strongly regular.
Now take an $\aleph_1$-saturated model $\mathcal{M}$ of T. $\mathcal{M}_{<>}$ (the model prime over $\emptyset$) is an elementary substructure of $\mathcal{M}$ up to isomorphism. As $\mathcal{M}_{<>}$ is countable, $\mathcal{M}$ contains a realization x of $p_{\mathcal{M}_{<>}}(1,a)$ for every $a \in M$. Also $\mathcal{M}' = \mathcal{M}_{<>}(x)$ is an elementary substructure of $\mathcal{M}$. Now consider $p_{\mathcal{M}'}(1,x)$, which is a strongly regular type, and again can be realized in $\mathcal{M}$ because $\mathcal{M}$ is $\aleph_1$-saturated and $\mathcal{M}'$ is countable. Furthermore $p_{\mathcal{M}'}(1,x)$ is orthogonal (equivalently is not $\sim_{RK}$-equivalent) to any non-algebraic 1-type p over $M_{<>}$; in other words $p_{\mathcal{M}'}(1,x) \perp p|_{\mathcal{M}'}$. In fact, if $p = p_{\mathcal{M}_{<>}}(n,a')$ for some $a' \in M_{<>}$ and positive integer n, then one easily observes $p|_{\mathcal{M}'} = p_{\mathcal{M}'}(n,a')$, while, if $p = q_{\mathcal{M}_{<>}}$, then $p|_{\mathcal{M}'} = q_{\mathcal{M}'}$. In any case, $p|_{\mathcal{M}'}$ is strongly regular. To conclude $p_{\mathcal{M}'}(1,x) \perp p|_{\mathcal{M}'}$, we can equivalently check $p|_{\mathcal{M}'} \not\sim_{RK} p_{\mathcal{M}'}(1,x)$, which is easily proved, as, for every $y \models p_{\mathcal{M}'}(1,x)$, $M'(y)$ contains no realization of $p|_{\mathcal{M}'}$ (and conversely). In conclusion there exist some strongly regular types over $\mathcal{M}'$ realized in $\mathcal{M}$ but orthogonal to the non-algebraic types over $\mathcal{M}_{<>}$. Moreover this procedure can be repeated as many times as you like (after realizing $p_{\mathcal{M}'}(1,x)$ by y in M, one can form $p_{\mathcal{M}'(y)}(1,y)$, and so on).

Let us come back to our general analysis. We wonder under which conditions any model $\mathcal{M}$ of our arbitrary ω-stable theory T can be reached at the top of the construction described before. To clarify this point we have to make our framework more precise. So let us first recall what follows.

Definition 7.6.2 *For any non-zero ordinal λ, let $\lambda^{<\omega}$ denote the set of the finite ordered sequences of elements of λ. $\lambda^{<\omega}$ is partially ordered by the relation $\leq$ according to which, for $s, t \in \lambda^{<\omega}$, $s \leq t$ if and only if s is an initial segment of t. $\lambda^{<\omega}$ has a least element in $\leq$ (the empty sequence $<>$); every $s \in \lambda^{<\omega}$ different from $<>$ has a (unique)* **predecessor** *(a greatest element $< s$), which will be denoted by s^-; s is called a* **successor** *of s^-.*

A **tree** *of* $\lambda^{<\omega}$ *is a downward closed subset* C*: if* $s \in C$, $t \in \lambda^{<\omega}$ *and* $t \leq s$*, then* $t \in C$ *as well.*

The following definition describes in detail the models $\mathcal{M}$ which can be built in the way sketched before: they are exactly those having a *presentation* of the form we are going to explain.

Definition 7.6.3 *Let* $\mathcal{M}$ *be a model of* T *of power* $\lambda > \aleph_0$*. A* **presentation** *of* $\mathcal{M}$ *is a pair* $(C, ((\mathcal{M}_s, a_s) : s \in C))$ *such that* C *is a tree of* $\lambda^{<\omega}$*, for every* $s \in C$ $\mathcal{M}_s$ *is a model of* T *and* $a_s \in M_s$*, and:*

(i) $\mathcal{M}_{<>}$ *is prime over* $\emptyset$*;*

(ii) *if* $s, t \in C$ *and* $s = t^-$*, then* $\mathcal{M}_t = \mathcal{M}_s(a_t)$ *and* $p_t = tp(a_t/M_s)$ *is strongly regular;*

(iii) *if* $s, t, t' \in C$ *and* $s = t^- = t'^-$*, then either* $p_t = p_{t'}$ *or* $p_t \perp p_{t'}$*;*

(iv) *for every* $s \in C$*,* $(a_t : t \in C, t^- = s)$ *is independent over* M_s*, namely, for every* $t \in C$ *such that* $t^- = s$*,* $a_t \downarrow_{M_s} \{a_{t'} : t' \in C, t' \neq t, t'^- = s\}$*;*

(v) *if* $s, t \in C$ *and* $s = t^{--}$*, then* p_t *is orthogonal to all the non-algebraic types over* M_s*;*

(vi) $\mathcal{M}$ *is prime over* $\bigcup_{s \in C} M_s$*.*

Of course we are interested in those theories T such that every model gets such a presentation.

Definition 7.6.4 *An* ω*-stable theory* T *is called* **presentable** *when, for all models* $\mathcal{M}_0$*,* $\mathcal{M}_1$*,* $\mathcal{M}_2$ *and* $\mathcal{M}$ *of* T *such that*

- $\mathcal{M}_0$ *is an elementary substructure of both* $\mathcal{M}_1$ *and* $\mathcal{M}_2$*,*
- $M_1 \downarrow_{M_0} M_2$ *(in other words* $\vec{a_1} \downarrow_{M_0} M_2$ *for every* $\vec{a_1} \in M_1$*, or, equivalently,* $\vec{a_2} \downarrow_{M_0} M_1$ *for every* $\vec{a_2}$ *in* M_2*),*
- $\mathcal{M}$ *is prime over* $M_1 \cup M_2$*,*

for every non-algebraic type $p \in S(M)$ *there exists some type over* $\mathcal{M}_1$ *or over* $\mathcal{M}_2$ *that is not orthogonal to* p*.*

Presentability is a new dichotomy within the classification problem. In fact, on the one side, one shows:

Theorem 7.6.5 (Shelah) *If T is an ω-stable theory and T is not presentable, then $I(T, \lambda) = 2^\lambda$ for every uncountable cardinal λ. In particular, T is not classificable.*

On the other hand

Theorem 7.6.6 (Shelah) *If T is ω-stable and presentable, then every uncountable model $\mathcal{M}$ of T has a presentation.*

A model $\mathcal{M}$ may admit several presentations. Actually there is a "quasi uniqueness" theorem stating under which conditions two "different" presentations yield isomorphic models; but it is impossible to discuss it here shortly, so we omit its treatment, and we conclude our report about presentability and presentations by proposing some examples and, in particular, an ω-stable theory which is not presentable.

Examples 7.6.7 1. Let T be the theory of two equivalence relations E_1, E_2 such that any E_1-class and any E_2-class share infinitely many common elements. One checks that T is complete. Here are the non-algebraic 1-types over a model $\mathcal{M}$ of T:

- for $a \in M$, the formulas $E_1(v, a)$, $E_2(v, a)$ and $\neg(v = b)$ for all $b \in M$ determine a type $p_{\mathcal{M}}(a)$;
- for $a \in M$ and $i = 1, 2$, the formulas $E_i(v, a)$ and $\neg E_{3-i}(v, b)$ for all $b \in M$ give a new type $p_{\mathcal{M}}(a, i)$;
- finally there is a type $q_{\mathcal{M}}$ determined by the formulas $\neg E_1(v, b)$ and $\neg E_2(v, b)$ for all $b \in M$.

In particular, if $\mathcal{M}$ is countable, then $S_1(M)$ is. Hence T is ω-stable. Moreover it is straightforward to check that all the types listed above are strongly regular and, for all $a \in M$, $p_{\mathcal{M}}(a, 1) \perp p_{\mathcal{M}}(a, 2)$. Now fix a model $\mathcal{M}_0$ of T and $a \in M_0$. For $i = 1, 2$, let x_i realize $p_{\mathcal{M}_0}(a, i)$ and $\mathcal{M}_i$ denote $\mathcal{M}_0(x_i)$: so $\mathcal{M}_i$ is built by taking $\mathcal{M}_0$ and adding a new E_{3-i}-class (the class of x_i) having countably many common elements with any E_i-class in $\mathcal{M}_0$. One can see that both $\mathcal{M}_1$ and $\mathcal{M}_2$ are elementary extensions of $\mathcal{M}_0$; furthermore $p_{\mathcal{M}}(a, 1) \perp p_{\mathcal{M}}(a, 2)$ implies $x_1 \downarrow_{M_0} x_2$, whence $M_1 \downarrow_{M_0} M_2$ (see the proof of Theorem 7.5.19). Now form the model $\mathcal{M}$ of T prime over $M_1 \cup M_2$; M is obtained just by adding countably many elements to $M_1 \cup M_2$ in the intersection between the E_2-class of x_1 and the E_1-class of x_2. Let x be an element in this intersection, and consider $p = p_{\mathcal{M}}(x)$: p is a non-algebraic type over M, and one can check that p is orthogonal (equivalently, is not $\sim_{RK}$-equivalent) to any type over M_1 or M_2. So T is not presentable.

Let us also check that, for every uncountable cardinal λ, $I(T, \lambda)$ just equals 2^λ. In fact take the disjoint union I of two sets I_1 and I_2 of power λ; let R be an irreflexive symmetric binary relation in I such that, if $s_1, s_2 \in I$ and $(s_1, s_2) \in R$, then $s_1 \in I_1$ and $s_2 \in I_2$ or conversely (so (I, R) is a *bipartite graph*). Now build a model $\mathcal{M}_R$ of T where the E_i-classes correspond to the elements of I_i for every $i = 1, 2$ and, for every E_1-class X_1 and E_2-class X_2,

$$|X_1 \cap X_2| = \lambda \quad \text{if} \quad (X_1, X_2) \in R$$

and

$$|X_1 \cap X_2| = \aleph_0 \quad \text{otherwise.}$$

It is clear that $\mathcal{M}_R$ has power λ for every R. Furthermore, for $R \neq R'$, $\mathcal{M}_R \not\simeq \mathcal{M}_{R'}$. So $I(T, \lambda) \geq 2^\lambda$ and consequently $I(T, \lambda) = 2^\lambda$: in fact, there exist 2^λ many relations R as before.

2. On the contrary the theory T in Example 7.6.1 is presentable. In fact take three models $\mathcal{M}_0$, $\mathcal{M}_1$, $\mathcal{M}_2$ of T such that $\mathcal{M}_0$ is an elementary substructure of both $\mathcal{M}_1$ and $\mathcal{M}_2$ and $M_1 \downarrow_{M_0} M_2$. We observed that $M_1 \cup M_2$ is the domain of a model $\mathcal{M}$ of T extending $\mathcal{M}_1$ and $\mathcal{M}_2$; of course, $\mathcal{M}$ is prime over $M_1 \cup M_2$. Let p be a non-algebraic type over M (and keep the same notation as in 7.6.1). If $p = p_{\mathcal{M}}(n, a)$ for some $a \in M$ and some positive integer n, then $p \sim_{RK} p_{\mathcal{M}_i}(n, a)$ where $i = 1, 2$ satisfies $a \in M_i$. If $p = q_{\mathcal{M}}$, then p is $\sim_{RK}$-equivalent to both $q_{\mathcal{M}_1}$ and $q_{\mathcal{M}_2}$. So T is presentable, as claimed.

   Nevertheless T is not classifiable, as it has 2^λ many pairwise non isomorphic models in every uncountable power λ. Let us see why. Indeed, for every λ, we can build 2^λ non isomorphic models satisfying the further assumption

$$\text{for every } a \in M \text{ and for some natural } n \ f^n(a) = 0.$$

   Notice that such a model $\mathcal{M}$ can be viewed as a tree of $\lambda^{<\omega}$ with respect to the relation $\leq$ defined as follows: $\forall a, b \in M$, $a \leq b$ if and only if $f^n(b) = a$ for some natural n; in particular $a = b^-$ if and only if $f(b) = a$, and $<>= 0$. The isomorphism class of $\mathcal{M}$ clearly determines the isomorphism type of $(M, \leq)$ as a tree. Moreover every point in $(M, \leq)$ has infinitely many successors, and $|M| = \lambda$. So it suffices to show that there exist 2^λ pairwise non isomorphic trees of $\lambda^{<\omega}$ satisfying the last additional properties.

Associate a tree $C(\nu)$ of $\lambda^{<\omega}$ with any ordinal $\nu < \lambda$ as follows.

(a) First let $\nu = 0$: in $C(0)$ the root $<>$ has λ successors, while any further vertex has $\aleph_0$ successors.

(b) Now let $\nu = \mu + 1$: in $C(\nu)$ $<>$ has $\aleph_0$ successors, and each of them is the root of a tree isomorphic to $C(\mu)$.

(c) Finally let ν be a limit ordinal: in $C(\nu)$, $<>$ has a successor s_μ for every $\mu < \nu$, and s_μ is the root of a tree isomorphic to $C(\mu)$.

At this point, let us build for every $S \subseteq \lambda$ a tree $C(S)$ of $\lambda^{<\omega}$ as follows: $<>$ has a successor s_ν for every $\nu < \lambda$ and, for every $\nu < \lambda$, s_ν is in its turn the root of a tree isomorphic to $C(\nu)$ if $\nu \in S$ and to $\omega^{<\omega}$ otherwise. It is clear that $|C(S)| = \lambda$ for every $S \supseteq \lambda$ and that different subsets $S \neq S'$ yield non isomorphic trees $C(S) \not\simeq C(S')$.

At this point one may wonder what is so wrong in the last example to exclude any classification of models. More generally one may ask why a presentable theory may be non-classifiable. Recall that, if T is presentable, then any uncountable model $\mathcal{M}$ of T has a presentation, whose "skeleton" is a tree C of $|M|^{<\omega}$. Of course this is a good feature towards a general classification of models. But the point is that some involved tree might be non well founded. Let us recall what this means.
Let C be any tree (say in $\lambda^{<\omega}$). One associates with any point s in C a *rank* $r(s)$ (an ordinal, or ∞) in the following way. First we define $r(s) \geq \alpha$ for any ordinal α. We proceed by induction on α:

1. $r(s) \geq \alpha$ if $\alpha = 0$;

2. if α is limit, then $r(s) \geq \alpha$ means $r(s) \geq \beta$ for every ordinal $\beta < \alpha$;

3. if $\alpha = \beta + 1$, then $r(s) \geq \alpha$ means that, for some $t \in S$ with $s = t^-$, $r(t) \geq \beta$.

If there is an ordinal α for which $r(s) \not\geq \alpha$, then the least ordinal with this property is necessarily a successor $\alpha_0 + 1$, and we put $r(s) = \alpha_0$. Otherwise (when $r(s) \geq \alpha$ for every ordinal α) we put $r(s) = \infty$. We say that C is *well founded* if $r(<>)$ is an ordinal.
Now let us arrange these definitions in our setting, when T is any ω-stable presentable theory. So every uncountable model of T has a presentation,

where the points of the involved tree correspond to strongly regular types. Take any model $\mathcal{M}$ of T and any strongly regular type p over M. We want to associate with p a *depth* $Dp(p)$ (an ordinal or ∞) in a way inspired by the above assignment of a rank to the points of a tree. First we define $Dp(p) \geq \alpha$ for every ordinal α. As usual, we proceed by induction on α.

Definition 7.6.8 *1. If $\alpha = 0$, then $Dp(p) \geq \alpha$.*

2. *If α is limit, then we put $Dp(p) \geq \alpha$ when $Dp(p) \geq \beta$ for every ordinal $\beta < \alpha$.*

3. *If $\alpha = \beta + 1$ is a successor β, then we put $Dp(p) \geq \alpha$ exactly when, for every x realizing p, there is some strongly regular type q over $\mathcal{M}(x)$ such that $Dp(q) \geq \beta$ and q is orthogonal to all the types over M.*

Now we can introduce $Dp(p)$.

Definition 7.6.9 *If there is an ordinal α such that $Dp(p) \not\geq \alpha$, then the least ordinal with this property is a successor $\alpha_0 + 1$, and we put $Dp(p) = \alpha_0$. Otherwise (when $Dp(p) \geq \alpha$ for every ordinal α) we put $Dp(p) = \infty$.*

Now we define the *depth* of the whole theory T $Dp(T)$. As usual, we agree that ∞ is greater than any ordinal.

Definition 7.6.10 *The* **depth** *of T $Dp(T)$ is the least upper bound of the depths of the strongly regular types over models of T. T is called* **deep** *if $Dp(T) = \infty$, and* **shallow** *if $Dp(T)$ is an ordinal.*

Actually the original definition of depth requires some rearrangements due to a technical convenience in the proofs. But this does not concern our sketched treatment here.
At this point it is easy to realize that the (presentable) theory T in 7.6.1 (and later in 7.6.7.2) is deep. In fact, let $\mathcal{M}$ be a model of T, $a \in M_0$, p be the strongly regular type $p_{\mathcal{M}}(1, a)$ (according to the notation introduced in 7.6.1). We have seen that, for every $x \models p$, $S(M(x))$ contains a strongly regular type $q = p_{\mathcal{M}(x)}(1, x)$ having depth ≥ 0 and orthogonal to all the non-algebraic types over M. Consequently $Dp(p) \geq 1$. But then even q has depth ≥ 1, and so $Dp(p) \geq 2$. Iterating this procedure, one eventually gets $Dp(p) = \infty$. Recall that T is not classifiable, because it gets too many models in any uncountable power. But this is a general fact, as the following theorem clarifies.

Theorem 7.6.11 (Shelah) *If T is an ω-stable, presentable and deep theory, then $I(T,\lambda) = 2^\lambda$ for every uncountable cardinal λ. In particular T is not classifiable.*

This is not the case for a shallow T. In fact, first one shows that $Dp(T) < \omega_1$ (and indeed, for every ordinal $\alpha < \omega_1$, there is some ω-stable presentable shallow theory T_α having depth α). Moreover, for every uncountable λ, one can upperly and underly bound $I(T, \lambda)$ with respect to $Dp(T)$ in an effective way, implying, among other things, $I(T, \lambda) < 2^\lambda$ for some λ. In this sense, we can conclude

Theorem 7.6.12 *Let T be an ω-stable theory. Then T is classifiable if and only if T is presentable and shallow.*

Of course, this statement makes sense provided that we agree with Shelah's proposal that T is not classifiable if and only if $I(T,\lambda) = 2^\lambda$ for every uncountable λ; should we think that a classifiability proof requires to show the existence of an invariant system of some ordinal rank, we might reasonably doubt that Theorem 7.6.12 is the last word about classification, and wonder if Shelah's classification program is fully reached in this theorem. This is a subtle and delicate matter, and may be discussed as long as one likes. So here we limit ourselves to a few comments which just aim at introducing a possible debate and do not claim to suggest any conclusion.

First one should undoubtedly acknowledge how formidable Shelah's work was; in some sense the "depth" itself of the new notions and techniques proposed by Shelah witness its validity and authoritativeness. Not surprisingly, Shelah himself celebrated its conclusion even in the title of a preliminary version of the final paper, which just stated "Why am I so happy". Actually Shelah's happiness is quite easy to understand and to share.

It should be also mentioned that Melles proposed some years ago an alternative approach to classifiability of a more recursion theoretic flavour, so looking for *effective* classification invariants; however Melles himself proved that his perspective eventually agrees with Shelah's point of view. It should be also said that Appenzeller (and others) observed some intriguing connections between the main notions arising in Shelah's classification, and some concepts coming from Stationary Logic and Descriptive Set Theory. This is quite interesting, and provides a further corroboration of Shelah's perspective.

Finally let us discuss what happens within superstable (and possibly non ω-stable) theories T. The main trouble in this enlarged framework is that

prime models may fail. However prime models (and existence and uniqueness properties) still make sense provided we restrict our attention to a suitable subclass of $Mod(T)$, formed by sufficiently saturated models (called a-models). This allows to define what T presentable, or T deep, or T shallow means, and to deduce that, when T is not presentable, or is presentable and deep, T is not classifiable: in fact, in these cases T has too many a-models, and consequently too many models, to get a classification.
But the point is whether a presentable shallow superstable T is classifiable. According to Shelah's perspective, this requires to upperly bound the number of models of a given uncountable power (warning: we have said *models*, and not *a-models*, bounding the latter ones does not imply a priori restricting the number of the former ones). Again the main difficulty in handling this setting is the possible lack of prime models, especially in the framework proposed in the definition of presentable theory. And actually Shelah singled out the following key property, called the *existence property*.

Definition 7.6.13 *A superstable T is said to satisfy the* **existence property** *if and only if, for every choice of models $\mathcal{M}_0$, $\mathcal{M}_1$, $\mathcal{M}_2$ of T such that*

- *$\mathcal{M}_0$ is an elementary substructure of both $\mathcal{M}_1$ and $\mathcal{M}_2$,*
- *$M_1 \downarrow_{M_0} M_2$,*

there is a model $\mathcal{M}$ of T prime and atomic over $M_1 \cup M_2$.

This is the last dichotomy in the superstable case, and provides the final dividing line between classifiable and non-classificable theories. In fact the following results hold.

Theorem 7.6.14 (Shelah) *If T is superstable but fails to have the existence property, then $I(T, \lambda) = 2^\lambda$ for every uncountable cardinal λ.*

Theorem 7.6.15 (Shelah) *If T is a superstable, presentable, shallow theory and has the existence property, then T is classifiable.*

Of course, the comments following Theorem 7.6.12 are still valid and do concern also Theorem 7.6.15.

7.7 Shelah's Uniqueness Theorem

In the forthcoming sections we discharge two debts we contracted in Chapter 6: the proofs of Shelah's Uniqueness Theorem for prime models in ω-stable theories and Morley's Theorem on uncountably categorical theories. In fact both require, in addition to the techniques developed in Chapters 5 and 6, the more sophisticated tools introduced in this Chapter, in particular the material of Section 7.5 on ω-stability. First we deal with Shelah's Uniqueness Theorem. We recall its framework: we are concerned with an ω-stable theory T and a small subset X of its universe Ω. We have to show that, if $\mathcal{A}_0$ and $\mathcal{A}_1$ are two models of T both elementarily including X and prime over X, then there is an isomorphism between $\mathcal{A}_0$ and $\mathcal{A}_1$ fixing X pointwise. As we saw in 6.4, Ressayre's Uniqueness Theorem for constructible sets reduces the whole problem to prove:

Theorem 7.7.1 (Shelah) *Let T be an ω-stable theory, X be a small subset of Ω, $\mathcal{A}$ be a model of T elementarily including X. If $\mathcal{A}$ is prime over X, then $\mathcal{A}$ is constructible over X.*

Proof. According to Morley's Existence Theorem (6.4.17) there is some model $\mathcal{B}$ of T constructible over X. Fix such a model $\mathcal{B}$. We can freely assume that X is a subset of B and that the corresponding inclusion is elementary. As $\mathcal{A}$ is prime over X, there is an elementary embedding of $\mathcal{A}$ into $\mathcal{B}$ fixing X pointwise. Again, without loss of generality we can assume that $\mathcal{A}$ is an elementary substructure of $\mathcal{B}$. In particular $X \subseteq A \subseteq B$, and all these inclusions are elementary. We claim that these assumptions

$$X \subseteq A \subseteq B, \ B \text{ constructible over } X$$

(so forgetting the additional hypothesis that both $\mathcal{A}$ and $\mathcal{B}$ are models of T) are sufficient to imply, for T ω-stable, that A is constructible over X, as expected.

We proceed by induction on the power $|B - X|$ of $B - X$. If $|B - X|$ is countable, then $|A - X|$ is countable, too; furthermore B is atomic over X because B is constructible over X; hence we can apply Lemma 6.4.6 and conclude that A is constructible over X, as claimed.

So assume $|B - X| = \lambda$ uncountable. Fix any construction $(b_\nu : \nu < \lambda)$ of B over X: here λ is viewed as an initial ordinal. Correspondingly we build an increasing sequence of subsets C_ν ($\nu < \lambda$) of B satisfying

1. $B = X \cup \bigcup_{\nu<\lambda} C_\nu$

and, for every $\nu < \lambda$,

2. $|C_\nu| = |\nu| + \aleph_0$,

3. C_ν is closed (in the sense of Section 6.4),

4. for every tuple $\vec{c}$ in C_ν, the type of $\vec{c}$ over A has the same Morley rank as its extension $tp(\vec{c}/A \cup C_\nu)$.

To form this sequence, start by putting $C_0 = \emptyset$. Moreover, for a limit ν, let C_ν be the union of the preceding sets in the sequence. It is straightforward to check that 2, 3, and 4 are satisfied in these cases. Now take a successor ordinal $\nu + 1$. If $b_\nu \in C_\nu$, then put $C_{\nu+1} = C_\nu$. Otherwise form $C_\nu \cup \{b_\nu\}$ and call it $C_\nu(0)$. For every tuple $\vec{b}$ in $C_\nu(0)$ there is a finite subset $A(\vec{b})$ of A for which

$$RM(tp(\vec{b}/A)) = RM(tp(\vec{b}/A(\vec{b}))).$$

Take $C_\nu(0) \cup \bigcup_{\vec{b}} A(\vec{b})$ and enlarge it to a minimal closed extension $C_\nu(1)$ in B, in the way we saw in Section 6.4. Apply the same procedure used to enlarge $C_\nu(0)$ to $C_\nu(1)$, and then to $C_\nu(n)$ for every natural n, so building for every n a closed set $C_\nu(n+1) \supseteq C_\nu(n)$ in B such that, for each $\vec{b}$ in $C_\nu(n)$, there is a finite subset $A(\vec{b})$ of $A \cap C_\nu(n+1)$ for which

$$RM(tp(\vec{b}/A)) = RM(tp(\vec{b}/A(\vec{b}))).$$

Finally put $C_{\nu+1} = \bigcup_n C_\nu(n)$. Again, a straightforward check proves 2, 3 and 4 for $C_{\nu+1}$.
It is also clear that $B = X \cup \bigcup_{\nu<\lambda} C_\nu$ and hence 1 holds. Now notice that, for a given $\nu < \lambda$,

5. $C_{\nu+1}$ is constructible over $X \cup C_\nu$,

6. $|C_{\nu+1} - (X \cup C_\nu)| < \lambda$.

The latter claim follows directly from 2, as $|B - X| = \lambda > |\nu| + \aleph_0$. 5 requires some more work. First form a (possibly transfinite) list of the elements in $C_{\nu+1} - (X \cup C_\nu)$

$$(d_\mu \, : \, \mu < \alpha)$$

extracting them from the given construction of B over X. Notice that, for every $\beta < \alpha$, $C_\nu \cup \{d_\mu \, : \, \mu < \beta\}$ is closed because both C_ν and $C_{\nu+1}$ are closed. So use Lemma 6.4.9 and deduce that the type of d_β over $X \cup C_\nu \cup \{d_\mu \, : \, \mu < \beta\}$ is isolated. Hence $\{d_\mu \, : \, \mu < \alpha\}$ is a construction of $C_{\nu+1}$ over $X \cup C_\nu$, and 5 is proved.
Owing to 5 and 6, the induction hypothesis applies to

$$X \cup C_\nu \subseteq X \cup C_\nu \cup (C_{\nu+1} \cap A) \subseteq X \cup C_{\nu+1}$$

and shows that $X \cup C_\nu \cup (C_{\nu+1} \cap A)$ is constructible over $X \cup C_\nu$. For every $\nu < \lambda$, fix a construction of $X \cup C_\nu \cup (C_{\nu+1} \cap A)$ over $X \cup C_\nu$, and glue all these constructions to build a (transfinite) list of the elements of A

$$(\star) \quad (a_\mu : \mu < \cup_{\nu<\lambda}\mu_\nu)$$

where, for every ν, $(a_\mu : \mu_\nu \leq \mu < \mu_{\nu+1})$ constructs $C_{\nu+1} \cap A$ over $X \cup C_\nu$. We claim that $(\star)$ is actually a construction of A over X: in other words, for every μ, $tp(a_\mu / X \cup \{a_\eta : \eta < \mu\})$ is isolated. In fact, suppose $\mu_\nu \leq \mu < \mu_{\nu+1}$, it suffices to show

$$RM(tp(a_\mu / X \cup C_\nu \cup \{a_\eta : \eta < \mu\})) = RM(tp(a_\mu / X \cup \{a_\eta : \eta < \mu\})).$$

In fact, the former type is isolated and consequently the Open Mapping Theorem ensures that the latter is isolated, too. Now, for every tuple $\vec{c}$ in C_ν,

$$RM(tp(\vec{c}/A)) = RM(tp(\vec{c}/A \cap C_\nu))$$

by 4. Now we use what we saw in Section 7.5 about independence in ω-stable theories. First notice that transitivity (I6) implies

$$RM(tp(\vec{c}/X \cup \{a_\eta : \eta \leq \mu\})) = RM(tp(\vec{c}/X \cup \{a_\eta : \eta < \mu\})).$$

Then symmetry yields

$$RM(tp(a_\mu / X \cup \vec{c} \cup \{a_\eta : \eta < \mu\})) = RM(tp(a_\mu / X \cup \{a_\eta : \eta < \mu\})).$$

As this holds for every tuple $\vec{c}$ in C_ν, (I3) implies that the type of a_μ over $X \cup C_\nu \cup \{a_\eta : \eta < \mu\}$ does not fork over $X \cup \{a_\eta : \eta < \mu\}$, and this is just what we have claimed. So A is constructible over X, and we are done. ♣

7.8 Morley's Theorem

At last here are to show Morley's Theorem on uncountably categorical theories. Let us recall its framework: for a complete theory T without finite models in a countable language L, the theorem says that T is categorical either in all uncountable cardinals, or in none of them. In other words

Theorem 7.8.1 (Morley) *If T is categorical in some uncountable cardinal, the T is categorical in every uncountable cardinal.*

Just to prepare the proof of Morley's Theorem, let us premit a particular case as a warm-up.

Proposition 7.8.2 *If T is strongly minimal, then T is categorical in every uncountable power.*

Proof. For every model $\mathcal{B}$ of T, the isomorphism class of $\mathcal{B}$ is given by its dimension with respect to the *acl* dependence relation. When B is uncountable, this dimension just equals the power of B because $B = acl(X)$ forces $|X| = |B|$ for every $X \subseteq B$. ♣

So the point is: how far is a theory T categorical in some uncountable power μ from being strongly minimal? Can we recover inside T a possibly "weaker" form of "strong" minimality making the previous argument work?
Notice that there do exist some theories T which are categorical in any infinite cardinal but are not strongly minimal: for instance, consider the theory of two infinite disjoint sets A and A' with a bijection f between them.
Anyhow take a μ-categorical T where μ is a fixed uncountable cardinal. We keep this assumption all throughout this section, unless explicitly stated. First we observe:

Lemma 7.8.3 *T is ω-stable.*

Proof. The strategy is simple. We deny ω-stability and consequently we succeed in building two models of T of power μ which cannot be isomorphic, so contradicting the hypothesis that T is μ-categorical. To do this, first we recall that, if T is not ω-stable, then there is a countable $X \subseteq \Omega$ over which there are uncountably many 1-types. So there is some model of T extending X and realizing uncountably many types, and using the Löwenheim-Skolem Theorem we can get a model $\mathcal{B}$ of T satisfying these properties and having power μ. The second part of the proof uses a method due to Ehrenfeucht and Mostowski and valid for every theory T and every uncountable cardinal μ: it yields in this general framework a model $\mathcal{C}$ of T of power μ realizing at most $\aleph_0$ 1-types over any countable subset. Ehrenfeucht-Mostowski's method is quite sophisticated, and its interest for our purposes in this book is confined to this lemma; so we omit its detailed treatment, and we limit ourselves to apply it to our particular setting. In fact, for a μ-categorical T, the model $\mathcal{C}$ just built in this way cannot be isomorphic to the structure $\mathcal{B}$ obtained before. ♣

Of course, Lemma 7.8.3 cannot imply that T is strongly minimal. But, as a consequence of the ω-stability of T, indeed as a consequence of this only hypothesis (we do not need μ-categoricity here), we show that there is some strongly minimal formula -possibly involving parameters- in T.

Lemma 7.8.4 *There is some strongly minimal formula $\varphi(v)$ in T.*

Just to avoid any misunderstanding later, let us underline once again that here formula means formula with parameters.

Proof. Otherwise, for every formula $\varphi(v)$ for which $\varphi(\Omega)$ is infinite, one can find a formula $\varphi'(v)$ such that $\varphi'(\Omega) \subseteq \varphi(\Omega)$ and both $\varphi'(\Omega)$ and $\varphi(\Omega) - \varphi'(\Omega)$ are infinite. Using this fact and starting from $\varphi_{()}(v) \, : \, v = v$, one defines, for every finite ordered sequence s of 0 and 1, a formula $\varphi_s(v)$ in such a way that, for any s, $\varphi_s(\Omega)$ is infinite and $\varphi_{s0}(\Omega)$ and $\varphi_{s1}(\Omega)$ partition $\varphi_s(\Omega)$. The parameters involved in these formulas are as many as the sequences s, and so form a countable set X. On the other side, every branch t of the tree $\{0, 1\}^\omega$ determines a(n incomplete) 1-type p_t over X given by the formulas $\varphi_{t|n}(v)$ where n ranges over naturals and $t|n$ denotes the restriction of t to the first n naturals; furthermore different branches yield different types. This gives at least $2^{\aleph_0}$ 1-types over X and contradicts ω-stability. ♣

But μ-categoricity implies even more. In fact, for a μ-categorical T, the formula $\varphi(v)$ can be chosen with parameters in the model $\mathcal{A}$ of T prime over $\emptyset$ (owing to ω-stability this models exists and is unique up to isomorphism). This result requires the following premise, which is now just a technical lemma, but will play a key role later in the proof of Morley's Theorem.

Lemma 7.8.5 *Let $\alpha(v, \vec{a})$ be a formula (with parameters $\vec{a}$ in Ω) such that $\alpha(\Omega, \vec{a})$ is infinite, and let $\mathcal{B}_0$ and $\mathcal{B}_1$ be two models of T such that $\mathcal{B}_0$ is an elementary substructure of $\mathcal{B}_1$, $\mathcal{B}_0 \neq \mathcal{B}_1$ and both $\mathcal{B}_0$ and $\mathcal{B}_1$ elementarily include $\vec{a}$. Then $\alpha(\mathcal{B}_0, \vec{a})$ is properly included in $\alpha(\mathcal{B}_1, \vec{a})$.*

Proof. Suppose not. Accordingly there are two models $\mathcal{B}_0$ and $\mathcal{B}_1$ of T elementarily including $\vec{a}$ such that $\mathcal{B}_0$ is an elementary substructure of $\mathcal{B}_1$, $\mathcal{B}_0 \neq \mathcal{B}_1$ but $\alpha(\mathcal{B}_1, \vec{a})$ equals $\alpha(\mathcal{B}_0, \vec{a})$ (an infinite set). Apply the Löwenheim-Skolem Theorem to the theory of $(\mathcal{B}_1, \mathcal{B}_0)$ in a language extending L by a 1-ary relation symbol for $\mathcal{B}_0$ and new constants for the elements in $\vec{a}$, and get a countable model $(\mathcal{B}_1^*, \mathcal{B}_0^*)$ of this theory: so $\mathcal{B}_1^*$ and $\mathcal{B}_0^*$ are still models of T elementarily including $\vec{a}$, $\mathcal{B}_0^*$ is an elementary substructure of $\mathcal{B}_1^*$, $\mathcal{B}_0^* \neq \mathcal{B}_1^*$, $\alpha(\mathcal{B}_1^*, \vec{a}) = \alpha(\mathcal{B}_0^*, \vec{a})$, $\mathcal{B}_0^*$, $\mathcal{B}_1^*$ and consequently $\alpha(\mathcal{B}_1^*, \vec{a}) = \alpha(\mathcal{B}_0^*, \vec{a})$ are countable. Hence, unless replacing $\mathcal{B}_0$ and $\mathcal{B}_1$ by $\mathcal{B}_0^*$

and $\mathcal{B}_1^*$ respectively, we can assume that $\alpha(\mathcal{B}_1, \vec{a}) = \alpha(\mathcal{B}_0, \vec{a})$ is countable. Now we define for every ordinal $\nu > 0$ a model $\mathcal{B}_\nu$ of T elementarily extending $\mathcal{B}_0$ (and $\vec{a}$) and even the $\mathcal{B}_\rho$'s for $\rho < \nu$, properly including all of them and still satisfying $\alpha(\mathcal{B}_\nu, \vec{a}) = \alpha(\mathcal{B}_0, \vec{a})$. Notice that this is sufficient for our purposes because, proceeding in this way, we eventually build a model $\mathcal{B}$ of T of power μ elementarily extending $\mathcal{B}_0$ but satisfying $\alpha(\mathcal{B}, \vec{a}) = \alpha(\mathcal{B}_0, \vec{a})$, hence admitting a countable $\alpha(\mathcal{B}, \vec{a})$; on the other side, a simple use of Compactness Theorem yields another model $\mathcal{C}$ of T of power μ such that, for every tuple $\vec{c}$ in C admitting the same type as $\vec{a}$ over the empty set, $|\alpha(\mathcal{C}, \vec{c})| = \mu$. Consequently $\mathcal{B}$ and $\mathcal{C}$ cannot be isomorphic, and this contradicts the μ-categoricity of T.

So let us build the $\mathcal{B}_\nu$'s. We already introduced $\mathcal{B}_1$. For a limit ν, put $\mathcal{B}_\nu = \cup_{\rho<\nu}\mathcal{B}_\rho$: this is a model of T and satisfies our conditions above. So it remains to treat the case of a successor ordinal $\nu = \rho + 1$ with $\rho > 0$. For simplicity we limit ourselves to $\rho = 1$, $\nu = 2$; indeed what we are going to say in this case applies to any $\rho > 0$ as well, and so is generally valid. First use independence theory, more precisely (I3) and (I4), and find an isomorphic copy $\mathcal{B}_1'$ of $\mathcal{B}_1$ inside Ω, corresponding to $\mathcal{B}_1$ by an isomorphism fixing B_0 pointwise, and satisfying $B_1' \downarrow_{B_0} B_1$. To build $\mathcal{B}_1'$, consider a language L^* enlarging L by a constant b^* for every $b \in \mathcal{B}_1$, and in L^* the theory T^* saying that, for every $\vec{b}$ in $\mathcal{B}_1$,

$$\vec{b}^* \text{ satisfies the non-forking extension of } tp(\vec{b}/B_0) \text{ over } B_1.$$

Any finite portion T_0^* of T^* has a model; in fact, let $\vec{b}$ glue all the tuples from B_1 arising in the sentences of T_0^*, use (I4) and obtain a tuple $\vec{b}^*$ realizing the non-forking extension of $tp(\vec{b}/B_0)$ over B_1; recall that every subsequence of $\vec{b}^*$ has the same property. By compactness, T^* has a model. The elements b' embodying the constants b^* in this model form a structure $\mathcal{B}_1'$ isomorphic to $\mathcal{B}_1$ over $\mathcal{B}_0$ (as, for every $L(B_0)$-formula $\varphi(\vec{v})$ and $\vec{b}$ in B_1, $\models \varphi(\vec{b})$ if and only if $\models \varphi(\vec{b}'))$ and independent of B_1 over B_0. In particular $\alpha(\mathcal{B}_1', \vec{a}) = \alpha(\mathcal{B}_0, \vec{a})$. Let $\mathcal{B}_2$ be the model of T prime over $B_1 \cup B_1'$. We claim that $\mathcal{B}_2$ is just the model we are looking for. The key point to check is that $\alpha(\mathcal{B}_2, \vec{a}) = \alpha(\mathcal{B}_0, \vec{a})$. $\supseteq$ is clear. On the other hand, take $d \in \alpha(\mathcal{B}_2, \vec{a})$. Then the type of d over $B_1 \cup B_1'$ is isolated by some suitable formula $\psi(\vec{b_1}, \vec{b_1'}, z)$ (in the free variable z) with parameters $\vec{b_1}$ from B_1 and $\vec{b_1'}$ from B_1'. Recall that $tp(\vec{b_1}/B_0)$ is definable: for every L-formula $\chi(\vec{v}, \vec{w})$ there is an L_{B_0}-formula $d\chi(\vec{w})$ such that, for any $\vec{b}$ in B_0,

$$\Omega \models \chi(\vec{b_1}, \vec{b}) \quad \Leftrightarrow \quad \Omega \models d\chi(\vec{b}).$$

This remains true if we enlarge B_0 to $B_1' \cup \{d\}$: for every $\vec{b}$ in $B_1' \cup \{d\}$, so also for $\vec{b} = (\vec{b'}, d)$ with $\vec{b'}$ in B_1',

$$\Omega \models \chi(\vec{b_1}, \vec{b'}, d) \quad \Leftrightarrow \Omega \models d\chi(\vec{b'}, d).$$

Let us see why. Otherwise $\Omega \models \neg(\chi(\vec{b_1}, \vec{b'}, d) \leftrightarrow d\chi(\vec{b'}, d))$ and consequently

$$\Omega \models \exists z(\alpha(z, \vec{a}) \wedge \neg(\chi(\vec{b_1}, \vec{b'}, z) \leftrightarrow d\chi(\vec{b'}, z)));$$

in other words,

$$\exists z(\alpha(z, \vec{a}) \wedge \neg(\chi(\vec{v}, \vec{b'}, z) \leftrightarrow d\chi(\vec{b'}, z))) \in tp(\vec{b_1}/B_1').$$

As B_1 and B_1' are independent over B_0, there is some $\vec{b''}$ in B_0 for which

$$\Omega \models \exists z(\alpha(z, \vec{a}) \wedge \neg(\chi(\vec{b_1}, \vec{b''}, z) \leftrightarrow d\chi(\vec{b''}, z))).$$

Then there is some $d'' \in B_1$ (and indeed in $\alpha(\mathcal{B}_1, \vec{a})$) such that

$$\mathcal{B}_1 \models \neg(\chi(\vec{b_1}, \vec{b''}, d'') \leftrightarrow d\chi(\vec{b''}, d'')).$$

Recall $\alpha(\mathcal{B}_1, \vec{a}) = \alpha(\mathcal{B}_0, \vec{a})$, so $d'' \in B_0$. But this contradicts the choice of $d\chi$ over B_0. Now apply what we have just observed to the formula $\psi(\vec{v}, \vec{b_1'}, d)$. As $\Omega \models \psi(\vec{b_1}, \vec{b_1'}, d)$, it follows $\Omega \models d\psi(\vec{b_1'}, d)$, where $d\psi$ is the defining formula. So $\mathcal{B}_1' \models \exists z d\psi(\vec{b_1'}, z)$, and we find $d' \in B_1'$ such that $\mathcal{B}_1' \models d\psi(\vec{b_1'}, d')$, and hence $\Omega \models \psi(\vec{b_1}, \vec{b_1'}, d')$. This means that d' has the same type as d over $B_1 \cup B_1'$, and so $d = d' \in B_1'$; but $\alpha(\mathcal{B}_1', \vec{a}) = \alpha(\mathcal{B}_0, \vec{a})$, hence $d \in \alpha(\mathcal{B}_0, \vec{a})$. This accomplishes our proof. ♣

Now we are in a position to show, as promised

Lemma 7.8.6 *Let $\mathcal{A}$ be a model of T prime over $\emptyset$. Then there is a strongly minimal formula $\varphi(v)$ in T with parameters from A.*

Proof. Proceed as in Lemma 7.8.5, but work in $\mathcal{A}$ instead of Ω. Using ω-stability, find again a formula $\varphi(v)$ with parameters in A such that $\varphi(\mathcal{A})$ is infinite but has no partition into 2 A-definable infinite subsets. This does not mean a priori that $\varphi(v)$ is strongly minimal, as in general we cannot imagine what happens if we allow parameters out of A. However we claim that in our particular setting, for a μ-categorical T, $\varphi(v)$ is just strongly minimal. Let us see why. Suppose not, so, enlarging our perspective to Ω, we can find an L-formula $\psi(v, \vec{z})$ and parameters $\vec{b}$ in Ω such that both

$\varphi(v) \wedge \psi(v, \vec{b})$ and $\varphi(v) \wedge \neg\psi(v, \vec{b})$ define infinite sets. Now let us restrict our horizon to $\mathcal{A}$: for every natural n, we can pick a tuple $\vec{a}(n)$ in A for which

$$|\varphi(\mathcal{A}) \cap \psi(\mathcal{A}, \vec{a}(n)|, |\varphi(\mathcal{A}) - \psi(\mathcal{A}, \vec{a}(n)| \geq n.$$

However, just owing the choice of $\varphi(v)$, at least one of these sets is finite, and consequently equals $\varphi(\Omega) \cap \psi(\Omega, \vec{a}(n))$, or $\varphi(\Omega) - \psi(\Omega, \vec{a}(n))$. So either $\varphi(\mathcal{A}) \cap \psi(\mathcal{A}, \vec{a}(n))$ is finite and equals $\varphi(\Omega) - \psi(\Omega, \vec{a}(n))$ for infinitely many n, or $\varphi(\mathcal{A}) - \psi(\mathcal{A}, \vec{a}(n))$ satisfies the same condition. Assume for simplicity that the former option holds. Unless forgetting $\vec{a}(n)$ for every exceptional n, we can even assuming that $\varphi(\mathcal{A}) \cap \psi(\mathcal{A}, \vec{a}(n))$ is finite of size $\geq n$ and equals $\varphi(\Omega) - \psi(\Omega, \vec{a}(n))$ for each n. Now a simple application of Compactness Theorem, using these features of $\mathcal{A}$, Ω and the $\vec{a}(n)$'s, yields two models $\mathcal{B}$ and $\mathcal{C}$ of T and a tuple $\vec{a}$ in B such that $\mathcal{B}$ is an elementary substructure of $\mathcal{C}$, $B \neq C$ and

$$\varphi(\mathcal{B}) \cap \psi(\mathcal{B}, \vec{a}) = \varphi(\mathcal{C}) \cap \psi(\mathcal{C}, \vec{a})$$

is infinite. This clearly contradicts Lemma 7.8.5. ♣

But we can even assume that our strongly minimal $\varphi(v)$ is actually an L-formula, and needs no additional parameters. Let us explain why. Essentially what we have to do is to insert in the language new constants for the parameters $\vec{a} \in A$ involved in $\varphi(v)$ and then, in this extended framework, to consider $T' = Th(\mathcal{A}, \vec{a})$ instead of $T = Th(\mathcal{A})$. Notice that the type of $\vec{a}$ over $\emptyset$ is isolated - say by the L-formula $\theta(\vec{w})$ - because A is atomic over $\emptyset$, so $tp(\vec{a}/\emptyset)$ is realized in every model of T and indeed the models of T' are just the structures $(\mathcal{B}, \vec{b})$ where $\mathcal{B}$ is a model of T and $\vec{b} \in B$ has the same type as $\vec{a}$ over $\emptyset$; in other words, $\vec{b}$ satisfies $\theta(\vec{w})$ in $\mathcal{B}$. In this setting it is not prohibitive to show:

Lemma 7.8.7 *For every infinite cardinal λ, T is λ-categorical if and only if T' is.*

The reader may try to check this as an exercise. In particular, if T is μ-categorical, then T' is; and, if we succeed in proving that T' is categorical in every uncountable power λ, then we can say the same of T. So, with no loss of generality, we can replace T by T', in other words to assume that T is a μ-categorical theory with a strongly minimal L-formula $\varphi(v)$ (without parameters). This resembles the plainer framework outlined at the beginning of this section, when T itself is strongly minimal. So the point is: for a model $\mathcal{B}$ of our theory T, can the dimension of $\varphi(\mathcal{B})$ with respect

to *acl*, and so (at least in uncountable powers) the cardinality of $\varphi(\mathcal{B})$, play the same role as the dimension of B in the strongly minimal case?
At this point Lemma 7.8.5 applies once again, and ensures that $\varphi(\mathcal{B}_0)$ is properly included in $\varphi(\mathcal{B}_1)$ when $\mathcal{B}_0$ and $\mathcal{B}_1$ are models of T, $\mathcal{B}_0$ is an elementary substructure of $\mathcal{B}_1$ and $B_0 \neq B_1$. But now we can say much more. In fact we have what follows.

Lemma 7.8.8 *For every model $\mathcal{B}$ of T, $\mathcal{B}$ is prime over $\varphi(\mathcal{B})$. Moreover* $|\varphi(\mathcal{B})| = |B|$.

Proof. As T is ω-stable, there exists a model $\mathcal{B}'$ of T prime over $\varphi(\mathcal{B})$. $\mathcal{B}'$ elementarily embeds into $\mathcal{B}$ via a function fixing $\varphi(\mathcal{B})$ pointwise; again, we can assume that $\mathcal{B}'$ is an elementary substructure of $\mathcal{B}$. Consequently $\varphi(\mathcal{B}') = \varphi(\mathcal{B}) \cap B'$; on the other side $\varphi(\mathcal{B}') \subseteq \varphi(\mathcal{B})$, and so $\varphi(\mathcal{B}')$ is equal to $\varphi(\mathcal{B})$. Apply Lemma 7.8.5 and deduce $B = B'$. Hence $\mathcal{B}$ is prime over $\varphi(\mathcal{B})$. It remains to check that $|\varphi(\mathcal{B})| = |B|$. $\leq$ is trivial. On the other hand an easy application of the Löwenheim-Skolem Theorem to the theory of $\mathcal{B}_{\varphi(\mathcal{B})}$ yields a model $\mathcal{C}$ of T such that $|C| = |\varphi(\mathcal{B})|$ and $\varphi(\mathcal{B}) \subseteq C$ by an elementary inclusion. As $\mathcal{B}$ is prime over $\varphi(\mathcal{B})$, $\mathcal{B}$ elementarily embeds in $\mathcal{C}$. In particular $|\varphi(\mathcal{B})| = |C| \geq |B|$, as required. ♣

At last we are in a position to conclude our way.

Proof. (Morley's Theorem). Let $\mathcal{B}_0$ and $\mathcal{B}_1$ be two models of T having the same uncountable power λ. Owing to Lemma 7.8.8, each $\mathcal{B}_i$ $(i = 0, 1)$ is prime over $\varphi(\mathcal{B}_i)$ and satisfies $|\varphi(\mathcal{B}_i)| = \lambda$. Choose a basis X_i of $\varphi(\mathcal{B}_i)$ with respect to the *acl* dependence relation: X_i has again power λ because $|\varphi(\mathcal{B}_i)| = \lambda > \aleph_0$. Then X_0 and X_1 correspond to each other by some elementary bijection, which enlarges to an elementary bijection h between $\varphi(\mathcal{B}_0)$ and $\varphi(\mathcal{B}_1)$; h defines in its turn an isomorphism between the models $\mathcal{B}_0$ and $\mathcal{B}_1$ prime over $\varphi(\mathcal{B}_0)$ and $\varphi(\mathcal{B}_1)$ respectively. ♣

7.9 Biinterpretability and Zilber Conjecture

We have devoted several sections to the problem of classifying structures up to isomorphism and to Shelah's analysis of this question. But, needless to say, isomorphism is not the only possible classifying equivalence relation, even within structures. Another possible criterion, deeply related to Model Theory, is interpretability. The following two examples illustrate this alternative perspective and its underlying idea.

Examples 7.9.1 1. Natural numbers (viewed as non-negative integers) form a definable set in the ring $(\mathbf{Z}, +, \cdot)$ of integers: as recalled in Chapter 1, a celebrated Theorem of Lagrange says that they are exactly the sums of four squares in $(\mathbf{Z}, +, \cdot)$. So the whole structure $(\mathbf{N}, +, \cdot)$ is definable in $(\mathbf{Z}, +, \cdot)$, because the addition and multiplication in $\mathbf{N}$ are just the restrictions of the corresponding operations of $\mathbf{Z}$. This is a fundamental result: in fact, as $(\mathbf{N}, +, \cdot)$ lives in $(\mathbf{Z}, +, \cdot)$ as a definable structure, $(\mathbf{Z}, +, \cdot)$ inherits its undecidability phenomena related to Gödel Incompleteness Theorems, and in this sense is a "wild" structure.

2. In the same way $(\mathbf{N}, +, \cdot)$ lives in the rational field $(\mathbf{Q}, +, \cdot)$ as a definable structure. This is a deep theorem of Julia Robinson. So even the rational field inherits the complexity of $(\mathbf{N}, +, \cdot)$ and its undecidability.

So, generally speaking, when we meet a structure $\mathcal{A}$ in a language L and we realize that $\mathcal{A}$ defines, or also interprets another structure $\mathcal{A}'$ (possibly of a different language L'), then we can reasonably agree that $\mathcal{A}$ inherits the full complexity of $\mathcal{A}'$, and consequently is at least as difficult to dominate as $\mathcal{A}'$ is. Of course this can be extended to classes of structures. In this enlarged framework, we compare two classes of structures, $\mathcal{K}$ in a language L and $\mathcal{K}'$ in a language L' respectively. For simplicity, we can agree that both $\mathcal{K}$ and $\mathcal{K}'$ are elementary. We assume that there are suitable L-formulas defining, or also interpreting, in any structure $\mathcal{A} \in \mathcal{K}$ a structure $\mathcal{A}' \in \mathcal{K}'$ and that, conversely, every $\mathcal{A}' \in \mathcal{K}'$ can be recovered by some $\mathcal{A} \in \mathcal{K}$ in this way; we assume also that these formulas do not depend on the choice of $\mathcal{A}$ in $\mathcal{K}$. Then we say that $\mathcal{K}$ interprets $\mathcal{K}'$ and in this case we can agree that $\mathcal{K}$ inherits the complexity of $\mathcal{K}'$. Here are some further examples illustrating this point.

Examples 7.9.2 1. Recall that a *graph* is a structure (G, R) where R is a symmetric irreflexive binary relation, usually called *adjacency*. The (elementary) class of graphs interprets any class of structures, and so inherits in this way the full complexity of mathematics. The proof of this fact requires patience rather than ingenuousness. To avoid too many tedious details, let us illustrate its idea in a particular case, and see how graphs interpret arbitrary binary relations. So take any structure (A, R) where R is a binary relation on A, and form a graph (A', R') as follows. Let A' include A. Moreover, for every $a \in A$, add two new vertices a_0 and a_1 in A', both adjacent to a (so

(a, a_0), $(a, a_1) \in R'$). Finally, for every pair $e = (a, b) \in R$, add in A' two new vertices e_0, e_1 satisfying (a, e_0), (e_0, e_1), $(e_1, b) \in R'$, three more vertices adjacent to e_0, and four more vertices adjacent to e_1. For instance, here is a picture of (A', R') when $A = \{a, b, c\}$ and $R = \{(a, b)\}$.

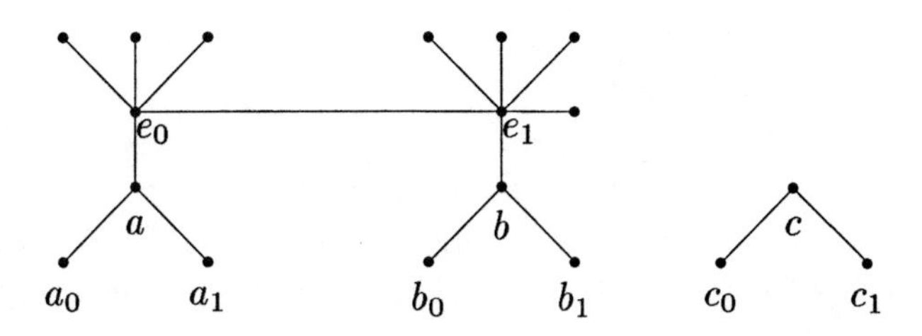

It is an easy exercise to realize that any (A, R) can be definably recovered inside the corresponding (A', R') in a way independent of the particular choice of (A, R). In fact, A is just the set of the vertices in A' having either 2 adjacent nodes, or 3 adjacent nodes such that 2 of them have no further adjacent node, while R is the set of the pairs $(a, b) \in R$ such that there are e_0, $e_1 \in A'$ for which (a, e_0), (e_0, e_1) $(e_1, b) \in R'$, e_0 has three adjacent nodes in addition to a and e_1 has four adjacent nodes besides e_0 and b.

2. The same can be said of the class of groups, and even of the class of nilpotent groups of class 2 (a comparatively slight generalization of abelian groups). This is a beautiful result of A. Mekler, showing that nilpotent groups of class 2 intepret graphs and so, through them, any class of structures. The proof uses brilliantly some non-trivial notions and tools from group theory. The conclusion is clear: groups, and even nilpotent groups of class 2, are a class as bad as possible, and inherits the full complexity of mathematics.

3. We said in 7.1 that, for a given (countable) field $\mathcal{K}$, $\mathcal{K}\langle x, y\rangle$-modules (i. e. $\mathcal{K}$-vectorspaces with two distinguished endomorphisms x and y) are an intractable class: no classification can be expected, even for finite dimensional objects, otherwise the word problem for groups would be solvable. Of course, every class interpreting $\mathcal{K}\langle x, y\rangle$-modules is at least as complicated as they are, and hence definitively a bad class. Several notable classes of modules share this negative feature. For instance,

this is the case of $\mathcal{K}[x, y]$-modules (with two commuting unknowns x and y), or $\mathbf{Z}[x]$-modules, and so on. The book of Prest quoted in the references at the end of the chapter includes a discussion of this point and a great deal of noteworthy examples.

So a possible way of classifying structures, or even classes of structures, is up to mutual interpretability (*biinterpretability*). Accordingly, one could try to characterize structures by looking at what is definable, or also interpretable, in them: groups, fields, and so on.
Incidentally notice that a relevant emphasis on the role of definability already arises within Shelah's classification analysis (for instance, think of the order property, looking at the orders definable in a given structure). However the study of mutual interpretability in mathematics did precede Model Theory, or, at least, modern Model Theory. For example, let us mention the celebrated Malcev correspondence between groups and rings, essentially showing that the class of unitary rings is biinterpretable with a suitable class of nilpotent groups of class 2, and confirming in this way how complicated these groups are.
But who mainly developed the biinterpretability program in Model Theory was Boris Zilber. Zilber's original project concerned the classification of uncountably categorical theories (those where Morley's Theorem applies) by looking at which groups, or fields, and so on, are definable in them.
As already observed, uncountably categorical theories include the strongly minimal ones, and the latter theories are the simplest possible (if we excludes finite structures). Accordingly their exam is a reasonable first step towards a general approach. So the question is: what strongly minimal structures look like? Can we reasonably classify them by looking at what they interpret?
Let us outline Zilber analysis in the strongly minimal setting. Recall that a structure $\mathcal{A}$ is said to be *strongly minimal* if its complete theory is: so, for every $\mathcal{B}$ elementarily equivalent to $\mathcal{A}$, the only definable subsets of B are those finite or cofinite. We introduced in Chapter 5 several examples of strongly minimal structures. To summarize them, let us work inside a fixed algebraically closed field $\mathcal{A}$.

Examples 7.9.3 1. Firstly view $\mathcal{A}$ as a structure in the language $\emptyset$, so as a mere infinite set. In this case, the theory of $\mathcal{A}$ equals that of infinite sets, and is strongly minimal. The structure of $\mathcal{A}$ is very poor, and no (infinite) group is definable here. Moreover

- for every subset X of A, $acl(X) = X$,

- for every positive integer n, the definable subsets of A^n are the finite Boolean combinations of

$$\{\vec{a} \in M \, : \, a_i = b\}, \quad \{\vec{a} \in M \, : \, a_i = a_j\}$$

with $i, j = 1, \ldots, n$, $b \in M$.

2. Let $\mathcal{A}_0$ be the prime subfield of $\mathcal{A}$, and look at $\mathcal{A}$ as a vectorspace over $\mathcal{A}_0$ in the appropriate language. Now the complete theory of $\mathcal{A}$ is that of infinite vectorspaces over $\mathcal{A}_0$, and is again strongly minimal. But this time

    - for every subset X of A, $acl(X)$ is the subspace of $\mathcal{A}$ spanned by X,
    - for every positive integer n, the definable subsets of A^n are the finite Boolean combinations of cosets of *pp*-definable subgroups of $\mathcal{A}$, and for $n > 1$ their class is larger than in Example 1. Notice that an infinite group is trivially definable in $\mathcal{A}$, but no field can be interpreted inside $\mathcal{A}$.

3. At last, view $\mathcal{A}$ just as an algebraically closed field. If p is its characteristic, then the complete theory of $\mathcal{A}$ is ACF_p and is strongly minimal. Moreover

    - for every subset X of A, $acl(X)$ is the algebraic closure of $\mathcal{A}$ in the field theoretic sense,
    - for every positive integer n, the definable subsets of A^n are the constructible ones, in other words the finite Boolean combinations of algebraic varieties of $\mathcal{A}^n$.

Now take any strongly minimal structure $\mathcal{A}$.

Definition 7.9.4 *$\mathcal{A}$ is called* **trivial** *if, for every $X \subseteq A$,*

$$acl(X) = \cup_{x \in X} acl(x).$$

Every (pure) infinite set is trivial. But, of course, vectorspaces and algebraically closed fields are not. Moreover no trivial strongly minimal structure can interpret an infinite group.

Definition 7.9.5 *$\mathcal{A}$ is called* **locally modular** *if, for every choice of X, $Y \subseteq A$ such that $X \cap Y \neq acl(\emptyset)$,*

$$(\star) \quad dim\,(X \cup Y) + dim\,(X \cap Y) = dim\,X + dim\,Y.$$

Every trivial structure $\mathcal{A}$ is locally modular (in fact, for every X and Y, $acl(X \cup Y) = acl(X) \cup acl(Y)$, so a basis of $X \cup Y$ can be formed by taking a basis a $X \cap Y$, extending it to a basis of X and a basis of Y, and gluing these bases together). But also vectorspaces are modular: in fact, in this case, $(\star)$ is just the Grassman formula (and does not need the assumption $X \cap Y \neq acl(\emptyset)$).

On the contrary, no algebraically closed field $\mathcal{A}$ (of transcendence degree ≥ 4) is locally modular. In fact, choose a_0, a_1, a_2, $a_3 \in A$ algebraically independent over the prime subfield $\mathcal{A}_0$, and form the extensions

$$X = A_0(a_0, a_1, a_2), \quad Y = A_0(a_0, a_3, a_1 a_3 + a_2).$$

Then $dim\, X = dim\, Y = 3$ but $dim(X \cap Y) = dim\, A_0(a_0) = 1$, whence $X \cap Y \neq acl(\emptyset)$, and $dim(X \cup Y) = dim(A_0(a_0, a_1, a_2, a_3)) = 4$. So $(\star)$ fails.

What is the significance of local modularity? Basically a locally modular $\mathcal{A}$ either is trivial or can define an infinite group. Furthermore one observes that any group $\mathcal{G}$ definable in a locally modular $\mathcal{A}$ is abelian-by-finite (in other words, it has a normal abelian subgroup of finite index); and every subset of any cartesian power G^n definable in $\mathcal{A}$ is a finite Boolean combination of cosets of definable subgroups of G^n. Accordingly, no infinite field is definable in $\mathcal{A}$.

In this setting Zilber raised in 1984 the following problem, generally called *Zilber Trichotomy Conjecture*.

Conjecture 7.9.6 (Zilber) *Let $\mathcal{A}$ be a strongly minimal non locally modular structure. Then $\mathcal{A}$ interprets an infinite field $\mathcal{K}$. Furthermore, for every positive integer n, the subsets of K^n definable in $\mathcal{A}$ are just those definable in $\mathcal{K}$ (and hence coincide with the constructible ones).*

Recall that, owing to Macintyre's Theorem, any infinite field interpretable in an ω-stable structure must be algebraically closed. Hence the importance of this conjecture is clear: according to it any strongly minimal structure $\mathcal{A}$ either is trivial, and so looks like an infinite set (as in Example 7.9.3 .1), or is locally modular and not trivial, and then resembles a module (as in Example 7.9.3.2), or looks like an algebraically closed field, because it interprets such a field (Example 7.9.3.3). Hence the conjecture would provide a quite satisfactory classification of strongly minimal structures (and theories) up to biinterpretability. But in 1993 Hrushovski showed that Zilber's Conjecture is false.

Theorem 7.9.7 (Hrushovski) *There do exist strongly minimal structures $\mathcal{A}$ which are not trivial but cannot interpret any infinite group.*

Clearly such a structure $\mathcal{A}$ is not locally modular and does not interpret any infinite field.
However Zilber Conjecture (more precisely, a suitable restatement) does hold in certain topological structures deeply related to strongly minimal models: the so called *Zariski geometries*. To introduce them, let us come back to Example 7.9.3.3, so dealing with algebraically closed fields $\mathcal{A}$.
We know that, for every positive integer n, the algebraic varieties of A^n are preserved under finite union and arbitrary intersection, and form the closed sets in the Zariski topology on $\mathcal{A}^n$. These topologies are Noetherian: none of them admits any infinite strictly decreasing sequence of closed sets. Moreover they satisfy the following properties (m and n denote below positive integers).

(Z1) Let $f = (f_1, \ldots, f_m)$ be a function from A^n in A^m. Assume that each component f_i $(1 \leq i \leq m)$, as a function from A^n in A, either projects A^n onto A or is constant. Then f is continuous.

(Z2) Every set $\{\vec{a} \in A : a_i = a_j\}$ with $1 \leq i, j \leq n$ is closed.

(Z3) The projection of a closed set of A^{n+1} onto A^n is a constructible set in A^n.

(Z4) A, as a closed set, is irreducible.

(Z5) Let X be a closed irreducible subset of A^n. For every $\vec{a} \in A^{n-1}$, let $X(\vec{a})$ denote the set of the elements $b \in A$ such that $(\vec{a}, b) \in X$. Then there is a natural N such that, for every $\vec{a} \in A^{n-1}$, either $|X(\vec{a})| \leq N$ or $X(\vec{a}) = A$. In particular, when $n = 1$, every closed proper subset of A must be finite.

(Z6) Let X be a closed irreducible subset of A^n, d denote the (topological) dimension of X. Then, for every i, j among $1, \ldots, n$, $X \cap \{\vec{a} \in A^n : a_i = a_j\}$ has dimension $\geq d - 1$.

(Z1) - (Z6) restate in a topological style some properties which are well known, or easy to check. For instance (Z5) follows directly from some simple algebraic facts and the strong minimality of $\mathcal{A}$ by a compactness argument: the reader may check this in detail as an exercise.
Now it is easy to realize that even infinite sets and vectorspaces over a countable field satisfy (Z1) - (Z6) provided one takes as closed subsets the finite Boolean combinations of the following sets:

- the sets of the tuples admitting a fixed coordinate in a given place, or equal coordinates in two different places, when dealing with pure infinite sets;
- the cosets of pp-definable subgroups when dealing with vectorspaces.

It is easy to control that in both cases these sets are actually the closed sets in a suitable topology.

Definition 7.9.8 *A* **Zariski structure** *(or* **geometry***) is a collection* $(A, \{T_n : n \text{ positive integer}\})$ *where* A *is a non-empty set, for every* n T_n *is a Noetherian topology on* A^n *and (Z1) - (Z6) hold.*

Hence the examples 7.9.3 produce Zariski structures. Conversely, let $(A, \{T_n : n \text{ positive integer}\})$ be any Zariski structure. Assume that A is the domain of some structure $\mathcal{A}$ (in a language L) such that, for every positive integer n, the subsets of A^n definable in $\mathcal{A}$ are just the finite Boolean combinations of closed sets in T_n (and so coincide with the constructible sets in T_n). Then it is easy to check that $\mathcal{A}$ is strongly minimal; moreover the possible triviality, or local modularity of $\mathcal{A}$ does not depend on L, or on the L-structure of $\mathcal{A}$, but only relies upon the characterization of the definable sets of $\mathcal{A}$ and so, after all, upon (Z1) - (Z6). More notably, in the restricted framework of Zariski structures, the Zilber Trichotomy Conjecture is true, as shown by Hrushoski and Zilber himself.

Theorem 7.9.9 (Hrushovski-Zilber) *Let* $(A, \{T_n : n \text{ positive integer}\})$ *be a Zariski structure,* $\mathcal{A}$ *be a strongly minimal structure with domain* A *such that, for every positive integer* n*, the subsets of* A^n *definable in* $\mathcal{A}$ *are just the constructible sets in* T_n*. If* $\mathcal{A}$ *is not locally modular, then* $\mathcal{A}$ *interprets an algebraically closed field* $\mathcal{K}$*, and* $\mathcal{K}$ *is unique up to definable isomorphism. Moreover, for every positive integer* n*, the subsets of* K^n *definable in* $\mathcal{A}$ *coincide with the ones definable in* $\mathcal{K}$*.*

7.10 Two algebraic examples

Let us summarize briefly some of the main notions introduced in this chapter by examining two relevant classes of algebraic structures, and their first order theories: differentially closed fields of characteristic 0, and existentially closed fields with an automorphism (again in characteristic 0). Both these examples play an important role in the model theoretic solution of some notable questions of Algebraic Geometry: we will describe these problems and their solution in the next chapter.

1. DCF_0. First let us deal with differentially closed fields of characteristic 0. Let us recall once again that their theory DCF_0 is complete and quantifier eliminable in its natural language L, containing the symbols $+$, $\cdot$, $-$, 0, 1, D and nothing more. So definable sets are easy to classify: as we saw in Chapter 3, they include the zero sets of (finite) systems of differential polynomials - in other words, the closed sets in the Kolchin topology - as well as their finite Boolean combinations - the constructible sets in this topology -, but nothing else. As a typical Kolchin closed set in a differentially closed field $(\mathcal{K}, D)$ let us mention the field of constants $C(\mathcal{K}) = \{a \in K : Da = 0\}$. This is an algebraically closed field - just as $\mathcal{K}$ -, and is strongly minimal even in L; in fact, D is identically 0 on $C(\mathcal{K})$ and so adds no further definable objects to the field structure on $C(\mathcal{K})$.

DCF_0 eliminates the imaginaries, too. Moreover DCF_0 is ω-stable with Morley rank ω, so independence makes sense in DCF_0, and indeed it is ruled by Morley rank: for $\vec{a}$, A and B as in Section 7.2,

$$\vec{a} \downarrow_A B \iff RM(tp(\vec{a}/B)) = RM(tp(\vec{a}/A)).$$

Of course this raises the question of characterizing algebraically Morley rank within differentially closed fields of characteristic 0. But there are also other ways of describing forking and independence in DCF_0, having a pretty algebraic flavour. For instance, one can preliminarily observe that, for every small A, $acl(A)$ - in the model theoretic sense - is just the (field theoretic) algebraic closure of the differential subfield generated by A $\mathbf{Q}(D^i a : a \in A, i \in \mathrm{N})$; at this point, one can realize that, for $\vec{a}$, A and B as usual, $\vec{a} \downarrow_A B$ just means that $acl(A \cup \vec{a})$ and $acl(B)$ are (algebraically) independent over $acl(A)$. Among other things, this characterization suggests an alternative rank notion, specifically concerning the differential framework: this is called *differential degree* or *D-degree* and denoted *D-dg*: for H a differential field and $\vec{a}$ a tuple of elements in Ω, $D\text{-}dg(\vec{a}/H)$ is the transcendence degre of the differential field generated by $H \cup \vec{a}$ over H. So, for $\vec{a}$, A and B as before, and A and B differential subfields for simplicity,

$$\vec{a} \downarrow_A B \iff D\text{-}dg(\vec{a}/B) = D\text{-}dg(\vec{a}/A).$$

However we have to be careful here: the last equivalence does not mean that RM and D-dg coincide. Their relationship, an algebraic characterization of RM in DCF_0 and the connection among differential degree, Morley rank and other possible ranks in DCF_0 are described in the references mentioned at the end of the chapter.

Now let us deal with biinterpretability, in particular let us consider strongly minimal sets in differentially closed fields $\mathcal{K}$ of characteristic 0. They include the constant subfield $C(\mathcal{K})$ (which also has differential degree 1, as it is easy to check). $C(\mathcal{K})$ is not locally modular, in fact the argument proposed in the last section for algebraically closed fields applies to $C(\mathcal{K})$ as well. But what is most remarkable in this setting is a theorem of Hrushovski and Sokolovic saying that Zilber Trichotomy Conjecture holds within strongly minimal sets in DCF_0. In fact all these sets are Zariski structures, and so obey the Hrushovski-Zilber Theorem. We can say even more: any strongly minimal set S which is not locally modular, and hence interprets an infinite field, does interpret the field of constants $C(\mathcal{K})$ up to a definable isomorphism. We will provide more details about these matters in Section 8.7.

2. $ACFA$. Now we deal with existentially closed fields with an automorphism. For simplicity, we still work in characteristic 0. Let $ACFA_0$ denote the corresponding theory in the natural language $L = \{+, \cdot, -, 0, 1, \sigma\}$ where σ is the symbol representing the automorphism. Recall that, this time, fixing the characteristic is not sufficient to ensure completeness: in order to characterize a model of $ACFA_0$ up to elementary equivalence, one has also to describe the action of the automorphism on the prime subfield $\mathbf{Q}$. Moreover $ACFA_0$ does not eliminate the quantifiers in L, although it is obviously model complete (as a model companion). Accordingly definable sets exhibit some more complications than in the differential case. In fact, they include the zero sets of (finite) systems of difference polynomials, as well as their finite Boolean combinations; the former are the closed sets, and the latter the constructible ones in a suitable topology. But now, as quantifier elimination fails, we have to consider also the projections of constructible sets - and nothing else, owing to model completeness - to capture the whole class of definable sets. An example of a closed set in a model $(\mathcal{K}, \sigma)$ of $ACFA_0$ is its fixed subfield $Fix(\sigma) = \{a \in K : \sigma(a) = a\}$. This is not algebraically closed (in particular it is not strongly minimal); but one can see that it is a pseudofinite field, so an infinite model of the theory of finite fields.

   $ACFA_0$ eliminates the imaginaries. This time no existentially closed field $(\mathcal{K}, \sigma)$ with an automorphism is ω-stable, or even stable. However $(\mathcal{K}, \sigma)$ is simple (as well as its fixed subfield, and any pseudofinite field). So independence makes sense in $ACFA_0$, and comes directly from forking, but cannot be ruled by Morley rank. Anyhow an explicit algebraic characterization can be done as follows. We work for simplicity in a big saturated model $(\mathcal{K}, \sigma)$ of $ACFA_0$. First one observes that, for every small A, $acl(A)$ coincides with the algebraic closure - in the field theoretic sense - of the difference subfield generated by A

$$\mathbf{Q}(\sigma^i(a) : a \in A, i \in \mathbf{Z})$$

(here we use the characteristic 0 assumption; prime characteristics cause some major trouble). At this point one shows that, for $\vec{a}$, A and B as usual, $\vec{a} \downarrow_A B$ just means that $acl(A \cup \vec{a})$ and $acl(B)$ are (algebraically) independent over $acl(A)$. This yields an appropriate notion of rank, of a pretty algebraic flavour, called *difference degree* or *σ-degree* and denoted *σ-dg*: for H a difference field and $\vec{a}$ a tuple of elements in Ω, $\sigma - dg(\vec{a}/H)$ is the transcendence degree of the difference field generated by $H \cup \vec{a}$ over H. When finite, the difference degree can reasonably replace Morley rank and provides a good notion of dimension in this unstable setting; on the other side, when $\sigma - dg(\vec{a}/H)$ is infinite, then clearly the $\sigma^i(a)$'s (when i ranges over integers) are algebraically independent over H. So, for $\vec{a}$, A and B as before, and A and B difference subfields for simplicity,

$$\vec{a} \downarrow_A B \quad \Longleftrightarrow \quad \sigma - dg(\vec{a}/B) = \sigma - dg(\vec{a}/A)$$

at least when the latter degrees are finite. When X is any definable set in K^n, the difference degree of X over H $\sigma - dg(X/H)$ is the maximal difference degree of a tuple $\vec{a}$ in X over H. In particular the fixed subfield of $\mathcal{K}$ gets difference degree 1. In this sense, $Fix(\sigma)$ is a "minimal" definable infinite set of $\mathcal{K}$. Notably, an adapted version of Zilber Trichotomy Conjecture holds for these "minimal" sets in $ACFA_0$, and even ensures in particular that, very roughly speaking, $Fix(\sigma)$ is the only non "locally modular" example among these structures.

7.11 References

The classification issue from the point of view of Descriptive Set Theory is treated in [68]; the particular and intriguing case of torsionfree abelian groups of finite rank is dealt with in [163]. Finite dimensional vectorspaces

with two distinguished endomorphisms over a fixed field, and the wildness of their classification problem are described in [136] or, more specifically, in [137].
The main references on Shelah's classification theory are just the Shelah book [149] and its revised and updated version [151]. Most of the topics of this chapter are treated there in detail. Another good and perhaps more accessible source on classification and stability theory is [8]. See also Makkai's paper [101]. Vaught's Conjecture is proposed in [174], and its solution in the ω-stable framework can be found in [152]. Lascar's paper [84] provides an enjoyable discussion of this matter.
The (uncountable) spectrum problem for complete countable first order theories is fully solved in [53].
Simple theories were introduced in [149], but it was Kim who showed, together with Pillay, their relevance within the classification program: see [71] and [73]. Kim again observed the key role of symmetry, transitivity and local character [72]. Wagner's book [175] provides a general and exhaustive report on this theme.
As already said, stability, superstability and the further dichotomies arising in the classification program are treated in Shelah books [149, 151], in [8] or also in [101]. [85] provides a nice and terse introduction to stability, using and emphasizing the notion of heir and coheir. [83] pursues this approach and deals in particular with Rudin-Keisler order and strong regularity.
The effectiveness aspects of Shelah's classification program is discussed in [114], while [2] examines its connections with Stationary Logic.
Turning our attention to the algebraic examples, let us mention [136] or also [56] for modules, and [19] for pseudofinite fields. [110] treats differentially closed and separably closed fields, and includes a wide bibliographical list on them. [62] and [155] provide other key references on DCF_0; see also [134]. Existentially closed fields with an automorphism are just the subject of [20]. An explicit example of such a field can be found in [58] and [88]; see also [95].
Shelah's uniqueness theorem is in [148], Morley's theorem in [117]. Another proof of Morley's Theorem is given in [9]; see also Sack's book [146], or [57]. The Ehrenfeucht - Mostowski models quoted in Section 7.8 are introduced in [38].
Finally, let us deal with biinterpretability. Malcev's correspondence is in [102] while Mekler's theorem on nilpotent groups of class 2 is in [113]. Zilber's program is developed in [182], where Zilber's Conjecture is also proposed. The negative solution of this conjecture is in [59], and the Hrushovski-Zilber theorem on Zariski structures in [64].

Chapter 8

Model Theory and Algebraic Geometry

8.1 Introduction

We have often emphasized in the past chapters the deep relationship between Model Theory and Algebraic Geometry: we have seen, and we are going to see also in this chapter that several relevant notions arising in Algebraic Geometry (like varieties, morphisms, manifolds, algebraic groups over a field $\mathcal{K}$) are definable objects and are consequently concerned with the model theoretic machinery developed in the previous pages. For instance, when $\mathcal{K}$ is algebraically closed, they are ω-stable structures.This connection can yield, and is actually yielding, significant fruits in both Model Theory and Algebraic Geometry. On the one hand, several techniques and ideas originated and employed within the specific setting of Algebraic Geometry can inspire a more abstract model theoretic treatment, applying to arbitrary classes of structures. In this sense Algebraic Geometry over algebraically closed fields can suggest new directions in the study of ω-stability: we will describe this connection in many sections of this chapter. However a parallel analysis can be developed inside other relevant areas, like differentially closed fields (and Differential Algebraic Geometry), or existentially closed fields with an automorphism, and so on.

On the other hand, it is right to observe that the benefits of this relationship regard not only Model Theory, but also, and relevantly, Algebraic Geometry. In particular, we will propose some prominent problems in Algebraic Geometry, whose solution does profit by Model Theory and its techniques. This will be the aim of the final section of this chapter.

8.2 Algebraic varieties, ideals, types

Let $\mathcal{K}$ be a field, n be a positive integer. We already introduced in the past chapters the (algebraic) varieties of $\mathcal{K}^n$. They are the zero sets of finite systems of polynomials of $\mathcal{K}[\vec{x}]$ (where $\vec{x}$ abbreviates, as usual, $(x_1, \ldots, x_n)$), and so are definable in $\mathcal{K}$. Moreover they form the closed sets in the Zariski topology of $\mathcal{K}^n$; accordingly even the Zariski open or constructible sets are definable in $\mathcal{K}$.
But the varieties of $\mathcal{K}^n$ are also closely related to the ideals of $\mathcal{K}[\vec{x}]$. In fact one can define a function $\mathcal{I}$ from varieties to ideals mapping any variety V of $\mathcal{K}^n$ into the ideal $\mathcal{I}(V)$ of the polynomials $f(\vec{x})$ in $\mathcal{K}[\vec{x}]$ such that $f(\vec{a}) = 0$ for every $\vec{a}$ in V. Checking that $\mathcal{I}(V)$ is indeed an ideal is straightforward; $\mathcal{I}(V)$ is even a *radical* ideal, in other words it coincides with its radical $rad\,\mathcal{I}(V)$: if $f(\vec{x}) \in K[\vec{x}]$ and, for some positive integer k, $f^k(\vec{x}) \in \mathcal{I}(V)$, then $f(\vec{x})$ already occurs in $\mathcal{I}(V)$. In particular $\mathcal{I}$ is not onto.
But there is also another function $\mathcal{V}$ in the other direction, from ideals to varieties, mapping any ideal I of $\mathcal{K}[\vec{x}]$ (in particular any radical ideal) into the set $\mathcal{V}(I)$ of those elements $\vec{a} \in \mathcal{K}^n$ annihilating all the polynomials of I

$$\mathcal{V}(I) = \{\vec{a} \in \mathcal{K}^n : f(\vec{a}) = 0 \quad \forall f(\vec{x}) \in I\}.$$

Due to the Hilbert Basis Theorem, I is finitely generated, and so $\mathcal{V}(I)$ is a variety. Indeed, any variety V can be obtained in this way by definition; in other words $\mathcal{V}$ is onto. Notice also that $V(I) = V(rad\,I)$ for every ideal I.
The definition of $\mathcal{I}$ and $\mathcal{V}$ trivially implies that, for every ideal I of $\mathcal{K}[\vec{x}]$, $\mathcal{I}(\mathcal{V}(I)) \supseteq I$. As $\mathcal{I}(\mathcal{V}(I))$ is a radical ideal, $\mathcal{I}(\mathcal{V}(I)) \supseteq rad\,I$. Hilbert's Nullstellensatz (see Chapter 3) ensures that, when $\mathcal{K}$ is an algebraically closed field, equality holds: for every ideal I of $\mathcal{K}[\vec{x}]$,

$$\mathcal{I}(\mathcal{V}(I)) = rad\,I.$$

It is easy to deduce that, if $\mathcal{K}$ is algebraically closed, then $\mathcal{I}$ and $\mathcal{V}$ determine two bijections, the one inverse of the other, between *varieties of* $\mathcal{K}^n$ and *radical ideals of* $\mathcal{K}[\vec{x}]$. We will still denote these restricted bijections by $\mathcal{I}$, $\mathcal{V}$ respectively. Notice that both reverse inclusion: for instance if V, W are two varieties of $\mathcal{K}^n$, then

$$\mathcal{I}(V) \supseteq \mathcal{I}(W) \quad \Longleftrightarrow \quad V \subseteq W.$$

We assume from now on that $\mathcal{K}$ is an algebraically closed field. By the way recall that, under this condition, what is definable in $\mathcal{K}$ is ω-stable of finite Morley rank, because $\mathcal{K}$ is strongly minimal.

Let us restrict our attention from radical ideals of $\mathcal{K}[\vec{x}]$ to *prime* ideals (those ideals I in $\mathcal{K}[\vec{x}]$ such that, if a product of two polynomials of $\mathcal{K}[\vec{x}]$ lies in I, then at least one factor polynomial is in I as well). Parallely we consider, among varieties of $\mathcal{K}^n$, the *irreducible* ones, so the non-empty varieties V that cannot decompose as a union of two proper subvarieties. It is easy to check that the previous bijections $\mathcal{I}$ and $\mathcal{V}$ between varieties of $\mathcal{K}^n$ and radical ideals of $\mathcal{K}[\vec{x}]$ link *irreducible varieties of* $\mathcal{K}^n$ and *prime ideals of* $\mathcal{K}[\vec{x}]$: for every variety V of $\mathcal{K}^n$, V is irreducible if and only if $\mathcal{I}(V)$ is prime. Notice also that $\mathcal{I}(\emptyset) = \mathcal{K}[\vec{x}]$.
A closer relationship links irreducible varieties and prime ideals. For instance, it is known that every non-empty variety V of $\mathcal{K}^n$ can be expressed as a finite irredundant union of irreducible varieties, and that this decomposition is unique up to permuting the involved irreducible varieties (which are consequently called the *irreducible components* of V); the irredundancy of the decomposition just means that no irreducible component of V is included in the union of the other components.
Specularly, every proper radical ideal I of $\mathcal{K}[\vec{x}]$ can be expressed as a finite intersection of prime ideals minimal with respect to inclusion; even this representation is unique up to permuting the involved minimal prime ideals. One can also realize that, under this point of view, for every non-empty variety V in $\mathcal{K}^n$, the irreducible components of V correspond to the minimal prime ideals occurring in the decomposition of $\mathcal{I}(V)$.
So far we have summarized some very familiar topics of basic Algebraic Geometry. Now let Model Theory intervene. As we saw in Section 5.3, prime ideals of $\mathcal{K}[\vec{x}]$ - and hence, through them, irreducible varieties of $\mathcal{K}^n$ - directly and naturally correspond to n-types over K. In fact, for an algebraically closed field $\mathcal{K}$, there are two bijections $\mathbf{i}$ and $\mathbf{p}$, one inverse of the other, between n-types over K and prime ideals of $\mathcal{K}[\vec{x}]$. Basically, for every n-type p over K,

$\mathbf{i}(p)$ is the ideal of the polynomials $f(\vec{x}) \in K[\vec{x}]$ such that the formula $f(\vec{v}) = 0$ is in p,

and, conversely, for every prime ideal I of $\mathcal{K}[\vec{x}]$,

$$\{f(\vec{v}) = 0 \,:\, f(\vec{x}) \in I\} \cup \{\neg(g(\vec{v}) = 0) \,:\, g(\vec{x}) \in K[\vec{x}] - I\}$$

enlarges to a unique n-type $\mathbf{p}(I)$ over K. Accordingly, for every irreducible variety V of $\mathcal{K}^n$ and polynomial $f(\vec{x}) \in K[\vec{x}]$,

$$f(\vec{a}) = 0, \forall \vec{a} \in V \Leftrightarrow f(\vec{x}) \in \mathcal{I}(V) \Leftrightarrow \text{“}f(\vec{v}) = 0\text{”} \in \mathbf{p}(\mathcal{I}(V));$$

$\mathbf{p}(\mathcal{I}(V))$ is called a *generic type* of the variety V. More generally a *generic type* of an arbitrary non-empty variety V of $\mathcal{K}^n$ is a generic type of an irreducible component of V. This model theoretic notion of generic type just corresponds to the idea of a generic point of an irreducible variety V introduced in Algebraic Geometry: the latter is a point just annihilating the polynomials in $\mathcal{I}(V)$ and nothing more, and so can be equivalently defined as a realization of the generic type $\mathbf{p}(\mathcal{I}(V))$. It can be obtained as follows. As V is irreducible, $\mathcal{I}(V)$ is prime and consequently the quotient ring $\mathcal{K}[\vec{x}]/\mathcal{I}(V)$ is an integral domain; $\mathcal{K}[\vec{x}]/\mathcal{I}(V)$ contains a subring $\{k+\mathcal{I}(V) : k \in K\}$ isomorphic to $\mathcal{K}$ and a tuple $\vec{x}+\mathcal{I}(V)$ annihilating just the polynomials of $\mathcal{I}(V)$; so the field of fractions $\mathcal{K}(V)$ of $\mathcal{K}[\vec{x}]/\mathcal{I}(V)$ extends $\mathcal{K}$ - up to isomorphism - and includes a generic point $\vec{a}(V) = \vec{x}+\mathcal{I}(V)$ of V.
To conclude this section let us observe what follows.

Proposition 8.2.1 *Let n be a positive integer, $\mathcal{K}$ be an algebraically closed field, V be an irreducible variety of $\mathcal{K}^n$, O be a Zariski open set of $\mathcal{K}^n$ satisfying $V \cap O \neq \emptyset$. Then the generic type p of V contains the formula "$\vec{v} \in 0$".*

Proof. In fact $V - O$ is a variety properly included in V; consequently $\mathcal{I}(V-O) \supset \mathcal{I}(V)$. If p does not contain "$\vec{v} \in O$", then p has to include "$\vec{v} \in V - O$", and hence every formula "$f(\vec{v}) = 0$" when $f(\vec{x})$ ranges over $I(V-O)$; accordingly, for some polynomial $f(\vec{x}) \notin I(V)$, "$f(\vec{v}) = 0$" belongs to p: a contradiction. ♣

8.3 Dimension and Morley rank

We maintain throughout this section our assumption that $\mathcal{K}$ is an algebraically closed field. Let V be an irreducible variety of $\mathcal{K}^n$. Algebraic Geometry equips V with a *dimension* in the followig way. As we saw in the last section, there is a minimal field $\mathcal{K}(V)$ extending $\mathcal{K}$ by a realization $\vec{a}(V)$ of the generic type of V, in other words by a generic point of V. The *dimension* of V ($dim(V)$) is just the transcendence degree of $\mathcal{K}(V)$ over $\mathcal{K}$. This makes sense for an irreducible V, but can be easily extended to any non-empty variety V of $\mathcal{K}^n$. In this enlarged setting the *dimension* of V ($dim(V)$) is the maximal dimension of an irreducible component of V. Finally, the *dimension* of a constructible non-empty set X $dim(X)$ is the dimension of the closure of X in the Zariski topology (a non-empty variety).

On the other side every variety V of $\mathcal{K}^n$ (irreducible or not) and, more generally, every constructible set $X \subseteq K^n$ is definable in $\mathcal{K}$; $\mathcal{K}$ is algebraically closed, hence ω-stable; accordingly V and X have their Morley rank. We want to compare the dimension and the Morley rank of V, or X, and to show that they coincide. We start by examining an irreducible variety V.

Lemma 8.3.1 *Let V be an irreducible variety of $\mathcal{K}^n$, p be its generic type. Then $dim(V) = RM(p)$.*

Proof. p is realized by $\vec{a}(V)$ in $\mathcal{K}(V)$. So the Morley rank of p equals the transcendence degree of $\mathcal{K}(V)$ over $\mathcal{K}$ (see Section 6.1), and hence the dimension of V. ♣

Lemma 8.3.2 *Let V, W be two irreducible varieties of $\mathcal{K}^n$, p, q denote their generic types. If $V \subset W$, then $RM(p) > RM(q)$.*

Proof. $V \subset W$ implies $\mathcal{I}(W) \supseteq \mathcal{I}(V)$ and hence

($\star$) for every $f(\vec{x}) \in \mathcal{K}[\vec{x}]$, if $f(\vec{a}(V)) = 0$, then $f(\vec{a}(W)) = 0$ as well.

In particular the transcendence degree of $\mathcal{K}(V)$ over $\mathcal{K}$ is not smaller than that of $\mathcal{K}(W)$ over $\mathcal{K}$, and so $RM(p) \geq RM(q)$. Now assume $RM(p) = RM(q)$, then $\mathcal{K}(V)$, $\mathcal{K}(W)$ have the same transcendence degree over $\mathcal{K}$. Recall that $\mathcal{K}(V) = \mathcal{K}(\vec{a}(V))$, and similarly for W. ($\star$) implies that, if $i_1, \ldots, i_m \leq n$ and $a_{i_1}(W), \ldots, a_{i_m}(W)$ form a transcendence basis of $\mathcal{K}(W)$ over $\mathcal{K}$, then $a_{i_1}(V), \ldots, a_{i_m}(V)$ are algebraically independent over $\mathcal{K}$ and so form a transcendence basis of $\mathcal{K}(V)$ over $\mathcal{K}$. Accordingly one can define an isomorphism of $\mathcal{K}_W = \mathcal{K}(a_{i_1}(W), \ldots, a_{i_m}(W))$ onto $\mathcal{K}_V = \mathcal{K}(a_{i_1}(V), \ldots, a_{i_m}(V))$ fixing K pointwise and mapping $a_i(W)$ into $a_i(V)$ for every $i = i_1, \ldots, i_m$. This isomorphism can be enlarged in the usual way to an isomorphism between $\mathcal{K}_W[\vec{x}]$ and $\mathcal{K}_V[\vec{x}]$. By using ($\star$) once again, one sees that, for every $j = 1, \ldots, n$ with $j \neq i_1, \ldots, i_m$, the minimum polynomial of $a_j(W)$ over $\mathcal{K}_W$ must correspond to the minimum polynomial of $a_j(V)$ over $\mathcal{K}_V$ in this isomorphism. Accordingly we obtain an isomorphism of $\mathcal{K}(W)$ onto $\mathcal{K}(V)$ fixing K pointwise and mapping $a_j(W)$ into $a_j(V)$ for every $j = 1, \ldots, n$. But $\vec{a}(V)$ realizes p and $\vec{a}(W)$ realizes q, whence $p = q$ and, in conclusion, $V = W$. ♣

Now we can show that dimension and Morley rank coincide for an irreducible variety V.

Theorem 8.3.3 *Let V be an irreducible variety of $\mathcal{K}^n$. Then $RM(V) = dim(V)$, $GM(V) = 1$.*

Proof. Let p be the generic type of V. p is the only n-type over $\mathcal{K}$ containing the formula "$\vec{v} \in V$" that defines V, and satisfying $\mathcal{V}(\mathbf{i}(p)) = V$. If q is another n-type over $\mathcal{K}$ containing "$\vec{v} \in V$", then $\mathbf{i}(q) \supset \mathbf{i}(p)$, whence $\mathcal{V}(\mathbf{i}(q)) \subset \mathcal{V}(\mathbf{i}(p)) = V$. By Lemma 8.3.2, $RM(q) < RM(p)$. Then $RM(V) = RM(p)$ and $GM(V) = 1$. By Lemma 8.3.1, $RM(V) = RM(p) = dim(V)$. ♣

Corollary 8.3.4 *Let V be an irreducible variety of $\mathcal{K}^n$ (with the relative topology of the Zariski topology). If O is a non-empty open set of V, then $RM(O) = RM(V)$. If $W \subset V$ is a closed set of V, then $RM(W) < RM(V)$.*

Proof. The former claim follows from Proposition 8.2.1 and the fact that, if p is the generic point of V, then $RM(p) = RM(V)$. At this point the latter claim is a consequence of the fact that V has Morley degree 1. ♣

Corollary 8.3.5 *Let V be a non-empty variety of $\mathcal{K}^n$. Then $RM(V) = dim(V)$, furthermore $GM(V)$ equals the number of the irreducible components of V having the same dimension as V.*

Proof. V is the (finite) union of its irreducible components. Then the Morley rank of V coincides with the maximal rank of its components. So by Theorem 8.3.3 and the definition of $dim(V)$ $RM(V) = dim(V)$. Moreover, if V_0 and V_1 are two different irreducible components of maximal rank of V, then $V_0 \cap V_1$ is a closed subset of V_0 properly included in V_0. By Corollary 8.3.4, $RM(V_0 \cap V_1) < RM(V_0)$. This implies that $GM(V)$ equals the number of the irreducible components of maximal rank in V. ♣

Dimension and Morley rank coincide even for constructible sets $X \subseteq K^n$. Recall that, owing to Tarski's quantifier elimination Theorem, constructible just means definable in $\mathcal{K}$. Furthermore a constructible $X \subseteq K^n$ can be represented as a union of finitely many sets, which are in their turn the intersection of a variety W - so a Zariski closed set, the zero set of a finite system of equations - and a Zariski open set O - the set of the elements of K^n satisfying finitely many inequations

$$g_0(\vec{x}) \neq 0, \ldots, g_s(\vec{x}) \neq 0$$

with $g_0(\vec{x}), \ldots, g_s(\vec{x}) \in K[\vec{x}]$, or also, equivalently, a single inequation

$$\prod_{j \leq s} g_j(\vec{x}) \neq 0 \ - .$$

An open set defined by a unique inequation is called *principal.*

Corollary 8.3.6 *Let $X \subseteq K^n$ be constructible (hence definable). Then $RM(X) = dim(X)$. Moreover, if $\bar{X}$ denotes the Zariski closure of X, then $RM(\bar{X}) = RM(X)$ and $RM(\bar{X} - X) < RM(X)$.*

Proof. We know $dim(X) = dim(\bar{X})$, and $dim(\bar{X}) = RM(\bar{X})$ (by Corollary 8.3.5). So the former claim is done if $RM(X) = RM(\bar{X})$. As observed before, X a finite union of sets of the form $W \cap O$ where W is a non-empty variety of $\mathcal{K}^n$ and O is open in $\mathcal{K}^n$. As any non-empty variety of $\mathcal{K}^n$ decomposes in its turn as the union of its irreducible components, we can assume that every variety W occurring in the above representation of X is irreducible. By Corollary 8.3.4, $RM(W \cap O) = RM(W) = dim(W)$. Consequently the Morley rank of X (that equals the maximal rank of the sets $W \cap O$) coincides with the Morley rank of $\bar{X}$. This proves the former claim. But, by Corollary 8.3.4 once again, $RM(W \cap O) < RM(W)$ for W and O as before and W irreducible. So, in conclusion, $RM(\bar{X} - X) < RM(X)$. ♣

8.4 Morphisms and definable functions

$\mathcal{K}$ still denotes an algebraically closed field. Let n, m be positive integers, V, W be two algebraic varieties in $\mathcal{K}^n$, $\mathcal{K}^m$ respectively. Algebraic Geometry defines what a *morphism* from V to W is: it is a function f from V into W such that, for every $i = 1, \ldots, m$, the composition f_i of f and the i-th projection of K^m onto K is a polynomial map. One easily sees that a morphism is a continuous function with respect to the topology induced on V and W by the Zariski topology.
But what is remarkable for our purposes is that a variety morphism is always definable in $\mathcal{K}$. For instance, if f is, as above, a morphism from V to W, then "$f(\vec{v}) = \vec{w}$" is defined by the formula

$$\text{“}\vec{v} \in V\text{”} \wedge \text{“}\vec{w} \in W\text{”} \wedge \bigwedge_{1 \leq i \leq m} \text{“}f_i(\vec{v}) = w_i\text{”}.$$

Conversely what can we say about an arbitrary definable function f in $\mathcal{K}$? Certainly both the domain and the image of f are definable sets. Furthermore, if the image of f is a subset of K^m, then, for every $i = 1, \ldots, m$, the composition f_i of f and the i-th projection of K^m onto K is definable. So we are restricted to characterize definable functions from a subset of K^n into K, for some positive integer n. In this framework one can observe what follows.

Theorem 8.4.1 *Let f be a definable function from $\mathcal{K}^n$ into $\mathcal{K}$. Then there are a non-empty open subset O of $\mathcal{K}^n$, a rational function r and (when $\mathcal{K}$ has a prime characteristic) a positive integer h such that*

- *if car $\mathcal{K} = 0$, then f equals r on O;*
- *if car $\mathcal{K}$ is a prime p and Fr denotes the Frobenius map from $\mathcal{K}$ to $\mathcal{K}$ (that mapping any $a \in K$ into $Fr(a) = a^p$), then f coincides with $Fr^{-h} \cdot r$ on O.*

Just to underline the power of this result, let us recall that, owing to Corollary 8.3.4, every non-empty open subset O of $\mathcal{K}^n$ has the same Morley rank n as $\mathcal{K}^n$, while the Morley rank of $\mathcal{K}^n - O$ is smaller (for, $\mathcal{K}^n$ is an irreducible variety of rank n). Secondly, it is worth recalling the general fact that, if f is a function from a variety V of $\mathcal{K}^n$ into $\mathcal{K}$ and, for every $\vec{a} \in V$, there is an open neighbourhood O of $\vec{a}$ such that f equals some rational function in $O \cap V$, then f can be globally expressed as a polynomial function.

Now let us show Theorem 8.4.1.

Proof. Let Ω be the universe of the theory of $\mathcal{K}$, $t_1, \ldots, t_n \in \Omega$ be algebraically independent over $\mathcal{K}$. As f is K-definable, any automorphism of Ω fixing $\mathcal{K}$ and $t_1, \ldots, t_n$ pointwise acts identically also on $f(\vec{t})$ ($\vec{t}$ denotes here the tuple $(t_1, \ldots, t_n)$). Whence $f(\vec{t})$ is in $dcl(K \cap \vec{t})$. So, if $\mathcal{K}$ has characteristic 0, then

$$f(\vec{t}) = r(\vec{t})$$

for some suitable rational function r with coefficients in K, while, if $\mathcal{K}$ has prime characteristic p, then

$$f(\vec{t}) = Fr^{-h_p}(r(\vec{t}))$$

where r is as before and h_p is a positive integer. Put $s = r$ when $car\,\mathcal{K} = 0$, $s = Fr^{-h_p} \cdot r$ otherwise. The elements in K^n where f coincides with s form a set X definable in $\mathcal{K}$ (just as f and s), and the formula "$s(\vec{v}) = f(\vec{v})$" defining it is in the type of $\vec{t}$ over K. As $t_1, \ldots, t_n$ are algebraically independent over $\mathcal{K}$, this type has Morley rank n. Hence $RM(X) \geq n$. As $X \subseteq K^n$, the Morley rank of X must equal n, and $RM(K^n - X) < n$. It follows that the

Zariski closure of $K^n - X$ has Morley rank $< n$, and so its complement is an open set O in X of Morley rank n where $f = s$. ♣

8.5 Manifolds

Throughout this section $\mathcal{K}$ still denotes an algebraically closed field and n a positive integer. We deal here with (abstract) *manifolds* in $\mathcal{K}^n$ and we show that they are definable objects. First let us recall their definition.

Definition 8.5.1 *A* **manifold** *of $\mathcal{K}^n$ is a structure $\mathbf{V} = (V, (V_i, f_i)_{i \leq m})$ where m is a natural and*

- *V is a subset of K^n (called* **atlas***);*
- *$V_0, \ldots, V_m$ are subsets of V, and V is the union $\bigcup_{i \leq m} V_i$;*
- *for every $i \leq m$, f_i is a bijection of V_i onto a Zariski closed set U_i (a* **coordinate chart** *of the atlas V);*
- *for $i, j \leq m$ and $i \neq j$, $f_i(V_i \cap V_j) = U_{ij}$ is an open subset of U_i;*
- *for $i, j \leq m$ and $i \neq j$, $f_{ij} = f_i \cdot f_j^{-1}$ (a bijection between U_{ji} and U_{ij}) can be locally expressed as a tuple of rational functions.*

Manifolds include several familiar examples.

Examples 8.5.2 1. Every algebraic variety (so every Zariski closed set) V of $\mathcal{K}^n$ is a manifold, provided we set $m = 0$ and choose $U_0 = V_0 = V$ as the only coordinate chart of the atlas; the resulting manifold is called *affine*.

2. Let O be an open principal set of $\mathcal{K}^n$; O is defined by a single inequation $\neg$"$g(\vec{v}) = 0$"; notice that the formula "$g(\vec{v}) \cdot v_{n+1} = 1$" defines a closed V in $\mathcal{K}^{n+1}$, and it is easy to control that the projection of K^{n+1} onto K^n by the first n coordinates determines a bijection f of V onto O. Accordingly $(V, (V, f^{-1}))$ is a manifold with the only chart O. Such a manifold is called *semiaffine*.

3. Also the projective space $\mathbf{P}^n(\mathcal{K})$ can be equipped with a manifold structure. In fact, view $\mathbf{P}^n(\mathcal{K})$ as the quotient set of $K^{n+1} - \{\vec{0}\}$ with respect to the equivalence relation $\sim$ linking two non zero tuples $\vec{x} = (x_0, x_1, \ldots, x_n)$ and $\vec{y} = (y_0, y_1, \ldots, y_n)$ in K^{n+1} if and only

if there is some $k \in K$ such that $x_i = ky_i$ for every $i \leq n$. For $\vec{x} \in K^{n+1} - \{0\}$, let $(x_0 : x_1 : \ldots : x_n)$ be the class of $\vec{x}$ with respect to this relation. Moreover, for $i \leq n$, let

- $\mathbf{A}_i$ denote the set of the elements $(x_0 : x_1 : \ldots : x_n)$ in $\mathbf{P}^n(\mathcal{K})$ such that $x_i \neq 0$,
- f_i be the function of $\mathbf{A}_i$ into K^n mapping any $(x_0 : x_1 : \ldots : x_n)$ in $\left(\frac{x_0}{x_i}, \ldots, \frac{x_{i-1}}{x_i}, \frac{x_{i+1}}{x_i}, \ldots, \frac{x_n}{x_i}\right)$.

It is straightforward to check that $(\mathbf{P}^n(\mathcal{K}), (\mathbf{A}_i, f_i)_{i \leq n})$ is a manifold.

Notice that affine and semiaffine varieties are definable -even as structures- in $\mathcal{K}$. Moreover $\mathbf{P}^n(\mathcal{K})$ is interpretable in $\mathcal{K}$ both as a set (since the relation $\sim$ is $\emptyset$-definable) and as a manifold. But algebraically closed fields uniformly eliminate the imaginaries, so we can view $\mathbf{P}^n(\mathcal{K})$ even as a definable structure in $\mathcal{K}$.
More generally one can show

Theorem 8.5.3 *Let $\mathbf{V} = (V, (V_i, f_i)_{i \leq m})$ be a manifold of $\mathcal{K}^n$. Then $\mathbf{V}$ is a structure definable in $\mathcal{K}$.*

Proof. As algebraically closed fields have the uniform elimination of imaginaries, it is sufficient to show that $\mathbf{V}$ is a structure interpretable in $\mathcal{K}$. In fact, every map U_i of the atlas V $(i \leq m)$ is definable in $\mathcal{K}$. V can be regarded as the quotient set of the disjoint union of the charts U_i (with $i \leq m$) with respect to the equivalence relation identifying U_{ij} and U_{ji} via f_{ij} for every $i < j \leq m$; moreover, for every $i \leq m$, V_i can be definably recovered as the image of U_i by the projection into the quotient set V, and f_i is given by the inverse function of this projection (restricted to U_i). So our claim is proved if we show that, for every $i, j \leq m$ with $i \neq j$, f_{ij} is definable. But f_{ij} can be locally expressed as a rational function, and its domain U_{ij} is an open subset of U_i and accordingly can be written as a finite union of principal open sets. So the theorem is a direct consequence of the next result.

Lemma 8.5.4 *Let O be a principal open of $\mathcal{K}^n$, and let $q(\vec{x}) \in K[\vec{x}]$ be a polynomial satisfying $O = \{\vec{a} \in K^n : q(\vec{a}) \neq 0\}$. Let f be a function of O into $\mathcal{K}^n$ which can be locally expressed as a rational function. Then there are a polynomial $r(\vec{x}) \in K[\vec{x}]$ and a positive integer m such that $f(\vec{a}) = r(\vec{a})/q^m(\vec{a})$ for every $\vec{a} \in O$. In particular f is definable.*

Proof. We know that O is canonically homeomorphic to the closed subset V of $\mathcal{K}^{n+1}$ defined by $q(\vec{v}) \cdot v_{n+1} = 1$. Under this perspective f can be replaced by the function $f^\star$ of V into K mapping any tuple $(\vec{a}, a_{n+1}) \in V$ in $f(\vec{a})$. Even $f^\star$ can be locally expressed as a rational function, and hence as a polynomial function in $(\vec{x}, x_{n+1})$ (by the general fact we recalled before the proof of Theorem 8.4.1). So there is some polynomial $s(\vec{x}, x_{n+1}) \in K[\vec{x}, x_{n+1}]$ such that

$$f^\star(\vec{a}, (q(\vec{a}))^{-1}) = s(\vec{a}, (q(\vec{a}))^{-1}) \quad \forall \vec{a} \in \Omega.$$

Let m denote the degree of $s(\vec{x}, x_{n+1})$ with respect to x_{n+1}. Then there is some polynomial $r(\vec{x}) \in K[\vec{x}]$ such that, for every $\vec{a} \in O$,

$$f(\vec{a}) = f^\star(\vec{a}, (q(\vec{a}))^{-1}) = \frac{r(\vec{a})}{q^m(\vec{a})}. \quad \clubsuit$$

Notice that a manifold, when regarded as a definable structure, may lose part of its geometric features. For instance the Zariski closed subset of $\mathcal{K}^2$ defined by

$$x_1 \cdot (x_2 - 1) = x_2 \cdot (x_2 - 1) = 0$$

is an affine manifold, formed by the line $x_2 = 1$ and the point $(0, 0)$, and hence is the disjoint union of two closed sets. However, consider the manifold given by the projective line $\mathbf{P}^1(\mathcal{K})$ (as seen in Example 8.5.2, 3). From the definable point of view, its atlas has two charts, each of them is a line and these lines coincide except a single point. So the resulting manifold is again the union of a line $\{(1 : x_1) : x_1 \in K\}$ and a point $(0 : 1)$, and as a definable object is quite similar to the previous one. But $\mathbf{P}^1(\mathcal{K})$ is not the union of two distinct closed sets.
On the other side, it is noteworthy that every definable subset $X \subseteq K^n$ can be easily equipped with a manifold structure. In fact X decomposes as a union $\bigcup_{i \leq m}(V_i \cap O_i)$ where m is a natural and, for every $i \leq m$, V_i is a Zariski closed of $\mathcal{K}^n$ and O_i is a principal open set, so that $V_i \cap O_i$ is canonically homeomorphic to a closed U_i of $\mathcal{K}^{n+1}$, as observed before. On this basis, it is easy to build a manifold structure on X (with $U_0, \ldots, U_m$ as atlas maps).
Furthermore a manifold $\mathbf{V}$, viewed as a definable structure, is ω-stable.

8.6 Algebraic groups

A basic example of algebraic group over a field $\mathcal{K}$ is the linear group of degree n over $\mathcal{K}$ $GL(n, \mathcal{K})$, where n is a positive integer. Observe:

- $GL(n, \mathcal{K})$ is a principal open of $\mathcal{K}^{n^2}$, because it is defined by the disequation $\neg(det(\vec{v}) = 0)$; hence $GL(n, \mathcal{K})$ is canonically homeomorphic to the Zariski closed of $\mathcal{K}^{n^2+1}$ given by the equation $det(\vec{v}) \cdot v_{n^2+1} = 1$;
- the product operation $\cdot$ in $GL(n, \mathcal{K})$ is a morphism of varieties and so is definable.

Accordingly the group $GL(n, \mathcal{K})$ is a structure definable in $\mathcal{K}$. Moreover, if we assume $\mathcal{K}$ algebraically closed, $GL(n, \mathcal{K})$ is an ω-stable group of finite Morley rank.
Also linear algebraic groups are examples of algebraic groups. Recall that a *linear algebraic group* over a field $\mathcal{K}$ is a closed subgroup $\mathcal{G}$ of some linear group $GL(n, \mathcal{K})$. In particular a linear algebraic group $\mathcal{G}$ is a variety over $\mathcal{K}$, hence is definable (as a group) in $\mathcal{K}$ and is ω-stable of finite Morley rank when $\mathcal{K}$ is algebraically closed. Under the last assumption, we can say even more: indeed, for an algebraically closed $\mathcal{K}$, the linear algebraic groups are just those subgroups of the linear groups $GL(n, \mathcal{K})$ which are definable in $\mathcal{K}$. Let us see why.

Theorem 8.6.1 *Let $\mathcal{K}$ be an algebraically closed field, n be a positive integer, $\mathcal{G}$ be a subgroup of $GL(n, \mathcal{K})$. Then $\mathcal{G}$ is closed if and only if $\mathcal{G}$ is definable (in $\mathcal{K}$).*

Proof. Clearly, if $\mathcal{G}$ is closed -in other words $\mathcal{G}$ is a variety-, then $\mathcal{G}$ is definable. Conversely suppose that $\mathcal{G}$ is a definable group; let $\bar{G}$ be the closure of G with respect to the Zariski topology, and let a be an element of $\bar{G}$. Every open set of $\mathcal{K}^{n^2}$ containing a overlaps G. Consequently, for $b \in G$, every open O including ba overlaps G; in fact, if $ba \in O$, then $a \in b^{-1}O$; $b^{-1}O$ is open because the left multiplication by b is a morphism of varieties, and so is continuous; whence $(b^{-1}O) \cap G \neq \emptyset$ and so $O \cap bG \neq \emptyset$; as $b \in G$, bG is just G, hence $O \cap G \neq \emptyset$. In conclusion, for every $a \in \bar{G}$, $Ga \subseteq \bar{G}$. If there is any $a \in \bar{G} - G$, then $Ga \subseteq \bar{G}$ is disjoint from G and has the same Morley rank as G. So $RM(\bar{G} - G) \geq RM(G)$, which contradicts Corollary 8.3.6. Hence $\bar{G} \subseteq G$ and G is closed. ♣

As already said, linear groups and linear algebraic groups exemplify algebraic groups. In fact an *algebraic group* over a field $\mathcal{K}$ is defined as a manifold over $\mathcal{K}$ carrying a group structure whose product and inverse operations are (manifold) morphisms.
Of course, understanding this definition preliminarily requires to state what a product of two manifolds is, and to realize that this product is a manifold

as well; moreover we should specify what a manifold morphism is. We omit here the details. The former point is comparatively simple and natural to clarify, while the concept of morphism is more complex to introduce; but, once this is done, one can easily show that manifold morphisms are definable (as it is reasonable to expect).
Hence, if $\mathcal{K}$ is algebraically closed, then every algebraic group $\mathcal{G}$ over $\mathcal{K}$ is a structure definable in $\mathcal{K}$ and so ω-stable of finite Morley rank.
As already said, linear algebraic groups are algebraic groups; in fact they just correspond to affine (or also semiaffine) manifolds. And indeed Theorem 8.6.1 can be regarded as a particular case of a general result ensuring that every group definable in an algebraically closed field $\mathcal{K}$ is an algebraic group over $\mathcal{K}$ up to definable isomorphism. This fact was shown by Hrushovski (and Van den Dries), who observed that it is implicitly contained in some results of A. Weil. So it is commonly quoted as the Hrushovski-Weil Theorem.

Theorem 8.6.2 (Hrushovski-Weil) *Let $\mathcal{K}$ be an algebraically closed field and $\mathcal{G}$ be a group definable in $\mathcal{K}$. Then $\mathcal{G}$ is isomorphic to an algebraic group over $\mathcal{K}$ by a function definable in $\mathcal{K}$*

Let us spend some more words about the connection between algebraic groups and ω-stable groups. We have seen that every algebraic group over an algebraically closed field $\mathcal{K}$ is ω-stable of finite Morley rank. Of course ω-stable groups include further relevant examples: for istance, any divisible torsionfree abelian group - so basically any vectorspace over $\mathbf{Q}$ - is ω-stable, and even strongly minimal. More generally, it was shown by Angus Macintyre that the ω-stable (pure) abelian groups are just the direct sums of a divisible abelian group and an abelian group of finite exponent.
However the techniques used in the investigation of abelian groups do not seem appropriate to handle ω-stable groups. On the contrary, algebraic groups and their theory fit very well for ω-stable groups. This is not surprising. In fact, even neglecting the similarities we emphasized in the last sections between varieties and definable sets, or dimension and Morley rank, and so on, we can recall that several notions introduced in Chapter 6 for studying ω-stable groups clearly come from Algebraic Geometry. In this setting, it is worth mentioning the following conjecture, proposed by Cherlin in 1979.

Conjecture 8.6.3 (Cherlin) *Let $\mathcal{G}$ be an ω-stable group of finite Morley rank. If $\mathcal{G}$ is simple, then $\mathcal{G}$ is an algebraic group over an algebraically closed field.*

Cherlin's Conjecture is still an open question. We should also point out that some remarkable progress in studying ω-stable groups of finite Morley rank has been obtained by using ideas and techniques coming from finite groups, in particular from the classification program of finite simple groups.

8.7 The Mordell-Lang Conjecture

Every promise must be honoured. And consequently, after emphasizing so many times that Model Theory does significantly apply to Algebraic Geometry, here we are to propose one application (indeed a beautiful and deep application, in our opinion): Hrushovski's proof of a question of Lang usually called Mordell-Lang Conjecture. Why is this solution noteworthy? Basically because it is the very first proof of this conjecture, at least in the general form we will state in 8.7.6 below; but also, and mainly, because it largely involves Model Theory (strongly minimal sets, Zariski structures, differentially closed and separably closed fields, as well as the material of this chapter).

So let us introduce the *Mordell-Lang Conjecture*, and briefly sketch its history. We assume some acquaintance with Algebraic Geometry. The question originally rose within Diophantine Geometry, which deals with the roots of systems of polynomials over the rational field $\mathbf{Q}$ or also over a *number field* $\mathcal{F}$ (that is an extension $\mathcal{F}$ of $\mathbf{Q}$ of finite degree). This was the setting where Mordell raised in 1922 the following problem.

Conjecture 8.7.1 (Mordell) *Let $\mathcal{F}$ be a number field, X be a curve of genus > 1 over $\mathcal{F}$. Then X has only finitely many F-rationals points.*

Incidentally recall that a curve X of genus 1 is an elliptic curve, and so is naturally equipped with a group structure.

In a more abstract perspective, on can observe what follows.

Remarks 8.7.2 1. A curve X of genus ≥ 1 over $\mathcal{F}$ is a Zariski closed subset of its Jacobian $J(X)$.

2. A theorem of Riemann says that the Jacobian $J(X)$ is an *abelian variety* (that is a connected complete algebraic group); by the way, every abelian variety is actually an **abelian** group.

3. A theorem of Mordell and Weil ensures that, if $\mathcal{A}$ is an abelian variety in the complex field defined over our number field $\mathcal{F}$, then the set $\mathcal{G}$ of the F-rational points in A is a finitely generated subgroup.

So Mordell's Conjecture can be restated more generally as follows.

Conjecture 8.7.3 *Let $\mathcal{A}$ be an abelian variety of* $\mathbf{C}$*, X be a curve of* $\mathbf{C}$ *embedded in $\mathcal{A}$, $\mathcal{G}$ be a finitely generated subgroup of $\mathcal{A}$. Then either X is an elliptic curve, or $X \cap G$ is finite.*

A similar question was raised by Manin and Mumford in 1963.

Conjecture 8.7.4 (Manin-Mumford) *Let $\mathcal{A}$ be an abelian variety of* $\mathbf{C}$*, X be a curve of* $\mathbf{C}$ *embedded in $\mathcal{A}$, $\mathcal{G} = Tor\, \mathcal{A}$ be the torsion subgroup of $\mathcal{A}$. Then either X is an elliptic curve, or $X \cap G$ is finite.*

Actually the original form of Manin-Mumford Conjecture said that, if X is a curve of genus > 1 and $\mathcal{A} = J(X)$ is its Jacobian, then $X \cap Tor\, \mathcal{A}$ is finite. But the more general statement given in 8.7.4 is easily obtained as in the Mordell case.
A possible unifying approach covering both the Mordell and the Manin-Mumford problem uses the notion of *group of finite type*: in our characteristic 0 framework, this can be introduced as an abelian group $\mathcal{G}$ with a finitely generated subgroup S such that, for every $g \in G$, there is some positive integer m for which mg is in S. In fact, every finitely generated abelian group $\mathcal{G}$ is of finite type (just take $S = G$), and every torsion group is of finite type as well (via $S = \{0\}$). Accordingly the conjectures of Mordell and Manin-Mumford can be regarded as two particular cases of the following more general question.

Conjecture 8.7.5 *Let $\mathcal{A}$ be an abelian variety over* $\mathbf{C}$*, X be a Zariski closed subset of $\mathcal{A}$, $\mathcal{G}$ be a subgroup of finite type of $\mathcal{A}$. Then $X \cap G$ is a (possibly empty) finite union of cosets of subgroups of $\mathcal{G}$.*

This statement was formulated by Lang in the sixties, and is usually called the *Mordell-Lang Conjecture*. As underlined before, it implies a positive answer to Mordell's Conjecture: to see this, just take a curve X_0 of genus > 1 over a number field $\mathcal{F}$, embed $X = X_0(\mathbf{C})$ into its Jacobian $\mathcal{A} = J(X)$ and apply the Mordell-Weil Theorem ensuring that the group $\mathcal{G}$ of F-rational points in A is finitely generated. Accordingly decompose $X_0 = X \cap G$ as a finite union of cosets $a + H$ where $a \in G$ and H is a subgroup of $\mathcal{G}$. Take a coset $a + H$. Its closure $a + \overline{H}$ is included in X_0 - an irreducible set of dimension 1 -. Consequently, if $a + H$ is infinite, then X_0 just equals $a + \overline{H}$ and so inherits a group structure, and genus ≤ 1. This means that, if the genus of X_0 is > 1, then every coset $a + H$ must be finite, whence X_0 itself is finite.

These are the questions we wish to deal with. Now let us report about their solution. Mordell's Conjecture was proved by Faltings in 1983. The echoes of this result spread far and wide, also because it implied an asymptotic solution of Fermat's Last Theorem: in fact, by applying Mordell's Conjecture (or, more precisely, Falting's Theorem) to the projective curve over **Q**

$$x^n + y^n = z^n$$

for $n \geq 3$, one gets only finitely many zeros for every n.
Also the Manin-Mumford Conjecture was positively answered by Raynaud in 1983. The Mordell-Lang Conjecture (as stated before) was solved just a few years ago: first Faltings handled the particular case when the group $\mathcal{G}$ is finitely generated, and then McQuillan provided a general positive solution, using Falting's work and other contributions of Hindry.
So far we have limited our analysis essentially to the characteristic 0, and to number fields. What can we say when passing to *function fields*, or prime characteristics? First let us deal with function fields. Still working in characteristic 0, Manin had proved in the sixties the following analogue of Mordell's Conjecture in this setting: if $\mathcal{K}$ is a function field over an algebraically closed field $\mathcal{K}_0$ (of characteristic 0) and X is a curve of genus > 1 over $\mathcal{K}$, then either X does not descend to $\mathcal{K}_0$ (in which case $X(\mathcal{K})$ is finite), or X is isomorphic to a curve X_0 defined over $\mathcal{K}_0$ (and all but finitely many points of $X(\mathcal{K})$ come from elements of $X(\mathcal{K}_0)$).
When considering prime characteristics p, even the notion of group $\mathcal{G}$ of finite rank must be rearranged. In fact, what we have to require now is that $\mathcal{G}$ has some finitely generated subgroup S such that, for every $g \in G$, there exists a positive integer m **prime to** p satisfying $mg \in S$.
However 8.7.5 - as it was stated before - does not hold any more. In fact $\mathcal{A}$ itself is a torsion group without elements of period p; but there may exist some curves of $\mathcal{A}$ which are not finite unions of cosets of subgroups of $\mathcal{A}$. A reasonable restatement of 8.7.5 in the general setting, for an arbitrary characteristic (0 or prime), is the following.

Conjecture 8.7.6 *Let $\mathcal{K}_0 \prec \mathcal{K}$ be algebraically closed fields, $\mathcal{A}$ be an abelian variety over $\mathcal{K}$ having trace 0 over $\mathcal{K}_0$ (this means that $\mathcal{A}$ has no non-zero abelian subvarieties isomorphic to abelian varieties over $\mathcal{K}_0$). If X is a Zariski closed subset of $\mathcal{A}$ and $\mathcal{G}$ is a subgroup of $\mathcal{A}$ of finite rank, then $X \cap G$ is a (possibly empty) finite union of cosets of subgroups of $\mathcal{G}$.*

In 1994 A. Buium proved this form of the Mordell-Lang Conjecture in characteristic 0.

Theorem 8.7.7 (Buium) *8.7.6 is a true statement when $\mathcal{K}_0 \prec \mathcal{K}$ are algebraically closed fields* **of characteristic** 0.

What is noteworthy for our purposes in Buium's line of proof is his use of Differential Algebraic Geometry; indeed Differential Algebra promptly recalls Model Theory and its treatment of differentially closed fields. So it is right to spend a few words to describe Buium's strategy: one equips $\mathcal{K}$ with a derivation D whose constant field is just $\mathcal{K}_0$, one embeds $\mathcal{G}$ in a differential algebraic group $\mathcal{G}_1$ and, finally, one shows by analytic arguments that $X \cap G_1$ is a finite union of cosets of $\mathcal{G}_1$ and one transfers this property to $\mathcal{G}$.
But it was Ehud Hrushovski who first proved the Mordell-Lang Conjecture in its more general form, in any characteristic, following the initial Buium approach and then using model theoretic methods and, above all, Zariski geometries, differentially closed fields in characteristic 0 and separably closed fields in prime characteristic, in addition to Morley rank, elimination of imaginaries and the definability results of this chapter. It should be emphasized that no alternative general proof of the conjecture is known; and indeed Hrushovski proposed, some time later, a new model theoretic proof of the Manin-Mumford Conjecture, based on a crucial use of existentially closed fields with an automorphism (in particular Zilber's Trichotomy in $ACFA_0$), and getting in this way nice effective bounds of the number of involved cosets in a decomposition of $X \cap G$. Coming back to the Mordell-Lang conjecture, we can say

Theorem 8.7.8 (Hrushovski) *8.7.6 is a true statement in any characteristic.*

This concludes our short and lacunose history of Mordell-Lang, Mordell and Manin-Mumford Conjectures. Which is our purpose now? Certainly we do not aim at providing a complete report of Hrushovski's proof: this would require many pages and serious efforts; moreover there do exist several nice expository papers and books wholly devoted to a detailed exposition (some of them are mentioned among the references at the end of this chapter). On the other hand, we would like to spend a few words about Hrushovski's approach, just to explain where Model Theory intervenes and why it plays a decisive role. With this in mind, we will sketch Hrushovski's proof in the characteristic 0 case, where some old friends of ours - differentially closed fields - are involved. Then we will shortly comment the prime characteristic case, where differentially closed fields are profitably replaced by the separably closed ones.

So take two algebraically closed fields $\mathcal{K}_0 \prec \mathcal{K}$ of characteristic 0. Let $\mathcal{A}$ be an abelian variety over $\mathcal{K}$ with trace 0 over $\mathcal{K}_0$, X be a Zariski closed set in $\mathcal{A}$, $\mathcal{G}$ be a subgroup of $\mathcal{A}$ of finite rank. Our claim is that $X \cap G$ is a (possibly empty) finite union of cosets of subgroups of $\mathcal{G}$.

(a) Without loss of generality one assumes that $\mathcal{K}$ has infinite transcendence degree over $\mathcal{K}_0$. Then one equips $\mathcal{K}$ with a derivation D making $\mathcal{K}$ a differential field, and even a differentially closed field, whose constant subfield $C(\mathcal{K})$ coincides with $\mathcal{K}_0$. Just to fix our symbols, let L denote from now on the usual language for fields, and $L' = L \cup \{D\}$ that of differential fields. So $\mathcal{K}_0$ is strongly minimal both as a structure of L and L': in fact D is identically 0 on $\mathcal{K}_0$ and so adds no definable objects to the pure field $\mathcal{K}_0$. On the contrary, $\mathcal{K}$ is a strongly minimal structure in L, as an algebraically closed field, but it is not any more as a differentially closed field; indeed $\mathcal{K}$, although ω-stable, has Morley rank ω in L'. Notice also that, owing to what we saw in the past sections, the abelian variety A is definable (even in L) in $\mathcal{K}$.

(b) At this point one recalls a general result of Manin on differential fields: the derivation D yields a group homomorphism μ (definable in L') from A onto $(\mathcal{K}^+)^d$, where $\mathcal{K}^+$ is the additive group of $\mathcal{K}$ and d is the dimension of A. The kernel of μ is definable in L' and has a finite Morley rank. Now we deal with $\mathcal{G}$. As $\mathcal{G}$ has finite rank and $\mathcal{K}^+$ has no nonzero torsion elements, the group $\mu(G)$ is finitely generated and there are $g_0, \ldots, g_m \in K$ such that

$$\mu(G) \subseteq \sum_{i \leq m} \mathbf{Q} \cdot g_i \subseteq \sum_{i \leq m} K_0 \cdot g_i.$$

Let H denote $\sum_{i \leq m} K_0 \cdot g_i$. H is definable (in L') and has finite Morley rank. Hence $\mu^{-1}(H)$ is a subgroup of $\mathcal{A}$ extending $\mathcal{G}$; moreover $\mu^{-1}(H)$ is definable (in L') and has finite Morley rank because both H and the kernel of μ are definable of finite Morley rank. Without loss of generality for our purposes, we can replace $\mathcal{G}$ by $\mu^{-1}(H)$. In fact, if $X \cap \mu^{-1}(H)$ is a finite union of cosets of subgroups of $\mu^{-1}(H)$, then the same can be said about $X \cap G$ and G. So we can assume that $\mathcal{G}$ itself is definable and has a finite Morley rank.

Now, just to explain Hrushovski's approach in a more accessible way, let us restrict a little more our framework to the particular case when X is an irreducible curve (the setting of the original Mordell Conjecture). If $X \cap G$ is finite, then we are done. Otherwise $X \cap G$ - as a definable set of finite Morley rank - contains a definable strongly minimal subset S. As usual, S can be viewed as a strongly minimal structure.

(c) Now we use the result of Hrushovski and Sokolovic saying that Zilber's Trichotomy Conjecture holds for strongly minimal sets definable in differentially closed fields of characteristic 0. In fact, these strongly minimal sets are Zariski structures, and so obey the Hrushovski-Zilber Theorem. This applies to S, of course. But what Hrushovski also points out is that S, as a Zariski structure, is locally modular. In fact, as S is strongly minimal, it suffices to exclude finitely many points from S to get an indecomposable set. So there is no loss of generality in assuming that S itself is indecomposable, and consequently each translate bS with $b \in G$ is also indecomposable. Up to replacing S by $b^{-1}S$ for a suitable $b \in G$ we can even assume that the identity element 1_G of G is in S. Hence we are just in a position to apply Zilber's Indecomposability Theorem; accordingly, one deduces that the subgroup generated by S in $\mathcal{G}$ is definable, and indeed every element in this subgroup can be expressed as $a \cdot c^{-1}$ with a and c in S. Hence S interprets an infinite group and so, as a Zariski structure, it cannot be trivial. This means that either S is locally modular or S interprets an infinite (algebraically closed) field. We have to exclude the latter option. To obtain this, one uses a result of Sokolovic already mentioned in 7.10 and saying what follows.

(d) An infinite field definable in a differentially closed field of characteristic 0 and having finite Morley rank is isomorphic to the constant subfield by a definable function.

Recall that, owing to the elimination of imaginaries, there is no difference between definable or interpretable within differentially closed fields. So, if S interprets any infinite field, then it defines even $C(\mathcal{K}) = \mathcal{K}_0$ up to an L'-definable isomorphism. Consequently the subgroup that S generates is isomorphic to some group $\mathcal{G}_0$ L'-definable in $\mathcal{K}_0$ by a function f also definable in L'. As $D = 0$ in K_0, $\mathcal{G}_0$ is definable even in L (just as f and f^{-1}). Then we can apply the Hrushovski-Weil Theorem and deduce that $\mathcal{G}_0$ is an algebraic group over $\mathcal{K}_0$ up to an L-definable isomorphism. At this point one checks that $\mathcal{G}_0$ defines an abelian subvariety of $\mathcal{A}$ in $\mathcal{K}$, which contradicts the hypothesis that $\mathcal{A}$ has trace 0 over $\mathcal{K}_0$. In conclusion, S must be locally modular.

Now S is of the form $(a + L) - \{b_0, \ldots, b_t\}$ for some strongly minimal subgroup L of $\mathcal{G}$ and suitable $a, b_0, \ldots, b_t \in G$. X is Zariski closed, and so contains the closure $a+L$ of S as well. Hence X equals $a+L$ and consequently inherits a group structure, as Mordell's Conjecture requires. This concludes our outline of Hrushovski's proof in characteristic 0 (when X is a curve).

But, as already said, the real novelty of Hrushovski's Theorem concerns prime characteristics p. So it is worth spending some words also on this case. The plan of the proof is similar, but requires some necessary rearrangements. In particular, one can avoid to refer to differentially closed fields and directly handle separably closed fields (with no derivation).

(a) First one replaces with no loss of generality $\mathcal{K}$ by a separably closed non algebraically closed extension having finite degree over $\mathcal{K}^p$. Observe that now the theory of $\mathcal{K}$ is not ω-stable, although it is stable.

(b) The role of the kernel of μ is now played by $\bigcap_n p^n A$, which is not a definable set, but is the intersection of infinitely many definable sets.

The other crucial points in the proof are:

(c) any strongly minimal structure definable in $\mathcal{K}$ is still a Zariski structure (so Zilber's Trichotomy holds);

(d) a field definable in $\mathcal{K}$ and having Morley rank 1 is isomorphic to $\mathcal{K}_0$ by a definable function (a result of M. Messmer).

For more details, look at the references quoted below.

8.8 References

The connection between Model Theory and Algebraic Geometry is clearly explained by Poizat's book [131] (recently translated in English [135]): this is a very rich reference on this matter. In particular it includes a proof of Hrushovski-Weil Theorem 8.6.2. Cherlin's Conjecture was raised in [24]. The classification of ω-stable groups was given by Macintyre in [89]. Groups of finite Morley rank are examined in [15].
Now let us deal with Mordell-Lang Conjecture and Manin-Mumford Conjecture. A geometrical introduction can be found in Lang's book [80]; a short but resonably exhaustive history of these two questions is also in the recent Pillay paper [128]. Pillay's book [126] explains the main model theoretic techniques involved in Hrushovski's approach. Hrushovski's original proof

of the Mordell-Lang Conjecture is in [60]. A detailed exposition is in the book [16]. [50] and [127] are shorter surveys, both very readable.
As already said, Hrushovski's proof in characteristic 0 case is based on some preliminaries concerning differentially closed fields: they can be found in [155] and [62]. For a prime characteristic, separably closed fields are enough: the basic preliminaries are described in [115].
Finally, let us deal with Manin-Mumford Conjecture: Hrushovski's proof is now in [61]. It is based on the model theory of $ACFA$, as developed in [20] and [21].

Chapter 9

O-minimality

9.1 Introduction

The last part of this book is devoted to o-minimal structures. As we saw in the past chapters, they are the infinite expansions $\mathcal{M} = (M, \leq, \ldots)$ of linear orderings such that the subsets of M definable in $\mathcal{M}$ are as trivial as possible, and restrict to the finite unions of singletons and open intervals, possibly with infinite endpoints $\pm\infty$ (equivalently to the finite unions of open, closed, ... intervals in the broad sense including half-lines and the whole M).

O-minimal structures are not simple according to the definition provided in Chapter 7, just because they define and even expand infinite linear orders; consequently no good independence system can be developed inside them, and they are not classifiable in Shelah's sense. Despite this, and just owing the relative triviality of their 1-ary definable sets, one can see that they enjoy several relevant model theoretic properties and, among them, a satisfactory notion of independence with a related dimension partly resembling Morley rank .

Furthermore, they include a lot of noteworthy algebraic examples. Indeed we have already seen that the ordered field of reals $(\mathbf{R}, +, \cdot, -, 0, 1, \leq)$, as well as any real closed field, is o-minimal; discrete or dense infinite linear orders, like $(\mathbf{N}, \leq)$, or $(\mathbf{Z}, \leq)$, or $(\mathbf{Q}, \leq)$, or $(\mathbf{R}, \leq)$, are o-minimal as well; and one can show that even divisible ordered abelian groups, such as $(\mathbf{Q}, +, 0, \leq)$ and $(\mathbf{R}, +, 0, \leq)$, are o-minimal.

In particular, considering the order of the reals, or expanding it by addition, or addition and multiplication together, yield o-minimal structures. On the other side, there do exist some expansion of $(\mathbf{R}, \leq)$ which are not o-minimal.

For instance, extend the order of reals by the sinus function *sin*; in this enlarged structure, **Z** -a denumerable set of isolated points- gets definable (by $sin\, \pi v = 0$), and so o-minimality gets lost. Of course the same argument applies to *cos*.
This chapter has a twofold aim. On the one hand, we will provide an abstract structure theory for o-minimal models $\mathcal{M}$. The starting point will be just the definition of o-minimality, and the consequent classification of the definable subsets of M. But we will see that, on these seemingly poor grounds, one can develop a significant theory, including a nice characterization of definable subsets of M^n for every positive integer n, as well as of definable functions from M^n into M. This will also lead to the already recalled notion of dimension, satisfying several remarkable properties resembling those of Morley rank in ω-stable theories. Actually there is a good deal of similarity between the o-minimal framework and the ω-stable setting, just in these dimensions, but also in definable groups and in other matters. We will emphasize these connections in Sections 9.2-9.6. In particular Section 9.4 and Section 9.5 will prove, among other things, the already mentioned and noteworthy fact that o-minimality, unlike minimality, is preserved by elementary equivalence: if $\mathcal{M}$ is an o-minimal structure, then every model of the theory of $\mathcal{M}$ is o-minimal as well. Accordingly a complete theory T is said to be o-minimal when some (equivalently every) model of T is o-minimal. The subsequent section 9.6 will treat definable groups, definable manifolds, and so on, in o-minimal structures.
On the other side, we will propose other relevant examples of o-minimal structures (to which the previous general theorems apply). This will be the theme of Section 9.7, where we will see that certain expansions of the real field by familiar functions, such as exponentiation, or suitably restricted analytic functions, are o-minimal. Here o-minimality largely overlaps real algebraic geometry and real analytic geometry, both in acquiring techniques and constructions from the geometric framework towards a general and larger spectre of applications, and in providing a new light and in opening new perspectives within the geometrical setting itself.
The subsequent section 9.8 will deal with some variations on the o-minimal theme, most notably with a notion of weak o-minimality, enlarging the o-minimal setting and intensively studied in the latest years.
At last, the final section 9.9 will introduce very shortly the quite recent and attractive work of A. Onshuus about a notion of independence enlarging both forking independence in simple theories and algebraic independence in o-minimal theories towards a common general framework.
Now a few historical notes. O-minimality began its life in the eighties; its

origin refers to a classical problem of Tarski, asking whether the real field, expanded by the exponential function $x \mapsto e^x$, still has a decidable theory. As we know, decidability closely overlaps definability, so a deeply related question is what is definable in $(\mathbf{R}, +, \cdot, -, 0, 1, \leq, exp)$: is the theory of this structure quantifier eliminable, or model complete? This was the scenery where L. Van Den Dries introduced o-minimality at the beginning of the eighties. But who gave a considerable impulse to this notion were A. Pillay and C. Steinhorn, who proved the basic structure theorems on o-minimal structures and greatly developed their abstract theory. In 1991 A. Wilkie partly solved Tarski's Conjecture, showing that the theory of $(\mathbf{R}, +, \cdot, -, 0, 1, \leq, exp)$ is model complete, and even o-minimal (as we will see in 9.7, decidability is still an open question, involving a deep number theoretic problem, usually known as Schanuel's Conjecture, while quantifier elimination fails). This emphasized the connection with analytic geometry, mentioned some lines ago. And indeed o-minimality became, and still is, a matter of interest not only to model theorists, but to geometers and analysts as well.
To conclude this section, we give the proof that any o-minimal ordered field is real closed. This is the converse of a result we already know, ensuring that every real closed field is o-minimal, and can be viewed as the o-minimal analogue of Macintyre's theorem saying that any ω-stable field must be algebraically closed. The proof requires a very basic machinery from o-minimality -just the definition itself- in addition to the necessary algebraic grounds. Let us preliminarily examine o-minimal groups. Here (and later in this chapter) intervals possibly admit infinite endpoints, and so include half-lines, and the whole line, in case.

Lemma 9.1.1 *Let $\mathcal{A} = (A, 0, +, -, \leq, \ldots)$ be an o-minimal structure expanding an ordered group $(A, 0, +, -, \leq)$ and let H be a subgroup of $(A, 0, +, -, \leq)$ definable in $\mathcal{A}$. Then $H = \{0\}$ oppure $H = A$.*

Proof. Suppose towards a contradiction that there exists some subgroup $H \neq \{0\}$, A of $(A, 0, +, -, \leq)$ definable in $\mathcal{A}$. Owing to o-minimality, H decomposes as a finite union of pairwise disjoint intervals (possibly closed, or with infinite endpoints). Accordingly write

$$(\circ) \quad H = \bigcup_{j \leq s} I_j$$

where $I_0, \ldots, I_s$ are intervals and s is minimal. Notice that H is infinite, because it must contain all the multiples nh of any nonzero element $h \in H$

when n ranges over integers. Hence there is $j \leq s$ for which I_j is infinite. Without loss of generality put $j = 0$. Moreover we can even assume that I_0 contains 0 and is symmetric with respect to 0 (in the sense that $-c \in I_0$ for every $c \in I_0$). This is because H is a subgroup, and so includes 0 and is preserved under $+$ and $-$. As $H \neq A$, I_0 is $[-a, a]$ or $]-a, a[$ for some $a > 0$ in A. In the former case, every b in A satisfying $a < b \leq 2a$ decomposes as $b = a + (b - a)$ where $a \in I_0 \subseteq H$ and $0 < b - a \leq a$, so even $b - a$ is in I_0 and consequently in H; hence $b \in H$. Therefore $[-2a, 2a]$ is an interval in H properly including I_0. This implies that $[-2a, 2a]$ shares at least one element with some interval I_j where $0 < j \leq s$, say with I_s. Put

$$I'_0 = [-2a, 2a] \cup I_s.$$

So I'_0 is an interval including $I_0 \cup I_s$ and contained in H. Consequently

$$H = I'_0 \cup \bigcup_{0<j<s} I_j,$$

and this contradicts the choice of s in ($\circ$).
In the latter case, fix $b \in A$ such that $0 < b < a$, then $0 < a - b < a$, and consequently even $a - b$ is in I_0. It follows $a = b + (a - b) \in H$. We get in this way an interval $[-a, a]$ properly including I_0 and contained in H. But this contradicts as before the minimality of s in ($\circ$). ♣

As a consequence, one can give a full characterization of o-minimal ordered groups. They are exactly those listed before, namely the ordered divisible abelian groups.

Theorem 9.1.2 *An o-minimal ordered group $\mathcal{A} = (A, 0, +, -, \leq)$ is abelian and divisible.*

Proof. For every $a \in A$, the centralizer $C(a)$ of a is a definable subgroup of $\mathcal{A}$, and consequently equals either $\{0\}$ or A. If $a = 0$, then clearly $C(a) = A$. On the other side, when $a \neq 0$, $C(a) = A$ as well, because $a \in C(a)$ and this excludes $C(a) = \{0\}$. Hence $C(a) = A$ for every $a \in A$, and $\mathcal{A}$ is abelian. Now take a positive integer n: nA is a definable subgroup of $\mathcal{A}$, clearly equalling A when $A = \{0\}$; on the other side, if $A \neq \{0\}$, then $nA \neq \{0\}$ and consequently $nA = A$. Then $nA = A$ for every positive integer n, in other words $\mathcal{A}$ is divisible. ♣

Coming back to ordered fields, we can eventually prove

Theorem 9.1.3 *Let $\mathcal{K} = (K, 0, 1, +, \cdot, -, \leq)$ be an o-minimal ordered ring with identity* 1 *(in particular let $\mathcal{K}$ be an o-minimal ordered field). Then $\mathcal{K}$ is a real closed field.*

Proof. First we claim that $\mathcal{K}$ is an ordered field, in other words the set $\mathcal{K}^\star$ of the nonzero elements in K is an abelian group with respect to $\cdot$. It suffices to show that the set $K^{>0}$ of the elements > 0 of K is so. In fact, for every $a \in K^{>0}$, aK is a definable subgroup of $(K, 0, +, -, \leq)$, and $aK \neq \{0\}$ because $a > 0$. So, owing to Lemma 9.1.1, $aK = K$, and in particular there exists some $b \in K$ satisfying $ab = 1$. As $a > 0$, b is positive as well. This proves that $K^{>0}$ is a(n ordered) group with respect to $\cdot$. It remains to check that $K^{>0}$ is also abelian. To show this, it suffices to observe that $K^{>0}$ is o-minimal, and then to use Theorem 9.1.2. In fact $K^{>0}$ is definable (as a group) in $\mathcal{K}$, and consequently every subset X of $K^{>0}$ definable in $K^{>0}$ is also definable in $\mathcal{K}$, and hence is a finite union of non-empty intervals of K. All the endpoints of these intervals lie in $K^{>0} \cup \{+\infty\}$, with the only possible exception of the leftmost endpoint in the first interval, that might equal 0, but can be replaced in this case by $-\infty$ in $K^{>0}$. In conclusion, X is actually a finite union of intervals in $K^{>0}$. Hence $K^{>0}$ is o-minimal and consequently abelian, as claimed.

Now let us prove that $\mathcal{K}$ is real closed. Accordingly take a polynomial $f(x) \in K[x]$ and two elements $a < b$ in K satisfying $f(a) \cdot f(b) < 0$, for example $f(a) < 0 < f(b)$. We have to show that there is some $c \in K$ such that $a < c < b$ and $f(c) = 0$. Recall that $\mathcal{K}$ is an ordered field, and hence $\leq$ is dense in K and in $]a, b[$. The polynomial function that f defines is continuous (with respect to the order topology), so both

$$]a, b[^+ = \{d \in K \,:\, a < d < b,\, f(d) > 0\}$$

and

$$]a, b[^- = \{d \in K \,:\, a < d < b,\, f(d) < 0\}$$

are open sets. If $]a, b[^- = \emptyset$, then the continuity of f is contradicted in a. Hence $]a, b[^-$ and similarly $]a, b[^+$ are not empty. Moreover both $]a, b[^+$ and $]a, b[^-$ are definable, and accordingly decompose as finite unions of intervals, indeed of open intervals. As $]a, b[^+$ and $]a, b[^-$ are disjoint, there is some $c \in]a, b[$ out of both $]a, b[^+$ and $]a, b[^-$: so c is a root of f, $f(c) = 0$. ♣

9.2 The Monotonicity Theorem

Let $\mathcal{M} = (M, \leq, \ldots)$ be an o-minimal structure. First observe that M is a topological space with respect to the order topology and so, for every positive integer n, M^n is a topological space as well, with respect to the product topology.
As already recalled, the o-minimality of $\mathcal{M}$ just requires that the only definable subsets are the finite union of singletons and open intervals (possibly including half-lines and the whole M). But what can we say about the definable subsets of M^n when $n > 1$? We give here a very partial and preliminary answer, dealing with 1-ary definable functions f. We show that, if the domain of f is an open interval $]a, b[$ with $a < b$ in $M \cup \{\pm\infty\}$, then one can partition $]a, b[$ into finitely many intervals such that, in each of them, f is either constant, or strictly increasing, or strictly decreasing, and anyhow continuous according to the order topology. This is the so-called Monotonicity Theorem, saying in detail what follows.

Theorem 9.2.1 *Let $\mathcal{M}$ be an o-minimal structure, $X \subseteq M$, a, $b \in M \cup \{\pm\infty\}$, $a < b$, a and b be X-definable when belonging to M. Let f be an X-definable function of $]a, b[$ into M. Then there are a positive integer n and $a_0, a_1, \ldots, a_n \in M \cup \{\pm\infty\}$ such that*

1. *$a = a_0 < a_1 < \ldots < a_n = b$, and $a_1, \ldots, a_{n-1}$ are X-definable;*
2. *for every $i < n$, f is either constant or strictly monotonic in $]a_i, a_{i+1}[$;*
3. *for every $i < n$, if f is strictly monotonic in $]a_i, a_{i+1}[$, then $f(]a_i, a_{i+1}[)$ is also an interval, and f contains a bijection preserving or reversing $\leq$ between $]a_i, a_{i+1}[$ and $f(]a_i, a_{i+1}[)$.*

In particular f is continuous in every interval $]a_i, a_{i+1}[$ for $i < n$.

Notice that, when more generally f is an arbitrary definable function with both domain and image in M, then the domain of f is definable as well, and consequently is a finite union of singletons and open intervals. Each of these intervals satisfies the assumptions of Theorem 9.1.2, and hence inherits its conclusions. Notice also that Theorem 9.1.2 implies, as a simple consequence, that, if $\mathcal{M}$ is an o-minimal structure, a, $b \in M \cup \{\pm\infty\}$, $a < b$ and f is a definable function from $]a, b[$ into M, then $f(x)$ has a limit in $M \cup \{\pm\infty\}$ when $x \to a^+$ and $x \to b^-$.
A full proof of Theorem 9.1.2, as stated before, would require several technical details and would be quite long. We prefer to propose here a simpler

argument showing only the continuity result when $\mathcal{M}$ expands $(\mathbf{R}, \leq)$ (actually this is more or less what we will need later).

Proposition 9.2.2 *Let $\mathcal{M}$ be an o-minimal structure expanding* $(\mathbf{R}, \leq)$, *$X \subseteq M$, f be an X-definable function from $]a, b[$ into* $\mathbf{R}$. *Then there are $a_1, \ldots, a_n \in]a, b[$ such that $a_1 < \ldots < a_n$, $a_1, \ldots, a_n$ are X-definable and f is continuous in $]a, a_1[$, $]a_n, b[$ and each interval $]a_i, a_{i+1}[$ for $1 \leq i < n$.*

Proof. Let S denote the set of the points in $]a, b[$ where f is not continuous. It is an easy exercise to prove that S is X-definable. If S is finite, then we are done: for, $a_1, \ldots, a_n$ are just the elements of S. Hence suppose S infinite. So S contains an infinite open interval I. For every natural n, build two infinite open intervals I_n and J_n such that, for every n,

(i) $I_n \subseteq I$,

(ii) the (topological) closure $\overline{I_{n+1}}$ of I_{n+1} is included in I_n,

(iii) $f(I_n) \supseteq J_n \supseteq f(I_{n+1})$,

(iv) the length of J_n is smaller than $\frac{1}{n+1}$.

Let us see how to define these intervals. First put $I_0 = I$. Then take any natural n, suppose I_n given and form J_n and I_{n+1} as follows. If $f(I_n)$ is finite, then for some $d \in f(I_n)$ the preimage $\{c \in I_n : f(c) = d\}$ is infinite. But $\{c \in I_n : f(c) = d\}$ is definable, and hence includes some infinite interval; f is constant, hence continuous, on this interval, which contradicts $I_n \subseteq I \subseteq S$. Accordingly $f(I_n)$ must be infinite. But $f(I_n)$ is definable, too, and hence includes in its turn an infinite interval; let J_n be such an interval, notice that we can assume with no loss of generality that the length of J_n is $< \frac{1}{n+1}$. The preimage of J_n in I_n is also definable, and contains some infinite interval. Let I_{n+1} denote such an interval; we can assume $\overline{I_{n+1}} \subseteq I_n$. This determines the I_n's and the J_n's for every n. Now put $I' = \bigcap_{n \in \mathbf{N}} I_n$. Clearly $I' = \bigcap_{n \in \mathbf{N}} \overline{I_n}$, whence $I' \neq \emptyset$ because $\mathbf{R}$ is compact. Pick $d \in I'$, we claim that f is continuous in d (this contradicts $d \in S$ and so accomplishes our proof). Take any interval U containing $f(d)$. Owing to (iv), there is some natural n for which $U \supseteq J_n$. But this implies $U \supseteq f(I_{n+1})$ where I_{n+1} is an open neighbourhood of d. ♣

A final remark. When $\mathcal{M}$ expands the field of reals, we can say even more, and state a smooth version of the theorem: in fact, one can partition $]a, b[$ into finitely many intervals where f is of class C^m for every positive integer m.

9.3 Cells

Now we move to characterize in an o-minimal structure $\mathcal{M} = (M, \leq, \ldots)$ the definable subsets of M^n, in particular the definable n-ary functions, for every positive integer n. To make our life easier, we assume all throughout this section (and also in the following ones) that $\leq$ is dense without endpoints: in particular every interval $]a, b[$ with $-\infty < a < b < +\infty$ in M must be infinite. As already said, the reason of this restriction is just to make our treatment and our proofs simpler; in fact, all the results we will show below can be extended -by the appropriate arrangements- to any o-minimal structure $\mathcal{M}$. On the other side, the order of reals is just dense without endpoints, so our framework include all the expansions of $(\mathbf{R}, \leq)$, in particular all the structures enlarging the real field; as we said before, the notable o-minimal examples we will propose in Section 9.7 lie in this setting. We know that the basic definable subsets of M are the intervals and the singletons. More generally, the basic definable subsets of M^n are the *cells*. So let us define what a cell is, more precisely what a k-cell in M^n is.

Definition 9.3.1 *First suppose $n = 1$. A subset C of M is a 0-cell if and only if C is a singleton, and a 1-cell if and only if C is a non-empty open interval, possibly with infinite endpoints.*
Now let $n > 1$, and let k a natural number $\leq n$. A subset C of M^n is called a k-cell if and only one of the following conditions hold:

1. *there are a k-cell D of $\mathcal{M}^{n-1}$ and a continuous and definable function f of D into M such that C is the graph of f, namely*

$$C = \{(\vec{a}, b) \in M^n : \vec{a} \in D, \ b = f(\vec{a})\};$$

2. *$k \geq 1$ and there are a $(k-1)$-cell D of $\mathcal{M}^{n-1}$ and two functions f and g with domain D such that*

   (i) *either the image of f is a subset of M and f is both continuous and definable, or $f(\vec{a}) = -\infty \ \forall \vec{a} \in D$,*

   (ii) *either the image of g is a subset of M and g is both continuous and definable, or $g(\vec{a}) = +\infty \ \forall \vec{a} \in D$,*

   (iii) *$f(\vec{a}) < g(\vec{a}) \ \forall \vec{a} \in D$,*

   (iv) *$C = \{(\vec{a}, b) \in M^n : \vec{a} \in D, \ f(\vec{a}) < b < g(\vec{a})\}$.*

One easily sees that every k-cell of $\mathcal{M}^n$ is definable, and that a k-cell of $\mathcal{M}^n$ is open in M^n if and only if $k = n$. One can also observe

Proposition 9.3.2 *Let $\mathcal{M}$ be an o-minimal structure, $k \leq n$ be positive integers. For every k-cell C of M^n, there is a definable homeomorphism of C onto a k-cell of M^k.*

Proof. It suffices to see, for $n > k$ and $n > 1$, how to determine, for every k-cell C of M^n, a definable homeomorphism π_C onto some k-cell C' of M^{n-1}, and then to iterate this procedure as long as one needs. First assume that C is the graph of some continuos definable function from a k-cell D of $\mathcal{M}^{n-1}$ into M. In this case it suffices to put $C' = D$, and choose as π_C the projection of C onto C'. Now assume

$$C = \{(\vec{a}, b) \in M^n : \vec{a} \in D, \ f(\vec{a}) < b < g(\vec{a})\}$$

where D is a $(k-1)$-cell of $\mathcal{M}^{n-1}$, and f, g satisfy the conditions in 9.3.1, 2. We proceed by induction on n.
If $n = 2$, then $k - 1 = 0$, and D reduces to a single point a of M. Put $C' =]f(a), g(a)[$, $\pi_C(a, b) = b$ for every $b \in C'$. Now let $n > 2$. We know that there is some definable homeomorphism π_D of D onto a $(k-1)$-cell D' of $\mathcal{M}^{n-2}$. Consider the two functions one gets by composing f, g respectively and the inverse of π_D. They have domain D' and satisfy the assumptions in 9.3.1, 2. Accordingly they define a k-cell C' of $\mathcal{M}^{n-1}$; furthermore

$$\pi_C(\vec{a}, b) = (\pi_D(\vec{a}), b) \quad \forall (\vec{a}, b) \in C$$

determines a definable homeomorphism of C onto C'. ♣

When $\mathcal{M}$ expands a real closed field (for instance, when $\mathcal{M}$ is the real field, or even an expansion of it), the cells in $\mathcal{M}$ can be characterized as follows.

Proposition 9.3.3 *Let $\mathcal{M}$ be an o-minimal expansion of a real closed field, k, n be two natural numbers satisfying $n \geq k, 1$. If C is a k-cell of $\mathcal{M}^n$, then there exists a definable homeomorphism of C onto $]0, 1[^k$.*

Proof. When $k = 0$, our claim is trivial, because a 0-cell reduces to a singleton. So take $k > 0$. Owing to the previous proposition, we can assume $n = k$. We proceed by induction on n. If $n = 1$, then $C =]a, b[$ where $a, b \in M \cup \{\pm\infty\}$ and $a < b$. So the required homeomorphism between C and $]0, 1[$ is easily obtained: for instance, if both a and b are in M, then it suffices to map every x in $C =]a, b[$ into

$$\frac{x-a}{b-a};$$

in the remaining cases one proceeds in a similar way. Now suppose $n > 1$ (and our claim true for $n-1$). Let C be an n-cell of $\mathcal{M}^n$. There do exist an $(n-1)$-cell D of $\mathcal{M}^{n-1}$ and two functions f and g as in 9.3.1, 2 such that C is the set of those tuples $(\vec{a}, b)$ in M^n for which $\vec{a} \in D$ and $f(\vec{a}) < b < g(\vec{a})$. By induction, there is a definable homeomorphism h of D onto $]0, 1[^{n-1}$. If both f and g take their values in M, then

$$(\vec{a}, b) \longmapsto (h(\vec{a}), \frac{b - f(\vec{a})}{g(\vec{a}) - f(\vec{a})}) \quad \forall(\vec{a}, b) \in C$$

is the required homeomorphism. All the other cases can be handled in a similar way, just as when for $n = 1$. ♣

In particular, when $\mathcal{M}$ expands the real field $\mathbf{R}$ and $k > 0$, every k-cell of $\mathcal{M}$ is connected in the order topology; for, $]0, 1[^k$ is. This does not hold any longer when $\mathbf{R}$ is replaced by any real closed field. For instance, if $\mathbf{R}_0$ is the ordered field of real algebraic numbers, then $\mathbf{R}_0$ is real closed, and hence o-minimal; $\mathbf{R}_0$ is a 1-cell of itself, but is not connected because, for every real trascendental t, $\mathbf{R}_0$ partitions as

$$\mathbf{R}_0 = \{r \in \mathbf{R}_0 : r < t\} \cup \{r \in \mathbf{R}_0 : r > t\}.$$

Hence connection gets lost. However every cell in an o-minimal structure $\mathcal{M}$ satisfies a weak form of connection, with respect to open **definable** sets. In fact, consider the following notion.

Definition 9.3.4 *Let $\mathcal{M}$ be an o-minimal structure, n be a positive integer. A definable set $X \subseteq M^n$ is said to be* **definably connected** *if and only if X cannot partition as the disjoint union of two non-empty open* **definable** *subsets.*

What we will see now is that every cell in an o-minimal structure $\mathcal{M}$ is definably connected. First we give an equivalent characterization of definably connected sets.

An *open box* of $\mathcal{M}^n$ is the cartesian product of n open intervals of $\mathcal{M}$. Hence open boxes form a basis of open neighbourhoods of the product topology of $\mathcal{M}^n$. Furthermore, for $Y \subseteq X \subseteq M^n$, an element $\vec{a}$ of X is called a *boundary point* of Y in X if and only if every open box of $\mathcal{M}^n$ containing $\vec{a}$ overlaps both Y and $X - Y$.

Lemma 9.3.5 *Let $\mathcal{M}$ be an o-minimal structure, n be a positive integer, X be a definable subset of M^n. Then X is definably connected if and only if, for every proper non-empty definable subset Y di X, X contains at least one boundary point of Y in X.*

Proof. X is definably connected if and only if, for every definable $Y \subseteq X$ such that $Y \neq \emptyset, X$, either Y or $X - Y$ is not open, in other words there is either a point $\vec{a} \in Y$ such that every open box containing $\vec{a}$ overlaps $X - Y$, or a point $\vec{a} \in X - Y$ such that every open box containing $\vec{a}$ overlaps Y. Accordingly X is definably connected if and only if, for every Y as above, there is $\vec{a} \in X$ which is a boundary point of Y in X. ♣

Now we can show, as promised,

Theorem 9.3.6 *Let $\mathcal{M}$ be an o-minimal structure, n be a positive integer. Then every cell C of $\mathcal{M}^n$ is definably connected.*

Proof. We proceed by induction on n.
If $n = 1$, then the claim is trivial because the cells of $\mathcal{M}$ reduce to singletons and open intervals.
Hence assume $n > 1$. Let C be a cell of $\mathcal{M}^n$. If C is a 0-cell, so a singleton, then C is definably connected. If C is a k-cell for some positive integer $k < n$, then, owing to Corollary 9.3.3, C is definably homeomorphic to a cell C' of $\mathcal{M}^k$; by the induction hypothesis, C' is definably connected, whence C is definably connected, too. Finally suppose that C is an n-cell of $\mathcal{M}^n$. Then there exist a cell D of $\mathcal{M}^{n-1}$ and two functions f and g as in 9.3.1, 2 such that

$$C = \{(\vec{a}, b) \in M^n : \vec{a} \in D, f(\vec{a}) < b < g(\vec{a})\}.$$

D is definably connected by the induction hypothesis. To deduce that even C is definably connected we use Lemma 9.3.5. Accordingly take a definable subset Y of C such that $Y \neq \emptyset, C$. First suppose that, for some $\vec{a} \in D$, there are two elements b and b' in M such that $(\vec{a}, b) \in Y$ and $(\vec{a}, b') \in C - Y$. Then the interval $]f(\vec{a}), g(\vec{a})[$ contains at least one boundary point b_0 of the definable set $\{b \in M : (\vec{a}, b) \in Y\}$; this implies that $(\vec{a}, b_0)$ is a boundary point of Y in C. Now suppose that, for every $\vec{a} \in D$, either

$$\{(\vec{a}, b) \in C : b \in M\} \subseteq Y$$

or

$$\{(\vec{a}, b) \in C : b \in M\} \subseteq C - Y.$$

Let Z denote the set of the elements $\vec{a}$ of D satisfying the former condition. $Y \neq \emptyset, C$ implies $Z \neq \emptyset, D$. As D is definably connected, there is some boundary point $\vec{a}_0$ of Z in D. Let $b \in M$ satisfy $(\vec{a}_0, b) \in C$, we claim that $(\vec{a}_0, b)$ is a boundary point of Y in C. Let B an open box of $\mathcal{M}^n$ containing $(\vec{a}_0, b)$. As C is open, we can suppose $B \subseteq C$. Let B' denote the projection of B in M^{n-1}. Then $\vec{a}_0 \in B'$ and $B' \subseteq D$. Since $\vec{a}_0$ is a boundary point of Z in D, there are $\vec{a}_1$ and $\vec{a}_2$ in B' such that $\vec{a}_1 \in Z$ and $\vec{a}_2 \notin Z$. Consequently, if $b_1, b_2 \in M$ satisfy $(\vec{a}_1, b_1), (\vec{a}_2, b_2) \in C$ respectively (in particular $(\vec{a}_1, b_1), (\vec{a}_2, b_2) \in B$), then $(\vec{a}_1, b_1) \in Y$, $(\vec{a}_2, b_2) \notin Y$. Hence $(\vec{a}_0, b)$ is a boundary point of Y in C. In conclusion, C is definably connected. ♣

9.4 Cell decomposition and other theorems

The aim of this section is to introduce and to state the basic general theorems for o-minimal structures $\mathcal{M} = (M, \leq, \ldots)$: we want to characterize all the definable sets and functions in $\mathcal{M}$, and we want also to emphasize some relevant consequences and, among them, the already mentioned fact that o-minimality is preserved under elementary equivalence. As the proofs of these fundamental results are quite long and intricate, we will defer part of them, and the corresponding details, to the next section; here we provide just a basic outline, illustrating these central cores of the theory of o-minimality. So a reader simply interested in a general view may limit her, or his attention to this section, and to skip the next one. Let us remind once again that, for simplicity's sake, we are assuming that $(M, \leq)$ is dense without endpoints: this is tacitly accepted all throughout these sections, unless otherwise stated. The first result we propose just describes definable sets (and functions) in o-minimal structures. It is a beautiful and powerful characterization, called *Cell Decomposition Theorem*. In fact, it says that every definable set decomposes as a finite union of cells.

Theorem 9.4.1 *Let $\mathcal{M}$ be an o-minimal structure, n be a positive integer.*

1. *Every definable set $X \subseteq M^n$ can be expressed as a finite (disjoint) union of cells in $\mathcal{M}^n$.*

2. *Furthermore, if X is the domain of a definable function f with values in M, then one can decompose X as a finite disjoint union of cells, such that f is continuous on each of them.*

Notice that this generalizes what we know when $n = 1$; in fact, in that case every definable $X \subseteq M$ is a finite union of points and open intervals (in other words, of 0-cells and 1-cells respectively), and, when X is the domain of some definable function f, one can also suppose that f is continuous on each of these pieces, owing to the Monotonicity Theorem. But now we can extend these results to any n.

As already said, the proof of the Cell Decomposition Theorem will be given in detail in the next section. But Cell Decomposition, and its analysis of definable sets, is also the key tool in showing another basic fact in o-minimality, that is its preserving under elementary equivalence.

Theorem 9.4.2 *If $\mathcal{M}$ is an o-minimal structure, then the theory of $\mathcal{M}$ is o-minimal.*

In fact, which is the trouble in this claim? Actually we do know that, for an o-minimal $\mathcal{M}$, for every formula $\theta(v, \vec{w})$ in the language L of $\mathcal{M}$ and tuple $\vec{a}$ in M, $\theta(\mathcal{M}, \vec{a})$ is a finite union of (possibly closed) intervals. But suppose that, when $\vec{a}$ ranges over M, the minimal number of the intervals involved in these decompositions of $\theta(\mathcal{M}, \vec{a})$ cannot be upperly bounded, and so, for every natural N, one finds some tuple $\vec{a}(N)$ in M such that any decomposition of $\theta(\mathcal{M}, \vec{a}(N))$ requires at least N intervals. If this is the case, then it suffices a straightforward application of Compactness Theorem to provide some $\mathcal{M}' \equiv \mathcal{M}$ and some tuple $\vec{a'}$ in M' for which $\theta(\mathcal{M}', \vec{a'})$ cannot be expressed as a finite union of intervals, and consequently to conclude that the theory of $\mathcal{M}$ is not o-minimal.
Hence the crucial point in showing that the o-minimality of $\mathcal{M}$ is preserved by elementary equivalence is to uniformly bound, for every formula $\theta(v, \vec{w})$ as before, the minimal number of intervals necessary to decompose $\theta(\mathcal{M}, \vec{a})$ when $\vec{a}$ ranges over M. This is a definability question concerning formulas in arbitrarily many free variables, and so directly refers to Cell Decomposition. On the other hand, bounding the number of the involved intervals in $\theta(\mathcal{M}, \vec{a})$ is the same as bounding the total number of their endpoints (forming a definable, and finite set). So the key step towards the proof of Theorem 9.4.2 is

Theorem 9.4.3 *Let $\mathcal{M}$ be an o-minimal structure in L, $\varphi(v, \vec{w})$ be an L-formula such that, for all $\vec{a}$ in M, $\varphi(\mathcal{M}, \vec{a})$ is finite. Then there is a positive integer N such that, for every $\vec{a}$ in M, $|\varphi(\mathcal{M}, \vec{a})| \leq N$.*

The proof of Theorem 9.4.3 will be deferred until the next section. But, as we have just pointed out, Theorem 9.4.2 is an almost immediate consequence of Theorem 9.4.3. Let us see in detail why.

Proof. (Theorem 9.4.2) Let L be the language of $\mathcal{M}$, and let $\vartheta(v, \vec{w})$ be an L-formula; in particular, let n denote the length of $\vec{w}$. For every $\vec{a} \in M^n$, $\theta(\mathcal{M}, \vec{a})$ is a finite union of intervals. Let $\varphi(v, \vec{w})$ be the L-formula saying

$$v \text{ is an endpoint of } \theta(\ldots, \vec{w}).$$

Then, for every $\vec{a} \in M^n$, $\varphi(\mathcal{M}, \vec{a})$ is finite; hence, by Theorem 9.4.3, there exists a positive integer N such that, for every $\vec{a} \in M^n$, $\varphi(\mathcal{M}, \vec{a})$ contains at most N elements, whence $\theta(\mathcal{M}, \vec{a})$ has at most N endpoints. But the sentence

$$\forall \vec{w} \exists^{\leq N} v\, \varphi(v, \vec{w})$$

remains true in every structure $\mathcal{M}' \equiv \mathcal{M}$. Accordingly, for every $\mathcal{M}' \equiv \mathcal{M}$ and $\vec{b} \in M'^n$, $\theta(\mathcal{M}', \vec{b})$ has at most N endpoints, and hence is a finite union of intervals. In conclusion the theory of $\mathcal{M}$ is o-minimal. ♣

The bounds given by Theorem 9.4.3 on formulas $\theta(v, \vec{w})$ can be extended to formulas $\theta(\vec{v}, \vec{w})$ with an arbitrarily long $\vec{v}$. In fact the following result holds.

Theorem 9.4.4 *Let $\mathcal{M}$ be an o-minimal structure in L, $\theta(\vec{v}, \vec{w})$ an L-formula (n be the length of $\vec{v}$ and m be that of $\vec{w}$). Then there exists a positive integer N such that, for every $\mathcal{M}' \equiv \mathcal{M}$ and tuple $\vec{a}$ in M', $\theta(\mathcal{M}'^n, \vec{a})$ is the union of at most N cells in $\mathcal{M}'$.*

Before beginning the proof, and just for preparing it, let us premit a simple example. Assume $n = 2$. Let $\theta(v_1, v_2, \vec{w})$ be an L-formula, $\vec{a}$ be a tuple in M. Suppose that the definable set $\theta(\mathcal{M}^2, \vec{a})$ decomposes as a disjoint union of 2 cells in M: the former is a singleton, so a 0-cell, while the latter is a 1-cell, and more precisely is the graph of a continuous definable function f whose domain is the open interval $]a, b[$ with $a < b$ in M. Notice that there are an L-formula $\eta(v_1, v_2, \vec{z})$ and a sequence $\vec{e}$ in M such that, for every c_1 and c_2 in M with $a < c_1 < b$,

$$f(c_1) = c_2 \quad \Leftrightarrow \quad \mathcal{M} \models \eta(c_1, c_2, \vec{e}).$$

Now consider the L-formula

$$\vartheta_{\vec{a}}(\vec{w}) \quad : \quad \forall v_1 \forall v_2 (\vartheta(v_1, v_2, \vec{w}) \leftrightarrow$$

$$\leftrightarrow \exists u_1 \exists u_2 \exists u \exists w \exists \vec{z} (((v_1 = u_1 \wedge v_1 = u_2) \vee (u < v_1 \wedge v_1 < w \wedge$$

$$\wedge \text{"}\eta(\cdot, \cdot, \vec{z}) \text{ defines a continuous function of domain }]u, w[\text{"}$$

$$\wedge\, \eta(v_1, v_2, \vec{z}))) \wedge \neg(u < u_1 < w \wedge \eta(u_1, u_2, \vec{z})))).$$

It is clear that the tuples $\vec{b}$ in $\mathcal{M}$, or even in a model $\mathcal{M}'$ of the theory of $\mathcal{M}$, satisfying $\mathcal{N} \models \theta_{\vec{a}}(\vec{b})$ are just those for which $\theta(\mathcal{M}'^2, \vec{b})$ has a cell decomposition as $\vartheta(\mathcal{M}^2, \vec{a})$ (the disjoint union of a singleton and a graph of a definable continuous function). It is also obvious that this is quite general: given an L-formula $\theta(\vec{v}, \vec{w})$ as in the statement of 9.4.4, a tuple $\vec{a}$ of the same length as $\vec{w}$ in the universe Ω of the theory of $\mathcal{M}$ and a cell decomposition of $\vartheta(\Omega^n, \vec{a})$, one can build an L-formula $\theta_{\vec{a}}(\vec{w})$ such that a tuple $\vec{b}$ in Ω satisfies $\theta_{\vec{a}}$ if and only $\vartheta(\Omega^n, \vec{b})$ has a cell decomposition just as $\vartheta(\Omega^n, \vec{a})$.

Proof. (Theorem 9.4.4). Owing to Theorem 9.4.2, any structure $\mathcal{M}'$ elementarily equivalent to $\mathcal{M}$ is o-minimal, and hence, by Theorem 9.4.1, any definable subset of M'^n (in particular $\theta(\mathcal{M}'^n, \vec{a})$ for $\vec{a} \in M'^m$) is a finite union of cell. We have seen that there is an L-formula $\theta_{\vec{a}}(\vec{w})$ (without parameters) describing the form of a given cell decomposition of $\theta(\mathcal{M}'^n, \vec{a})$. Let Φ denote the (countable) set of all these formulas $\theta_{\vec{a}}(\vec{w})$ when $\vec{a}$ ranges over M'^n and $\mathcal{M}'$ is a model of the theory of $\mathcal{M}$. Use Compactness Theorem and get finitely many formulas

$$\vartheta_0(\vec{w}), \ldots, \vartheta_s(\vec{w}) \in \Phi$$

such that

$$\forall \vec{w} \bigvee_{i \leq s} \vartheta_i(\vec{w}) \in Th(\mathcal{M}).$$

Consequently there is a positive integer N such that, for every $\mathcal{M}' \equiv \mathcal{M}$ and $\vec{a} \in M'^m$, $\theta(\mathcal{M}'^n, \vec{a})$ decomposes as the union of at most N cells: N is just the maximal number of involved cells in the decompositions described by $\vartheta_0(\vec{w}), \ldots, \vartheta_s(\vec{w})$. ♣

Recall that, when $\mathcal{M}$ expands the real field $\mathbf{R}$, every cell of $\mathcal{M}$ is also connected. Hence in this case, for every formula $\theta(\vec{v}, \vec{w})$, there is a positive integer N such that, for a tuple $\vec{a}$ in M^m, $\theta(\mathcal{M}^n, \vec{a})$ is the union of $\leq N$ connected components. We will see in Section 9.7 several relevant examples of o-minimal expansions of the real field $\mathbf{R}$. In this framework it is worth stating the following result of Wilkie's.

Theorem 9.4.5 (Wilkie) *Let $\mathcal{M}$ expand the ordered field of reals by C^∞ functions from some cartesian powers $\mathbf{R}^t$ of $\mathbf{R}$ into $\mathbf{R}$. Assume that, for every* **quantifier free** *formula $\theta(\vec{v}, \vec{w})$ in the language L of $\mathcal{M}$, there is a positive integer N such that, for any tuple $\vec{a} \in M^m$, $\theta(\mathcal{M}^n, \vec{a})$ decomposes as the union of $\leq N$ connected components. Then the same is true for every formula of L. In particular, $\mathcal{M}$ is o-minimal.*

Cell Decomposition is an important tool also in developing a dimension theory inside o-minimal structures $\mathcal{M}$, and in equipping every definable X

in $\mathcal{M}$ with a natural number (its *dimension*). Recall that no o-minimal $\mathcal{M}$ is simple, and hence ω-stable; accordingly the Morley rank of X might be ∞. More generally, as simplicity fails, other possible rank notions arising within simple, or stable, or superstable settings and replacing RM in these enlarged frameworks lose their interest in o-minimal models. But Cell Decomposition does assign a dimension to X in a quite reasonable way. Let us see how.

Definition 9.4.6 *The* **dimension** *of a k-cell of $\mathcal{M}^n$ is k. The* **dimension** *of a definable X in $\mathcal{M}$ is the maximal dimension of a cell arising in a cell decomposition of X.*

Of course, a cell decomposition of X is not unique; but the maximal dimension of an involved cell is, and does depend only on X. So the previous definition makes sense for every X. There is another alternative way to introduce a dimension notion in $\mathcal{M}$. In fact, as we saw in Chapter 5, the algebraic closure *acl* determines a dependence relation $\prec$ in $\mathcal{M}$ (and in every model $\mathcal{M}'$ of the theory of $\mathcal{M}$). This relation generates in its turn an independence system as axiomatized in section 7.2, so satisfying the conditions **(I1)**-**(I6)** (but not the further requirement **(I7)**)). With respect to this independence notion, we can define the *dimension* of a tuple $\vec{a} = (a_1, \ldots, a_m)$ in M'^m as the size of a minimal subsequence $\vec{b}$ such that $\vec{a}$ lies in $acl(\vec{b})$. Then we can introduce, just as in the strongly minimal case, the dimension of a definable set $\varphi(\mathcal{M}'^m, \vec{b})$ as the maximal dimension of a tuple $\vec{a}$ in $\varphi(\mathcal{M}''^m, \vec{b})$ where $\mathcal{M}''$ is an elementary extension of $\mathcal{M}'$.
Notably these dimension notions (the former arising from topology, the latter more directly related to model theory) coincide and share good properties and, after all, a satisfactory behaviour.
To conclude this section, let us mention some other nice properties of o-minimal structures and theories, closely resembling what happens in the ω-stable setting.

Theorem 9.4.7 (Pillay-Steinhorn) *Let T be a (complete) o-minimal theory, A be a small subset of the universe Ω of T. Then there is a model of T prime over A, and this is unique up to isomorphism fixing A pointwise.*

Actually this result does not need the Cell Decomposition Theorem, but can be proved by referring directly to the definition of o-minimality and to the Monotonicity Theorem. The same ingredients yield a nice classification of $\aleph_0$-categorical o-minimal theories, again due to Pillay and Steinhorn.

The uncountable spectrum of an o-minimal T theory takes everywhere the maximal value

$$I(T, \lambda) = 2^\lambda \quad \forall \lambda > \aleph_0.$$

With respect to the countable framework, it is worth emphasizing that Vaught's Conjecture holds, in the following strong form.

Theorem 9.4.8 (Mayer) *Let T be a (complete) o-minimal theory. Then, up to isomorphism, either T has continuum many countable models, or there are two naturals n and m such that T has $3^n \cdot 6^m$ countable models.*

Of course, one might get curious in reading the statement of this theorem: why, and where, do n and m arise? Basically, they depend on a careful analysis of types in o-minimal structures. The interested reader may directly consult Laura Mayer's work, quoted below.

9.5 Their proofs

This section provides the proofs of Theorems 9.4.1 and 9.4.3, stated in Section 9.4. As said, they are long and intricate. In spite of this, we think it right to propose them for at least two reasons. The former (and the principal) is that we believe that Theorem 9.4.1 (the Cell Decomposition Theorem) and Theorem 9.4.2 (the one saying that o-minimality is preserved under elementary equivalence) are two beautiful and fundamental results and deserve a full report, including a technical preliminary like Theorem 9.4.3. The latter reason just concerns the intricacy of the proof; actually this is due to its length and ingenuity, but does not depend on a relevant and deep theory, indeed the premises it needs are rather elementary and accessible (they just include the definition of o-minimality, the Monotonicity Theorem, some properties of cells and an induction argument). So our exposition should require no particular efforts but a little attention and patience. And anyhow the reader who is not interested in these details may neglect this section and proceed directly to the next ones, that will not use these proofs.
So consider an o-minimal structure $\mathcal{M}$ in a language L; for simplicity, we keep our assumption that the order of $\mathcal{M}$ is dense without endpoints.
What we said in the last section in introducing Theorem 9.4.3 suggests the following definition.

Definition 9.5.1 *Let $\varphi(v, \vec{w})$ be a formula of $L(M)$, n be the length of $\vec{w}$, X be a subset of M^n. We say that:*

- $\varphi(v, \vec{w})$ *is* **finite** *in* X *if and only if, for every* $\vec{a} \in X$, $\varphi(\mathcal{M}, \vec{a})$ *is finite;*

- $\varphi(v, \vec{w})$ *is* **uniformly finite** *in* X *if and only if there is a positive integer* h *such that, for every* $\vec{a} \in X$, *the size of* $\varphi(\mathcal{M}, \vec{a})$ *is at most* h.

When $\varphi(v, \vec{w})$ is finite in X, we can introduce two partial functions φ_- and φ_+ mapping any $\vec{a} \in X$ into the minimal and the maximal element in $\varphi(\mathcal{M}, \vec{a})$ respectively -provided that $\varphi(\mathcal{M}, \vec{a})$ is not empty, of course-. If X is definable, then φ_- and φ_+ are definable as well.

Definition 9.5.2 *Let* n *be a positive integer,* X *be a definable subset of* M^n. *A* **decomposition** *of* X *is a partition of* X *into finitely many cells. If* $Y \subseteq X$ *is definable, we say that a decomposition* P *of* X **partitions** Y *when no cell in* P *overlaps both* Y *and* $X - Y$.

Here is another technical preliminary notion.

Definition 9.5.3 *Let* C *be an open cell of* $\mathcal{M}^n$ *(so an* n*-cell),* $\varphi(v, \vec{w})$ *be a formula finite in* C. *Call a point* $\vec{a} \in C$ **good** *for* $\varphi(v, \vec{w})$ *if and only if the following conditions hold:*

1. *for every* $b \in \varphi(\mathcal{M}, \vec{a})$, *there are an open box* $B \subseteq C$ *containing* $\vec{a}$ *and an open interval* I *containing* b *such that* $\varphi(\mathcal{M}^{n+1}) \cap (I \times B)$ *is the graph of some continuous function of* B *in* I;

2. *for every* $b \in M - \varphi(\mathcal{M}, \vec{a})$, *there is an open neighbourhood of* $(\vec{a}, b)$ *in* M^{n+1} *disjoint from* $\varphi(\mathcal{M}^{n+1})$.

A point $\vec{a} \in C$ which is not good for $\varphi(v, \vec{w})$ is called (with no particular imagination) **nasty** for $\varphi(v, \vec{w})$. Notice that both good and nasty points for $\varphi(v, \vec{w})$ form definable subsets of the cell C.
At this point we can begin our proof. The following lemma is its crucial step.

Lemma 9.5.4 *Let* $\mathcal{M}$ *be an o-minimal structure,* n *be a positive integer,* C *be a cell of* $\mathcal{M}^n$.

$(1)_n$ *For every element* i *in a finite set* I *of indexes, let* X_i *denote a definable subset of* C. *Then there exists a decomposition of* C *partitioning each* X_i.

$(2)_n$ *Let f be a definable function of C into M. Then there is a decomposition P of C such that f is continuous on any cell of P.*

$(3)_n$ *A formula $\varphi(v, \vec{w})$ of $L(M)$ finite in C is uniformly finite in C.*

$(4)_n$ *Let $\varphi(v, \vec{w})$ be a formula finite in C. If the functions φ_- e φ_+ are defined and continuous in C and every point of C is good for $\varphi(v, \vec{w})$, then the size of $\varphi(\mathcal{M}, \vec{a})$ is constant when $\vec{a}$ ranges over C.*

Lemma 9.5.4 immediately implies both Theorem 9.4.1 and Theorem 9.4.3.

Proof. (9.4.1.1) M^n is an n-cell. If X is a definable subset of M^n, then $(1)_n$ in Lemma 9.5.4 provides a decomposition P of M^n partitioning X. Hence X is the (finite) union of the cells of P it contains. ♣

Proof. (9.4.1.2) Just apply $(2)_n$ to the cells of the decomposition of X given by 9.4.1, 1. ♣

Proof. (9.4.3) This is just $(3)_n$ when $C = M^n$. ♣

Now let us show Lemma 9.5.4. We proceed by induction on n.

$(1)_1$ This just rephrases o-minimality.

$(2)_1$ This is the Monotonicity Theorem (in the weak form we saw in 9.2).

$(4)_1$ If C reduces to a singleton, then the claim is trivial. Hence assume that C is an open interval $]a, b[$ where $-\infty \leq a < b \leq +\infty$. Suppose that, for some positive integer h, the set Y of the points c in $]a, b[$ such that $|\varphi(\mathcal{M}, c)| = h$ is not empty and does not equal $]a, b[$. As φ_- and φ_+ are defined throughout the interval C, we can assume $h > 1$. Let c be an endpoint of Y in C, and put $\varphi(\mathcal{M}, c) = \{d_0, \ldots, d_L\}$ where L is a suitable natural and $d_0 < \ldots < d_L$. As c is good for φ, there are an interval $I \subseteq]a, b[$ containing c and $L+1$ pairwise disjoint intervals $J_0, \ldots, J_L$ such that, for every $i \leq L$, J_i includes d_i and $\varphi(\mathcal{M}^2) \cap (J_i \times I)$ is the graph of a continuous function g_i of I into J_i. Each function g_i is clearly definable. Furthermore, for every $c' \in I$,

$$g_0(c') < \ldots < g_L(c');$$

consequently $|\varphi(\mathcal{M}, c')| \geq L + 1$. But c is an endpoint of Y in C, and $|\varphi(\mathcal{M}, c)| = L + 1$. Therefore, for some open interval I' having c as a left or right endpoint,

$$|\varphi(\mathcal{M}, c')| > L + 1, \quad \forall c' \in I'.$$

Assume for simplicity that c is a left endpoint of I'. Define the following function g in I': for every $c' \in I'$,

$g(c')$ is the least element $d' \in \varphi(\mathcal{M}, c')$ such that $d' \neq g_i(c')$ for every $i \leq L$.

g is definable and its domain coincides with the whole interval I'. So, as observed in § 2, g has a limit in $M \cup \{\pm\infty\}$ when $x \to c^+$

$$d = lim_{x \to c^+} g(x).$$

Moreover

$$\varphi_-(c') < g(c') < \varphi_+(c') \quad \forall c' \in I',$$

and so d cannot equal $+\infty$ or $-\infty$, in other words $d \in M$. If $\models \varphi(d, c)$, then $d = d_i$ for some $i \leq L$, and consequently

$$d = lim_{x \to c^+} g_i(x);$$

on the other side, we know that, for every $c' \in I'$,

$$g(c') \neq g_i(c'), \quad g(c'), g_i(c') \in \varphi(\mathcal{M}, c').$$

But we contradict in this way the fact that c is good for φ (recall Definition 9.5.3, 1). Accordingly $\models \neg\varphi(d, c)$, which again excludes that c is good for φ (this time by 9.5.3, 2). This yields the required contradiction. Hence $(4)_1$ holds.

$(3)_1$ The claim is trivial when C is a singleton. Accordingly suppose $C =]a, b[$ where $-\infty \leq a < b \leq +\infty$. The set of the elements c of C for which $\varphi(\mathcal{M}, c) \neq \emptyset$ is definable, and consequently is a finite union of singletons and open intervals in C. Of course, it is enough to show that $\varphi(v, w)$ is uniformly finite on every interval in this decomposition. Accordingly we can even assume

$$\varphi(\mathcal{M}, c) \neq \emptyset \quad \forall c \in]a, b[;$$

in particular φ_- and φ_+ are defined throughout $]a, b[$. Owing to $(2)_1$, we can even suppose (up to replacing again $]a, b[$ with a suitable subinterval) that both φ_- and φ_+ are continuous in $]a, b[$. Now let Y denote the set of those points in $C =]a, b[$ that are nasty for φ. If Y is finite, say $Y = \{c_0, \ldots, c_t\}$ with $a < c_0 < \ldots < c_t < b$, then φ_- and φ_+ are continuous in each interval $]a, c_0[$, $]c_i, c_{i+1}[$ for $i < t$, $]c_t, b[$, and every point in these intervals is good for φ. So we are just in a position to apply $(4)_1$, and accordingly the size of $\varphi(\mathcal{M}, c)$ is constant throughout every interval, and, in conclusion, $\varphi(v, w)$ is uniformly finite in $]a, b[$.
Hence it suffices to show that Y is finite. Suppose not. Anyhow Y is definable, and hence, by o-minimality, it contains some infinite interval.

Without loss of generality, we can assume that this interval just equals $C =]a, b[$, and so that every point c of $]a, b[$ is nasty for φ. Then, for every $c \in]a, b[$, there exists $d \in M$ satisfying one of the following conditions:

(i) $d \in \varphi(\mathcal{M}, c)$ but, for no pair of open intervals I and J containing c and d respectively, $\varphi(\mathcal{M}^2) \cap (J \times I)$ is the graph of a function of I in J;

(ii) $d \notin \varphi(\mathcal{M}, c)$ but every open box B containing (d, c) overlaps $\varphi(\mathcal{M})^2$ as well.

In both cases we say that (d, c) is a black sheep of φ (of type (i) or (ii) according to whether (i), or (ii), holds).

Observe that, for every $c \in]a, b[$, there is a minimal $d \in M$ for which (d, c) is a black sheep of φ. In fact, as $\varphi(v, w)$ is finite in $]a, b[$, there are at most finitely many $d \in M$ such that (d, c) is a black sheep of type (i). Consequently it suffices to prove that, if $d_1, d_2 \in M$, $d_1 < d_2$ and, for every $d \in]d_1, d_2[$, (d, c) is a black sheep of type (ii) of φ, then also (d_1, c) is a black sheep of φ. This is certainly true when d_1 does not belong to $\varphi(\mathcal{M}, c)$ (in this case (d_1, c) is a black sheep of type (ii)). Accordingly assume $d_1 \in \varphi(\mathcal{M}, c)$ and fix two open intervals I and J containing c and d_1 respectively. If $d \in J \cap]d_1, d_2[$, then $d \notin \varphi(\mathcal{M}, c)$ and any open box including (d, c) -in particular, any open box in $J \times I$ including (d, c)- intersects $\varphi(\mathcal{M}^2)$. Then $\varphi(\mathcal{M}^2) \cap (J \times I)$ cannot be the graph of a continuous function of I in J, and (d_1, c) is a black sheep of type (i) of φ.
Therefore we can consider the function g mapping any $c \in]a, b[$ into the minimal $d \in M$ for which (d, c) is a black sheep of φ. g is definable, and so, owing to $(2)_1$, we can find an open interval I in $]a, b[$ such that g is continuous in I and one of the following conditions holds:

(iii) for every $c \in I$, $(g(c), c)$ is a black sheep of type (i) of φ;

(iv) for every $c \in I$, $(g(c), c)$ is a black sheep of type (ii) of φ.

As before, we can suppose that I is just $]a, b[$. Assume (iii). Introduce two functions g_1 and g_2 as follows. For every $c \in]a, b[$,

- $g_1(c)$ is the maximal element $d \in \varphi(\mathcal{M}, c)$ satisfying $d < g(c)$, if such an element exists, and is $-\infty$ otherwise;

- $g_2(c)$ is the minimal element $d \in \varphi(\mathcal{M}, c)$ satisfying $d > g(c)$, if such an elements exists, and is $+\infty$ otherwise.

Both g_1 and g_2 are definable (in the obvious sense); moreover we can suppose that g_1 is continuous or constantly $-\infty$ in $]a, b[$ and, similarly, g_2 is continuous or constantly $+\infty$ in $]a, b[$. Choose $c \in]a, b[$, $d_1, d_2 \in M$ such that

$$g_1(c) < d_1 < g(c) < d_2 < g_2(c).$$

By the definition of g_1, g_2 and the continuity of g_1, g_2, g, there exists an interval $I \subseteq]a, b[$ containing c such that, for every $c' \in I$,

$$g_1(c') < d_1 < g(c') < d_2 < g_2(c').$$

Hence again the definition of g_1 and g_2 implies that $\varphi(\mathcal{M})^2 \cap (]d_1, d_2[\times I)$ is the graph of the continuous function g, whence $(g(c), c)$ is not a black sheep of φ. So (iii) cannot hold.
Accordingly assume (iv). Let $c \in]a, b[$. Clearly $g(c) \neq \varphi_-(c)$. If $g(c) < \varphi_-(c)$, then one can use the continuity of φ_- and g and determine an open box containing $(g(c), c)$ and disjoint from $\varphi(\mathcal{M})$. But this contradicts the fact that $(g(c), c)$ is a black sheep of φ. So $\varphi_-(c) < g(c)$. In the same way one proves $g(c) < \varphi_+(c)$. Now let us introduce two functions g_1 and g_2 as follows. For every $c \in]a, b[$,

- $g_1(c)$ is the maximal element $d \in \varphi(\mathcal{M}, c)$ such that che $d < g(c)$;
- $g_2(c)$ is the minimal element $d \in \varphi(\mathcal{M}, c)$ such that $d > g(c)$.

Both g_1 and g_2 have domain $]a, b[$ because, for every $c \in]a, b[$, $\varphi_-(c) < g(c) < \varphi_+(c)$ and $\varphi(\mathcal{M}, c)$ is finite. We can even assume that g_1 and g_2 are continuous in $]a, b[$. The Monotonicity Theorem provides, for any $c \in]a, b[$, two intervals I, J containing c, $g(c)$ respectively and satisfying, for every $c' \in I$,

$$g(c') \in J, \quad g_1(c'), g_2(c') \notin J.$$

Hence $\varphi(\mathcal{M}^2) \cap (J \times I) = \emptyset$, which contradicts the fact that $(g(c), c)$ is a black sheep of type (ii). This excludes also (iv).
In conclusion Y cannot be infinite. As already pointed out, this implies our claim: $\varphi(v, w)$ is uniformly finite in $]a, b[$.

Now let $n > 1$. Assume $(1)_j$, $(2)_j$, $(3)_j$ and $(4)_j$ for $1 \leq j < n$.

$(1)_n$ Let C be a cell of $\mathcal{M}^n$. For every $i \in I$ let X_i be a definable subset of C. We are looking for a decomposition of C partitioning each X_i. If C is a k-cell for $k < n$, then there is some definable homeomorphism π_C of C onto a cell C' of $\mathcal{M}^{n-1}$. By $(1)_{n-1}$ there is a decomposition P' of C' partitioning

each $\pi_C(X_i)$. Through π_C^{-1}, P' can be lifted to a decomposition P of C partitioning every X_i.
So assume that C is just an n-cell. Consequently there are an $(n-1)$-cell C' of $\mathcal{M}^{n-1}$ and two functions f and g satisfying 9.3.1, 2; in particular

$$C = \{(\vec{a}, b) \in M^n : \vec{a} \in C', f(\vec{a}) < b < g(\vec{a})\}.$$

Let π_C be the projection of C onto C'. By $(1)_{n-1}$ there exists a decomposition P' of C' partitioning each $\pi_C(X_i)$. For every $Y' \in P'$, let

$$Y = \{(\vec{a}, b) \in C : \vec{a} \in Y'\}.$$

Clearly these sets Y partition C when Y' ranges over P'. So it suffices to show that, for every Y,

(v) there is a decomposition of Y partitioning each set X_i for which $\pi_C(X_i) \cap Y' \neq \emptyset$.

To simplify the notation, assume without loss of generality that $\pi_C(X_i) \cap Y' \neq \emptyset$ for every $i \in I$. Fix $i \in I$ and, for every $\vec{a} \in Y'$, consider

$$X_i(\vec{a}) = \{b \in M : (\vec{a}, b) \in X_i\}.$$

There is a formula $\varphi_i(v, \vec{w})$ (possibly with parameters from M) such that, for every $\vec{a} \in Y'$, $\varphi_i(\mathcal{M}, \vec{a}) = X_i(\vec{a})$. Moreover $X_i(\vec{a})$ is a non-empty subset of $]f(\vec{a}), g(\vec{a})[$. For every $\vec{a} \in Y'$, let $B_i(\vec{a})$ denote the set of the endpoints of $X_i(\vec{a})$ in $]f(\vec{a}), g(\vec{a})[$. There is a formula $\theta_i(v, \vec{w})$ (with parameters in M) such that, for every $\vec{a} \in Y'$, $\theta_i(\mathcal{M}, \vec{a})$ defines $B_i(\vec{a})$. $\theta_i(v, \vec{w})$ is finite in Y' and Y' is a cell. $(3)_{n-1}$ implies that $\theta_i(v, \vec{w})$ is uniformly finite in Y'. Accordingly there exists a positive integer h_i such that, for every $\vec{a} \in Y'$, $|B_i(\vec{a})| \leq h_i$. X_i is the finite union of the tuples $(\vec{a}, b)$ where $\vec{a} \in Y'$, $b \in X_i(\vec{a})$ and $|B_i(\vec{a})| = h$ for $h = 1, \ldots, h_i$. Without loss of generality, we can even suppose that $|B_i(\vec{a})| = h_i$ for every $\vec{a} \in Y'$. Consequently we can define h_i functions from Y' into M

$$f_1^i, \ldots, f_{h_i}^i,$$

mapping any $\vec{a} \in Y'$ into the first, ..., the h_i-th element of $B_i(\vec{a})$. All these functions are definable; by $(2)_{n-1}$, we can assume that they are continuous on Y'. Unless partitioning again each X_i, we can also suppose that, for every i and j in I, $h = 1, \ldots, h_i$ and $h' = 1, \ldots, h_j$, exactly one of the following cases holds:

$$\forall \vec{a} \in Y', \quad f_h^i(\vec{a}) = f_{h'}^j(\vec{a}),$$

$$\forall \vec{a} \in Y', \quad f_h^i(\vec{a}) < f_{h'}^j(\vec{a}),$$

$$\forall \vec{a} \in Y', \quad f_h^i(\vec{a}) > f_{h'}^j(\vec{a}).$$

Accordingly we can rearrange the functions f_h^i ($i \in I$, $1 \leq h \leq h_i$) and form a new sequence

$$g_0, \ldots, g_t$$

such that, for $r \leq s \leq t$, $g_r(\vec{a}) < g_s(\vec{a})$ for every $\vec{a} \in Y'$. But this implies that the sets of the tuples $(\vec{a}, b) \in Y$ such that $\vec{a} \in Y'$ and

$$f(\vec{a}) < b < g_0(\vec{a}),$$

$$g_s(\vec{a}) < b < g_{s+1}(\vec{a}) \quad (s < t),$$

$$g_t(\vec{a}) < b < g(\vec{a}),$$

$$b = g_s(\vec{a}) \quad (s \leq t)$$

respectively, form a decomposition of Y partitioning each X_i, as claimed. This concludes the proof of $(1)_n$.

Accordingly assume from now on also $(1)_n$.

$(2)_n$ Let C be a cell of $\mathcal{M}^n$ and let f be a definable function from C into M. What we have to find is a decomposition P of C in cells where f is continuous. If C is a k-cell for some $k < n$, then there exists a definable homeomorphism π_C of C onto a cell C' of $\mathcal{M}^{n-1}$. By $(2)_{n-1}$, there is a decomposition P' of C' in cells such that $f\pi_C^{-1}$ is continuous on each of them, and π_C^{-1} lifts P' to a decomposition P as required.
So assume that C is an n-cell, in other words an open cell in $\mathcal{M}^n$. Let

C_1 be the set of tuples $(\vec{a}, b) \in C$ such that

$$(x_1, \ldots, x_{n-1}) \mapsto f(x_1, \ldots, x_{n-1}, b)$$

defines a continuous function f_b on some open box B of M^{n-1} containing $\vec{a}$ and satisfying $B \times \{b\} \subseteq C$,

C_2 be the set of tuples $(\vec{a}, b)$ such that $x_n \mapsto f(\vec{a}, x_n)$ defines a function $f_{\vec{a}}$ either constant or strictly monotonic on some open interval I containing b and satisfying $\{\vec{a}\} \times I \subseteq X$, and, in the latter case, also $f(I)$ is an open interval and f is a bijection between I and $f(I)$.

Use $(1)_n$ and get a decomposition P of X partitioning both C_1 and C_2. So it is enough to show that f is continuous on every cell of P, and even on every open cell D of P, owing to what we observed at the beginning of this point. Then there are a cell D' of $\mathcal{M}^{n-1}$ and two functions f_1, f_2 satisfying 9.3.1, 2 and

$$D = \{(\vec{a}, b) \in M^n : \vec{a} \in D', f_1(\vec{a}) < b < f_2(\vec{a})\}.$$

Notice:

(vi) $D \subseteq C_1$.

In fact let $b \in M$ satisfy $(\vec{a}, b) \in D$ for some suitable $\vec{a} \in M^{n-1}$. As the domain of g includes D, f_b is defined in some open subset of D'. By $(1)_{n-1}$ and $(2)_{n-1}$, there is some open cell in D' where f_b is continuous. For $\vec{a}$ in this cell, $(\vec{a}, b) \in C_1$. Hence $D \cap C_1 \neq \emptyset$. As P partitions C_1, $D \subseteq C_1$.
Now we claim

(vii) $D \subseteq C_2$ and, for every $\vec{a} \in D'$, $f_{\vec{a}}$ is either constant or strictly monotonic in $]f_1(\vec{a}), f_2(\vec{a})[$ (and, in the latter case, the image of $]f_1(\vec{a}), f_2(\vec{a})[$ is an open interval and $f_{\vec{a}}$ is a bijection between these intervals).

$D \subseteq C_2$ can be shown by proceeding as for C_1. Now take $\vec{a} \in D'$. Owing to the Monotonicity Theorem, there are a natural m and $b_1, \ldots, b_m \in M$ such that

$$f_1(\vec{a}) < b_1 < \ldots < b_m < f_1(\vec{a}),$$

$f_{\vec{a}}$ is either constant or strictly monotonic in each interval J among $]f_1(\vec{a}), b_1[$, $]b_j, b_{j+1}[$ (for $1 \leq j < m$) and $]b_m, f_2(\vec{a})[$, and, in the latter case, even $f_{\vec{a}}(J)$ is an interval and $f_{\vec{a}}$ induces a bijection between J and $f_{\vec{a}}(J)$. Choose m minimal. If $m = 0$, then the only involved interval J is just $]f_1(\vec{a}), f_2(\vec{a})[$, and (vii) is trivial. On the other side, if $m > 0$, then $f_{\vec{a}}$ is neither constant nor strictly monotonic in any open interval containing b_1; as $(\vec{a}, b_1) \in D$ and $D \subseteq C_2$, $(\vec{a}, b_1) \in C_2$, which contradicts the definition of C_2. So $m = 0$, and we are done.
At this point, we are in a position to conclude the proof of $(2)_n$. In fact, let $(\vec{a}, b) \in D$, J be an open interval containing $f(\vec{a}, b)$; we are looking for an open box B of $\mathcal{M}^n$ including $(\vec{a}, b)$ and satisfying $f(B) \subseteq J$. Owing to (vii), there is a closed interval $I = [b_1, b_2]$ of M such that b is in the interior of I, $I \subseteq]f_1(\vec{a}), f_2(\vec{a})[$ and $f_{\vec{a}}(I) \subseteq J$. By (vi) there exist two open boxes B_1 and B_2 of $\mathcal{M}^{n-1}$ both containing $\vec{a}$, and satisfying the additional conditions

$$B_i \times \{b_i\} \subseteq D, \quad f(B_i, b_i) \subseteq J \quad \forall i = 1, 2.$$

Let B' be an open box of $\mathcal{M}^{n-1}$ such that $B' \subseteq B_1 \cap B_2$, $\vec{a} \in B'$. Then $f(B' \times I) \subseteq J$. In fact, let $\vec{a'} \in B'$, $b' \in I$, so $f(\vec{a'}) < b_1$ because $B' \times \{b_1\} \subseteq D$ and $b_2 < f_2(\vec{a'})$ because $B' \times \{b_2\} \subseteq D$. Consequently

$$f_1(\vec{a'}) < b_1 < b' < b_2 < f_2(\vec{a'}).$$

Furthermore $f(\vec{a'}, b_1) \in J$, $f(\vec{a'}, b_2) \in J$. As $f_{\vec{a'}}$ is either constant or strictly monotonic in $]f_1(\vec{a'}), f_2(\vec{a'})[$, $f(\vec{a'}, b') \in J$. Accordingly $f(B' \times I) \subseteq J$. This accomplishes our proof of $(2)_n$.

So we can assume even $(2)_n$ true from now on. At this point let us deal with $(3)_n$.

$(3)_n$ Let $\varphi(v, w_1, \ldots, w_n)$ be a formula, C be a cell of $\mathcal{M}^n$ such that $\varphi(v, w_1, \ldots, w_n)$ is finite in C. The case when the dimension of C is strictly smaller than n can be handled as before. So we can limit our analysis to the case when C is an n-cell, in other words is open. Let

C_1 be the set of those points $(\vec{a}, b) \in C$ such that $\vec{a}$ is good for $\varphi(v, w_1, \ldots, w_{n-1}, b)$,

C_2 be the set ot the points $(\vec{a}, b) \in C$ such that b is good for $\varphi(v, \vec{a}, w_n)$

respectively ($\vec{a}$ abbreviates here $(a_1, \ldots, a_{n-1})$). By $(1)_n$, there is a decomposition P of C partitioning both C_1 and C_2. We claim

(viii) for every open Y in P, $Y \subseteq C_1$ and $Y \subseteq C_2$.

In fact, let $(\vec{a}, b) \in Y$, and let B be an open box in Y including $(\vec{a}, b)$. The projection B' of B onto the first $n-1$ coordinates is an open box of $\mathcal{M}^{n-1}$. By $(3)_{n-1}$, $\varphi(v, w_1, \ldots, w_{n-1}, b)$ is uniformly finite in B'. By $(1)_{n-1}$ and $(2)_{n-1}$, there are an open cell $C' \subseteq B'$ and a positive integer h such that, for every $\vec{a'} \in C'$, $\varphi(\mathcal{M}, \vec{a'}, b)$ has size h and the functions mapping any $\vec{a'} \in C'$ into

the first, ..., the h-th element of $\varphi(\mathcal{M}, \vec{a'}, b)$ respectively

are continuous. Let $\vec{a'} \in C'$. It is easy to see that $\vec{a'}$ is good for $\varphi(v, w_1, \ldots, w_{n-1}, b)$.
Then $(\vec{a'}, b) \in C_1$, and $Y \cap C_1 \neq \emptyset$. Hence $Y \subseteq C_1$. $Y \subseteq C_2$ is shown in a similar way.
In conclusion, if Y is an open set of P, then for every $(\vec{a}, b) \in Y$, $\vec{a}$ is good for $\varphi(v, w_1, \ldots, w_{n-1}, b)$, b is good for $\varphi(v, \vec{a}, w_n)$. So it suffices to show what follows.

(ix) If Y is an open cell of $\mathcal{M}^n$, $\varphi(v, w_1, \ldots, w_n)$ is an $L(M)$-formula finite in Y and, for every $(\vec{a}, b) \in Y$, $\vec{a}$ is good for $\varphi(v, w_1, \ldots, w_{n-1}, b)$ and b is good for $\varphi(v, \vec{a}, w_n)$, then the size of $\varphi(\mathcal{M}, \vec{c})$ is constant when $\vec{c}$ ranges over Y.

Otherwise there is a natural h such that the set Y' of the elements $\vec{c} \in Y$ satisfying $|\varphi(\mathcal{M}, \vec{c})| = h$ is different from Y and $\emptyset$. But Y' is definable and so, as Y is definably connected, Y contains some boundary point $\vec{c}$ of Y': any open box $B \subseteq Y$ including $\vec{c}$ overlaps both Y' and its complement $Y - Y'$ in Y. Hence it is sufficient to prove (ix) when Y just coincides with some open box B. Let $(\vec{a}_1, b_1), (\vec{a}_2, b_2) \in B = Y$. Notice that every $\vec{a}$ in the projection B' of B onto the first $n-1$ coordinates is good for $\varphi(v, w_1, \ldots, w_{n-1}, b_1)$: in fact $(\vec{a}, b_1) \in Y$. Hence $(4)_{n-1}$ ensures

$$|\varphi(\mathcal{M}, \vec{a}_1, b_1)| = |\varphi(\mathcal{M}, \vec{a}_2, b_1)|.$$

Similarly $(4)_1$ implies

$$|\varphi(\mathcal{M}, \vec{a}_2, b_1)| = |\varphi(\mathcal{M}, \vec{a}_2, b_2)|.$$

In conclusion

$$|\varphi(\mathcal{M}, \vec{a}_1, b_1)| = |\varphi(\mathcal{M}, \vec{a}_2, b_2)|,$$

and this accomplishes the proof of $(3)_n$.

The last claim to be examined is $(4)_n$. But now the proof is a direct consequence of what we have just observed. In fact, recall that, if Y is an open box of $\mathcal{M}^n$ and the points $(\vec{a}, b)$ of Y are good for $\varphi(v, w_1, \ldots, w_n)$, then, for every $(\vec{a}, b) \in Y$, $\vec{a}$ is good for $\varphi(v, w_1, \ldots, w_{n-1}, b)$ and b is good for $\varphi(v, \vec{a}, w_n)$. Use this and (ix) and deduce $(4)_n$.

Hence the proof of Lemma 9.5.4 is concluded, and at last we can also end this section. ♣

9.6 Definable groups in o-minimal structures

Which structures are definable in o-minimal models? The aim of this section is just to measure how complicated o-minimal structures are up to biinterpretability, and so to answer the previous question, and to realize which groups, or rings, or manifolds are definable in them. In particular the interest in definable manifolds arises quite naturally from the connection between o-minimality and (analytic) geometry underlined at the beginning of this chapter. We saw what a manifold is (inside an algebraically closed field $\mathcal{K}$) in Section 8.5, where we observed that every manifold in $\mathcal{K}$ is definable in $\mathcal{K}$. In an arbitrary o-minimal structure $\mathcal{M}$ a manifold may not be definable. Accordingly, first we fix what *definable manifold* means. It is just a finite family $(V, (V_i, f_i)_{i \leq m})$ where m is a natural number and

- ⋆ $V = \cup_{i \leq m} V_i$ is the *atlas*,
- ⋆ each f_i is a bijection from V_i onto a **definable** open subset $U_i = f_i(V_i)$ of M^n for some natural n independent of i,
- ⋆ for $i, j \leq m$ and $i \neq j$, $U_{i,j} = f_i(V_i \cap V_j)$ is, again, **definable** and open in U_i;
- ⋆ for $i, j \leq m$ and $i \neq j$, $f_{i,j} = f_i f_j^{-1}$ is a **definable** homeomorphism between $U_{i,j}$ and $U_{j,i}$.

After fixing this definition, let us look for groups and manifolds definable in o-minimal structures. Even at a first superficial sight one can meet some non-trivial examples: for instance, it is quite obvious that, for a real closed field $\mathcal{K}$, the linear groups $GL(n, \mathcal{K})$ are definable in $\mathcal{K}$. Indeed, a sharp analysis displays some notable similarities with the ω-stable framework. In particular, by adapting the Hrushovski-Weil Theorem 8.6.2, Pillay showed

Theorem 9.6.1 *Let $\mathcal{M}$ be an o-minimal structure, and $\mathcal{G}$ be a group definable in $\mathcal{M}$. Then $\mathcal{G}$ can be equipped with a (unique) definable manifold structure making it into a topological group.*

When $\mathcal{M}$ expands the real field, the manifold topology makes $\mathcal{G}$ into a locally Euclidean topological group and in conclusion, owing to the Montgomery, Zippen and Gleason solution of Hilbert's Fifth Problem, into a Lie group. Definable groups have been intensively studied in o-minimal structures. In particular we would like to mention an o-minimal analogue of Cherlin's Conjecture, proved by Peterzil, Pillay and Starchenko.

Theorem 9.6.2 *Let $\mathcal{M}$ be an o-minimal structure, $\mathcal{G}$ be a connected group definable in $\mathcal{M}$ and having no definable non-trivial normal abelian subgroup. Then there is a definable isomorphism of $\mathcal{G}$ onto the connected component of an algebraic group over a real closed field.*

Notably, a local version of Zilber's Conjecture is true in the o-minimal setting, as shown by Peterzil and Starchenko. Let us discuss briefly this matter.

Given an o-minimal $\mathcal{M}$, call an element $a \in M$ *trivial* when there are no open interval I including a and no definable f from I^2 into I which is strictly monotone in each variable.

Theorem 9.6.3 (Trichotomy Theorem) *Let $\mathcal{M}$ be an $\aleph_0$-saturated o-minimal structure, $a \in M$. Then exactly one of the following conditions holds:*

(i) *a is trivial,*

(ii) *there is some convex neighbourhood of a where $\mathcal{M}$ induces a structure of an ordered vectorspace over an ordered division ring,*

(iii) *there is some open interval including a where $\mathcal{M}$ induces a structure of a real closed field.*

9.7 O-minimality and Real Analysis

In this section, we introduce some new examples of o-minimal structures. They concern some expansions of the real field closely related to Real Analysis and Geometry. Indeed Model Theory meets these areas within o-minimality, and provides new ideas, new tools and, definitively, new perspectives in studying the involved structures.

1. $\mathbf{R}_{exp}$

   The first example we wish to deal with is the most famous as well. It concerns the exponentiation in the real field. We have seen in Chapter 2 Tarski's Theorem showing that the theory of the real field $\mathbf{R}$ has the elimination of quantifiers in the language L of ordered fields: accordingly

   $$\textit{definable} = \textit{semialgebraic}$$

   in this setting. Tarski also gave an effective procedure reducing any formula $\varphi(\vec{v})$ di L into an equivalent quantifier free L-formula $\varphi'(\vec{v})$. When applying this reduction method to a sentence φ of L, we get explicitly a quantifier free sentence φ' of L equivalent to φ in RCF; φ' is a finite Boolean combination of sentences $n \geq m$ where n and m are integers; accordingly it is easy to check whether φ' (and so φ) is true in $\mathbf{R}$ or not. In conclusion, the theory of $\mathbf{R}$ -in other words RCF- is decidable.

Tarski also proposed the following question. Expand $\mathbf{R}$ to a structure

$$\mathbf{R}_{exp} = (\mathbf{R}, 0, 1, +, \cdot, -, \leq, exp)$$

where exp is the 1-ary function mapping any real x into e^x. Accordingly add a 1-ary operation symbol (for exp) to L and denote by L_{exp} the enlarged language: hence

$$L_{exp} = L \cup \{exp\} = \{0, 1, +, \cdot, -, \leq, exp\}.$$

Conjecture 9.7.1 (Tarski) *The theory of* $\mathbf{R}_{exp}$ *is decidable.*

One can observe that the theory of $\mathbf{R}_{exp}$ is not quantifier eliminable in L_{exp}: this was shown by Van den Dries in 1982. However in 1991 Wilkie proved its model completeness.

Theorem 9.7.2 (Wilkie) *The theory of* $\mathbf{R}_{exp}$ *is model complete.*

In particular, the definable sets in $\mathbf{R}_{exp}$ can be obtained as follows. For n any positive integer, call a subset E of $\mathbf{R}^n$ *exponential* when it has the form

$$E = \{(a_1, \ldots, a_n) \in \mathbf{R}^n : f(a_1, \ldots, a_n, e^{a_1}, \ldots, e^{a_n}) = 0\}$$

for a suitable real polynomial f with $2n$ unknowns. Notice that exponential sets are closed under finite union and intersection (as the points annihilating at least one of finitely many polynomials are just the zeros of their product, and the points annihilating a finite system of **real** polynomials are just the zeros of the sum of their squares). But, according to the Van den Dries remark on (the failure of) quantifier elimination, the definable sets in $\mathbf{R}_{exp}$ are something larger than the finite Boolean combinations of exponential sets. So let us introduce *subexponential* sets. A subexponential set in $\mathbf{R}^n$ is just the image of an exponential set of $\mathbf{R}^{n+m}$ (for some m) under the projection map of $\mathbf{R}^{n+m}$ onto the first n coordinates in $\mathbf{R}^n$. Clearly exponential and subexponential sets are definable. What Wilkie showed is that subexponential sets are closed under complement. This implies model completeness, and proves that in $\mathbf{R}_{exp}$

definable = *subexponential.*

At this point, one can use a theorem of Khovanskii saying that every exponential set, and consequently every subexponential set, has only finitely many connected components. Just apply this result to definable (equivalently subexponential) subsets of **R** and get

Corollary 9.7.3 $\mathbf{R}_{exp}$ *(and its theory) are o-minimal.*

Later Ressayre gave a nice axiomatization of $Th(\mathbf{R}_{exp})$, showing that its model theory requires very simple global information about exponentiation. But now let us come back to Tarski's Conjecture. What can we say about it? Well, there is a famous conjecture in transcendental number theory, due to Schanuel and saying:

Conjecture 9.7.4 (Schanuel) *Let n be a positive integer, $a_1, \ldots, a_n$ be complex numbers linearly independent over the rational field $\mathbf{Q}$. Then the transcendence degree of $\mathbf{Q}(a_1, \ldots, a_n, e^{a_1}, \ldots, e^{a_n})$ over $\mathbf{Q}$ is at least n.*

Remarks 9.7.5 (a) Schanuel's Conjecture has been proved in some particular cases, for example when $n = 1$, or $a_1, \ldots, a_n$ are algebraic (Lindemann).

(b) 1, e are linearly independent over $\mathbf{Q}$. Hence Schanuel's Conjecture would imply that $\mathbf{Q}(1, e, e^1, e^e)$, in other words $\mathbf{Q}(e, e^e)$, has transcendence degree 2 over $\mathbf{Q}$, and hence e, e^e are algebraically independent. Nevertheless, as far as one presently knows, it is still an open question whether e^e is irrational.

(c) 1, $i\pi$ are linearly independent over $\mathbf{Q}$. Hence Schanuel's Conjecture would imply that $\mathbf{Q}(1, i\pi, e^1, e^{i\pi})$, hence $\mathbf{Q}(e, i\pi)$ (as $e^{i\pi} + 1 = 0$), has transcendence degree 2 over $\mathbf{Q}$, and consequently that e and π are algebraically independent: but this is still an open question, as well known.

It is generally felt that a solution of Schanuel's Conjecture is vary far, and should go beyond the present knowledge in Mathematics. However a positive answer to the question of Schanuel would imply a solution of Tarski's Conjecture as well.

Theorem 9.7.6 (Macintyre-Wilkie) *If Schanuel's Conjecture holds, then $Th(\mathbf{R}_{exp})$ is decidable.*

2. $\mathbf{R}_{an}$

Now we deal with real analytic functions f. Here we have to be very careful in fixing our setting. In fact, we have to recall what happens when we expand the reals by *sin* (or *cos*): o-minimality gets lost. However one sees that o-minimality is preserved if we restrict the domain of the sinus function to a suitable interval $]-\frac{\pi}{2}, \frac{\pi}{2}[$. Accordingly one takes a language L_{an} enlarging the language L of ordered fields by a 1-ary operation symbol $\hat{f}$ for every function f analytic on some open subset U of $\mathbf{R}^n$ containing the cube $[0, 1]^n$ (n ranges, as usual, over positive integers, and the only reason to choose $[0, 1]$ instead of another interval is just to fix and normalize our setting); then one takes the L_{an}-structure $\mathbf{R}_{an}$ expanding the real field $\mathbf{R}$ and interpreting any symbol $\hat{f}$ in the function equalling f in $[0, 1]^n$ and assuming the constant value 0 elsewhere. Notice that L_{an} is uncountable. $\mathbf{R}_{an}$ is called the *real field with restricted analytic functions.*

Theorem 9.7.7 (Van den Dries) $\mathrm{Th}(\mathbf{R}_{an})$ *is model complete and o-minimal.*

Van den Dries' analysis also determines what is definable in $\mathbf{R}_{an}$. In fact, the definable sets are exactly the so called *globally subanalytic* sets. They are obtained as follows. Call a subset A of an analytic manifold X *semianalytic* in X if there is an open covering $\mathcal{U}$ of X such that, for every $U \in \mathcal{U}$, $A \cap U$ is a finite union of sets

$$\{a \in U \,:\, f(a) = 0,\, g_0(a), \ldots, g_k(a) > 0\}$$

where f and the g's are analytic functions on U. At this point call a subset B of X *subanalytic* in X if there is an open covering $\mathcal{U}$ of X such that, for every $U \in \mathcal{U}$, $B \cap U$ is the image of some A semianalytic in $U \times \mathbf{R}^m$ by the projection map from $U \times \mathbf{R}^m$ onto U (here m may depend on B and U). Finally call $S \subseteq \mathbf{R}^n$ *globally subanalytic* if it is subanalytic in the analytic manifold $(\mathbf{P}_1(\mathbf{R}))^n$ (where $\mathbf{P}_1(\mathbf{R})$ is the real projective line).

Gabrielov showed a "theorem of the complement" for subanalytic sets in an analytic manifold X, ensuring that they are just closed under complement. This is the key result in showing the model completeness and the o-minimality of the theory of $\mathbf{R}_{an}$, and also in proving that

definable = globally subanalytic

in $\mathbf{R}_{an}$.

Does $Th(\mathbf{R}_{an})$ admit quantifier elimination? Denef and Van den Dries showed that the answer is positive, provided one extends the language L_{an} by a symbol $^{-1}$ for the inverse function (with the usual convention $0^{-1} = 0$).

Theorem 9.7.8 (Denef-Van den Dries) *$Th(\mathbf{R}_{an})$ eliminates the quantifiers in $L_{an} \cup \{^{-1}\}$.*

3. $\mathbf{R}_{an,\,exp}$

Finally let us examine what happens when we expand the reals both by the exponential function and the restricted analytic functions. Let $L_{an,\,exp} = L_{an} \cup \{exp\}$ the corresponding language, $\mathbf{R}_{an,\,exp}$ the resulting structure in $L_{an,\,exp}$. First Van den Dries and Miller, adapting Wilkie's work on exponentiation, proved

Theorem 9.7.9 (Van den Dries-Miller) *The theory of $\mathbf{R}_{an,\,exp}$ is model complete and o-minimal.*

Subsequently, Van den Dries, Macintyre and Marker found a different proof providing a nice axiomatization of the theory of $\mathbf{R}_{an,\,exp}$ in Ressayre's style. They got also quantifier elimination in a language extending $L_{an,\,exp}$ by the logarithm function *log*.

Theorem 9.7.10 (Van den Dries-Macintyre-Marker) *The theory of $\mathbf{R}_{an,\,exp}$ eliminates the quantifiers in the language $L_{an,\,exp} \cup \{log\}$.*

Notably, the logarithm function cannot be ignored to obtain quantifier elimination. The Van den Dries-Macintyre-Marker approach also provides an explicit description of the definable sets in $\mathbf{R}_{an,\,exp}$, following the same lines as in the cases before.

Of course these examples are very far from exhausting a general display of the o-minimal expansions of the real field (a wider information can be found in the references quoted at the end of the chapter). But they can illustrate how rich and interesting this research field is.

Let us conclude this section with some final remarks partly exceeding the o-minimal limits. In fact, it is noteworthy that, although Model Theory and Real Analysis closely interact via o-minimality, Complex Analysis has raised a lot of difficulties to a model theoretic treatment. For instance, while expanding the reals by exponentiation gives an o-minimal structure (by Wilkie's Theorem), $(\mathbf{C}, +, \cdot, -, 0, 1, exp)$ defines the integers by the formula

$$exp(2\pi i v) = 1,$$

so that there is a very little hope to dominate its first order theory, its definable sets, and so on. Indeed, the zero sets of complex analytic functions can be quite pathological.
Some years ago, Boris Zilber proposed a satisfactory strategy to develop the model theory of the complex exponentiation, but his program needs some very strong conjectures on transcendental numbers (even beyond Schanuel's Problem). Zilber also followed a more successful approach, looking at analytic compact manifolds X rather than at analytic functions: in fact, these manifolds can be viewed as first order structures in a language with a relation for any subanalytic subset of every power of X. In this setting one shows

Theorem 9.7.11 (Zilber) *The theory of a compact complex manifold eliminates the quantifiers and has a finite Morley rank.*

9.8 Variants on the o-minimal theme

Strongly minimal theories have a natural enlargement to totally transcendental (i. e. ω-stable) theories via Morley rank. In the ordered setting nothing is known extending sistematically o-minimality in a parallel way. However just the ordered framework suggests several notions widening o-minimality: they have been intensively studied in the latest years.
In particular we want to discuss here briefly *weak o-minimality*. As said, we still work within linearly ordered structures $\mathcal{M} = (M, \leq, \ldots)$. Recall that $\mathcal{M}$ is o-minimal when every definable subset D of M is a finite union of intervals (possibly with infinite endoints); notice that intervals are convex.

Definition 9.8.1 *$\mathcal{M}$ is called* **weakly o-minimal** *when every definable subset D of M is a finite union of convex (definable) sets.*

Remark 9.8.2 Of course, o-minimality implies weak o-minimality. Moreover, among expansions of the real line $(\mathbf{R}, \leq)$, the converse is also true, and weakly o-minimal just means o-minimal. This is because $(\mathbf{R}, \leq)$ is Dedekind complete, and every bounded set has its own least upper bound and its own greatest under bound; in particular every convex set is an interval (in the broader sense recalled before).

However there do exist weakly o-minimal structures which are not o-minimal.

Example 9.8.3 Take the ordered field of real algebraic numbers $\mathbf{R}_0$. This is a real closed field, and so an o-minimal structure. Add a 1-ary relation selecting the elements of $\mathbf{R}_0$ lying between $-\pi$ and π (or, if you like, between any two reals $a < b$ with a or b transcendental). The resulting structure is not o-minimal any more, because $D = \{r \in \mathbf{R}_0 : -\pi < r < \pi\}$ is convex and definable, but cannot be expressed as a finite union of intervals with real algebraic endpoints. But actually D is convex, and indeed one can see that the new structure is weakly o-minimal.

Notably, every expansion of an o-minimal structure by convex subsets is weakly o-minimal. This is a beautiful result of Baisalov-Poizat, generalizing the last example and answering in this way a question of Cherlin. Other relevant examples, arising from several frameworks in Algebra, can be proposed. By the way, weak o-minimality was first introduced by M. Dickmann in 1985, dealing with certain ordered rings extending real closed fields.
Not surprisingly, weakly o-minimal structures do not behave so well as o-minimal do. In particular

weak o-minimality is not preserved by elementary equivalence

(so there are weakly o-minimal structures whose theory has some non weakly o-minimal models). Furthermore

Monotonicity and Cell Decomposition fail

as well as existence and uniqueness of prime models. However some "weaker" versions of these results can be recovered, and a relevant, although not so fluent, theory has been developed.

9.9 No rose without thorns

We have seen that o-minimal theories admit an independence notion related to algebraic closure and satisfying the same basic properties **(I1)**-**(I6)** forking independence has in simple theories. However these independence notions -forking indipendence in simple theories and algebraic independence in o-minimal theories- were introduced in a different way and were developed independently. So a natural question arises in Model Theory, i.e. to find a new concept of independence so convincing to satisfy **(I1)**-**(I6)** in most theories and so general to enlarge both the previous cases. This is the content of a recent work of Alf Onshuus who, following suggestions from Thomas Scanlon, introduced

- a new notion of independence (called *thorn-independence*)

and

- a related class of theories (named *rosy theories*)

where thorn-independence enjoys all the basic assumptions **(I1)**-**(I6)**, so local character, symmetry and so on. Rosy theories include both simple and o-minimal theories, as well as further relevant examples. Thorn-independence agrees with algebraic independence in the o-minimal case and with forking independence in *stable* theories: in fact, when these pages are written (at the end of 2002), it is not clear whether thorn-independence equals forking-independence even in the simple setting, although this has been checked to be true in all the known key examples of simple theories.
Notably symmetry, or also local character, is a key property towards rosiness. In fact, a theory T is rosy if and only if thorn-independence satisfies symmetry or local character.

9.10 References

O-minimal theories where introduced by Van den Dries [166] and extensively studied by Pillay and Steinhorn in [129] and (together with J. Knight) in [74]. Van den Dries' book [169] provides a nice and stimulating treatment of o-minimality, also describing its genesis and motivations, and emphasizing its connections with real analysis and real algebraic geometry. These interactions are illustrated in the more recent survey [170], where the o-minimal expansions of the real field are examined. Also [109] gives a short, but captivating introduction to o-minimality.

A general proof of Monotonicity Theorem 9.2.1 can be found in [129]. Wilkie's Complement Theorem 9.4.5 is shown in [178], and the Pillay-Steinhorn Theorem 9.4.7 on prime models in o-minimal theories is in [129] again. Laura Mayer's solution of Vaught Conjecture in the o-minimal setting is in [111]. Pillay's analysis of the groups definable in o-minimal structures (Theorem 9.6.1) is in [125], while the o-minimal analogue of Cherlin Conjecture shown by Peterzil, Pillay and Starchenko (9.6.2) is in [123] and the Peterzil-Starchenko Theorem 9.6.3 is in [124].

As already said [170] provides a general survey of the main o-minimal expansions of the real field, and a rich and detailed bibliography on this matter. In particular Wilkie's Theorem 9.7.2 on $\mathbf{R}_{exp}$ is in [177], and the Khovanskii's results on exponential sets in [70]; Ressayre's approach to the theory of $\mathbf{R}_{exp}$ can be found in [139]; the Macintyre-Wilkie Theorem on the decidability of the theory of $\mathbf{R}_{exp}$ and its relationship to Schanuel's Conjecture is in [99]. The o-minimality of the theory of $\mathbf{R}_{an}$ is proved in [167], using [49], while the Denef-Van den Dries treatment -including the quantifier elimination result in a language with the inverse operation- is in [30]. The o-minimality of the theory of $\mathbf{R}_{an,exp}$ is already shown in [171], but the subsequent analysis of Macintyre, Marker and Van den Dries is in [97].

Zilber's Theorem 9.7.11 on compact complex manifolds can be found in [183]; see also [128].

Now let us deal with weak o-minimality. This was introduced in [32], and extensively examined by Macpherson, Marker and Steinhorn in [100]. The nice theorem of Baisalov-Poizat (mentioned at the end of Section 9.8) is in [7].
Rosy theories and thorn-independence are just the matter of [122].

Bibliography

[1] J. W. Addison, L. Henkin, and A. Tarski, editors. *The theory of models.* North Holland, Amsterdam, 1965.

[2] F. Appenzeller. Classification Theory through Stationary Logic. *Ann. Pure Appl. Logic*, 102:27–68, (2000).

[3] J. Ax. The elementary theory of finite fields. *Ann. Math.*, 88:239–271, (1968).

[4] J. Ax and S. Kochen. Diophantine problems over local fields I. *Amer. J. Math.*, 87:605–630, (1965).

[5] J. Ax and S. Kochen. Diophantine problems over local fields II: A complete set of axioms for p-adic number theory. *Amer. J. Math.*, 87:631–648, (1965).

[6] T. Ax and S. Kochen. Diophantine problems over local fields III: Decidable fields. *Ann. Math.*, 83:437–456, (1966).

[7] Y. Baisalov and B. Poizat. Paires de structures o-minimales. *J. Symbolic Logic*, 63:570–578, (1998).

[8] J. Baldwin. *Fundamentals of Stability Theory.* Springer Verlag, Berlin, 1988.

[9] J. Baldwin and A. Lachlan. On strongly minimal sets. *J. Symbolic Logic*, 36:79–96, (1971).

[10] J. Barwise, editor. *Handbook of Mathematical Logic.* North Holland, Amsterdam, 1977.

[11] J. Barwise and S. Feferman, editors. *Model-theoretic logics.* Springer, New York, 1985.

[12] L. Blum. *Generalized Algebraic Theories.* PhD thesis, M.I.T., Cambridge, 1968.

[13] L. Blum, F. Cucker, M. Shub, and S. Smale. *Complexity and Real computation.* Springer, New York, 1998.

[14] L. Blum, M. Shub, and S. Smale. On a theory of computation and complexity over the real numbers. *Bull. Amer. Math. Soc.*, 21:1–46, (1989).

[15] A. Borovik and A. Nesin. *Groups of finite Morley rank.* Oxford University Press, Oxford, 1994.

[16] E. Bouscaren, editor. *Model Theory and Algebraic Geometry*, volume 1696 of *Lecture Notes in Mathematics.* Springer, Berlin, 1998.

[17] A. B. Carson. The model completion of the theory of commutative regular rings. *J. Algebra*, 27:136–146, (1973).

[18] C. Chang and H. Keisler. *Model Theory.* North Holland, Amsterdam, third edition, 1990.

[19] Z. Chatzidakis. Théorie des modèles des corps finis et pseudo-finis. Prépublications 59, Equipe de Logique Mathématique, Université Paris VII, (1996).

[20] Z. Chatzidakis and E. Hrushovski. The model theory of difference fields. *Trans. Amer. Math. Soc.*, 351:2997–3071, (1999).

[21] Z. Chatzidakis, E. Hrushovski, and Y. Peterzil. Model theory of difference fields, II: Periodic ideals and the trichotomy in all characteristics. Preprint.

[22] G. Cherlin. Algebraically closed commutative rings. *J. Symbolic Logic*, 38:493–499, (1973).

[23] G. Cherlin. *Model Theoretic Algebra*, volume 521 of *Lecture Notes in Mathematics.* Springer Verlag, Berlin, 1976.

[24] G. Cherlin. Groups of small Morley rank. *Ann. Math. Logic*, 17:1–28, (1979).

[25] G. Cherlin and S. Shelah. Superstable fields and rings. *Ann. Math. Logic*, 18:227–270, (1980).

[26] P. Clote and J. Krajicek, editors. *Arithmetic, Proof Theory and Computational Complexity.* Oxford University Press, Oxford, 1993.

[27] P. Cohen. Decision procedures for real and p-adic fields. *Comm. Pure Appl. Math.*, 22:131–151, (1969).

[28] R. M. Cohn. *Difference algebra*, volume 17 of *Tracts in Mathematics.* Interscience, New York, 1965.

[29] G. E. Collins. Quantifier elimination for real closed fields by cylindrical algebraic decomposition. In *Automata Theory and Formal Language*, volume 33 of *Lecture Notes in Computer Science*, pages 134–183, Berlin, 1975. Second GI Conference, Kaiserslautern, Springer.

[30] J. Denef and L. Van den Dries. p-adic and real subanalytic sets. *Ann. Math.*, 128:79–138, (1988).

[31] K. Devlin. *The joy of sets.* Undergraduate Texts in Math. Springer Verlag, Berlin, 1993.

[32] M. Dickmann. Elimination of quantifiers for ordered valuation rings. In *Proceedings 3rd Easter Model Theory Conference, Humboldt Universität*, pages 64–88, 1985.

[33] A. Dolzmann, T. Sturm, and V. Weispfenning. Real quantifier elimination in practice. In *Algorithmic Algebra and number theory*, pages 221–247. Springer, Berlin, 1999.

[34] J. Doner and W. Hodges. Alfred Tarski and decidable theories. *J. Symbolic Logic*, 53:20–35, (1988).

[35] J. L. Duret. Les corps pseudo-finis ont la proprieté d'indépendence. *C. R. Acad. Sci.*, 290:981–983, (1980).

[36] H. D. Ebbinghaus. *Numbers*, volume 123 of *Graduate Texts in Math.* Springer Verlag, New York, 1990.

[37] H. D. Ebbinghaus, J. Flum, and W. Thomas. *Mathematical Logic.* Undergraduate Texts in Math. Springer Verlag, Berlin, 1979.

[38] A. Ehrenfeucht and A. Mostowski. Models of axiomatic theories admitting automorphisms. *Fund. Math.*, 43:50–68, (1956).

[39] P. Eklof. *Ultraproducts for Algebraists*, pages 105–137. In Barwise [10], 1977.

[40] P. Eklof and E. Fisher. The elementary theory of abelian groups. *Ann. Math. Logic*, 4:115–171, (1972).

[41] P. Eklof and G. Sabbagh. Model completions and modules. *Ann. Math. Logic*, 2:251–295, (1970-71).

[42] H. Enderton. *Elements of recursion theory*, pages 527–566. In Barwise [10], 1977.

[43] E. Engeler. A characterization of theories with isomorphic denumerable models. *Notices Amer. Math. Soc.*, 6:161, (1959).

[44] Yu. L. Ershov. On the elementary theory of maximal normed fields. *Algebra i Logika*, 4:31–70, (1965).

[45] Yu. L. Ershov. On the elementary theory of maximal normed fields II. *Algebra i Logika*, 5:5–40, (1966).

[46] Yu. L. Ershov. Fields with a solvable theory. *Soviet Math. Dokl.*, 8:575–576, (1967).

[47] M. Fisher and M. Rabin. Super exponential complexity of Presburger's arithmetic. *SIAM-AMS Proc.*, 7:27–41, (1974).

[48] J. Flum. *First order Logic and its extensions.* Number 499 in Lecture Notes in Mathematics. Springer Verlag, Berlin, 1975.

[49] A. Gabrielov. Projections of semianalytic sets. *Functional Analysis and its Applications*, 2:282–291, (1968).

[50] J. B. Goode. H.L.M. (Hrushovski-Lang-Mordell). Technical Report 48me année, Séminaire Bourbaki, 1995-96.

[51] R. C. Gunning and H. Rossi. *Analytic functions of several complex variables.* Princeton Hall Inc., 1965.

[52] P. Halmos. *Lectures on Boolean Algebras.* Van Nostrand, Princeton, 1963.

[53] B. Hart, E. Hrushovski, and M. Laskowski. The uncountable spectrum of countable complete theories. *Ann. Math.*, 152:207–257, (2000).

[54] D. Haskell, A. Pillay, and C. Steinhorn, editors. *Model Theory, Algebra and Geometry.* MSRI Publ., Cambridge UP, 2000.

[55] L. Henkin. The completeness of the first order functional calculus. *J. Symbolic Logic*, 14:159–166, (1949).

[56] W. Hodges. *Model Theory.* Cambridge University Press, Cambridge, 1993.

[57] W. Hodges. *A shorter model theory.* Cambridge University Press, Cambridge, 1997.

[58] E. Hrushovski. The elementary theory of the Frobenius. Preprint.

[59] E. Hrushovski. A new strongly minimal set. *Ann. Pure Appl. Logic*, 62:147–166, (1993).

[60] E. Hrushovski. The Mordell-Lang Conjecture for function fields. *J. Amer. Math. Soc.*, 9:667–690, (1996).

[61] E. Hrushovski. The Manin - Mumford conjecture and the model theory of difference fields. *Ann. Pure Appl. Logic*, 112:43–115, (2001).

[62] E. Hrushovski and Z. Sokolovic. Minimal subsets of differentially closed fields. (to appear).

[63] E. Hrushovski and B. Zilber. Zariski's geometries. *Bull. Amer. Math. Soc.*, 28:315–324, (1993).

[64] E. Hrushovski and B. Zilber. Zariski's geometries. *J. Amer. Math. Soc.*, 9:1–56, (1996).

[65] N. Jacobson. *Basic Algebra.* Freeman, San Francisco, 1974-1980.

[66] T. Jech. *Set Theory.* Academic Press, New York - London, 1978.

[67] I. Kaplansky. *Differential Algebra.* Hermann, 1957.

[68] A. Kechris. New directions in Descriptive Set Theory. *Bull. Symbolic Logic*, 5:161–174, (1999).

[69] J. Ketonen. The structure of countable Boolean algebras. *Ann. Math.*, 108:41–89, (1978).

[70] A. Khovanskii. On a class of systems of transcendental equations. *Soviet Math. Doklady*, 22:762–765, (1980).

[71] B. Kim. Forking in simple unstable theories. *J. London Math. Soc.*, 57:257–267, (1998).

[72] B. Kim. Simplicity and stability in there. *J. Symbolic Logic*, 66:822–836, (2001).

[73] B. Kim and A. Pillay. Simple theories. *Ann. Pure Appl. Logic*, 88:149–164, (1997).

[74] J. Knight, A. Pillay, and C. Steinhorn. Definable sets in ordered structures II. *Trans. Amer. Math. Soc.*, 295:593–605, (1986).

[75] R. Knight. The Vaught Conjecture: a counterexample. Technical report, Oxford University, 2002.

[76] E. Kolchin. *Differential Algebra and Algebraic Groups.* Academic Press, 1973.

[77] T. Kucera and M. Prest. Imaginary modules. *J. Symbolic Logic*, 57:698–723, (1992).

[78] K. Kunen. *Set Theory. An introduction to independence proofs.* North Holland, Amsterdam, 1983.

[79] S. Lang. *Algebra.* Addison-Wesley, Reading, 1971.

[80] S. Lang. *Number Theory III: Diophantine Geometry.* Encyclopedia of Mathematical Sciences. Springer, Berlin, 1991.

[81] C. H. Langford. Some theorems on deducibility. *Ann. Math.*, 28:16–40, (1926).

[82] C. H. Langford. Theorems on deducibility. *Ann. Math.*, 28:459–471, (1926).

[83] D. Lascar. Ordre de Rudin-Keisler et poids dans les théories ω-stables. *Zeitschr. Math. Logik*, 28:413–430, (1982).

[84] D. Lascar. Why some people are excited by Vaught's Conjecture. *J. Symbolic Logic*, 35:973–982, (1985).

[85] D. Lascar and B. Poizat. An intoduction to forking. *J. Symbolic Logic*, 46:781–788, (1981).

[86] L. Lipschitz and D. Saracino. The model companion of the theory of commutative rings without nilpotent elements. *Proc. Amer. Math. Soc.*, 37:381–387, (1973).

[87] L. Löwenheim. Über Möglichkeiten in Relativ Kalkül. *Math. Ann.*, 76:447–470, (1915).

[88] A. Macintyre. Non-standard Frobenius and generic automorphisms. Preprint.

[89] A. Macintyre. On ω_1-categorical theories of abelian groups. *Fund. Math.*, 70:253–270, (1971).

[90] A. Macintyre. On ω_1-categorical theories of fields. *Fund. Math.*, 71:1–25, (1971).

[91] A. Macintyre. On algebraically closed groups. *Ann. Math.*, 95:53–97, (1972).

[92] A. Macintyre. Model completeness for sheaves of structures. *Fund. Math.*, 81:73–89, (1973).

[93] A. Macintyre. On definable subsets of p-adic fields. *J. Symbolic Logic*, 41:605–610, (1976).

[94] A. Macintyre. *Model-completeness*, pages 139–180. In Barwise [10], 1977.

[95] A. Macintyre. Generic automorphisms of fields. *Ann. Pure Appl. Logic*, 88:165–180, (1997).

[96] A. Macintyre, editor. *Connections between Model Theory and Algebraic and Analytic Geometry*, volume 6 of *Quaderni di Matematica*. Dipartimento di Matematica II Università di Napoli, 2000.

[97] A. Macintyre, D. Marker, and L. Van den Dries. The elementary theory of restricted analytic fields with exponentiation. *Ann. Math.*, 140:183–205, (1994).

[98] A. Macintyre, K. McKenna, and L. Van den Dries. Elimination of quantifiers in algebraic structures. *Adv. Math.*, 47:74–87, (1983).

[99] A. Macintyre and A. Wilkie. On the decidability of the real exponential field. In *Kreiseliana (P. Odifreddi ed.)*, pages 441–467. A. K. Peters, Wellesley, 1996.

[100] D. Macpherson, D. Marker, and C. Steinhorn. Weakly o-minimal structures and real closed fields. *Trans. Amer. Math. Soc.*, 352:5435–5483, (2000).

[101] M. Makkai. A survey of basic stability theory, with particular emphasis on orthogonality and regular types. *Israel J. Math.*, 49:181–238, (1984).

[102] A. Malcev. A correspondence between groups and rings. In *The metamathematics of algebraic systems*, pages 124–137. North Holland, Amsterdam, 1971.

[103] J. Malitz. *Introduction to Mathematical Logic*. Undergraduate Texts in Math. Springer Verlag, Berlin, 1979.

[104] A. Marcja and P. Mangani. Shelah rank for Boolean algebras and some applications to elementary theories. *Alg. Univ.*, 10:247–257, (1980).

[105] A. Marcja and C. Toffalori. On pseudo-$\aleph_0$-categorical theories. *Zeitschr. Math. Logik*, 30:533–540, (1984).

[106] A. Marcja and C. Toffalori. On the Cantor-Bendixson Spectra containing (1,1)-I. In *Logic Colloquium '83*, pages 331–350, Berlin, 1984. Springer.

[107] A. Marcja and C. Toffalori. On the Cantor-Bendixson Spectra containing (1,1)-II. *J. Symbolic Logic*, 50:611–618, (1985).

[108] A. Marcja and C. Toffalori. *Introduzione alla Teoria dei Modelli*, volume 43 of *Quaderni dell'Unione Matematica Italiana.* Pitagora Editrice, Bologna, 1998.

[109] D. Marker. Model theory and exponentiation. *Notices Amer. Math. Soc.*, 43:753–759, (1996).

[110] D. Marker, M. Messmer, and A. Pillay. *Model Theory of fields.* Lecture Notes in Logic. Springer, Berlin, 1996.

[111] L. Mayer. Vaught's conjecture for o-minimal theories. *J. Symbolic Logic*, 53:146–159, (1988).

[112] K. Meer. A note on a $P \neq NP$ result for a restricted class of real machines. *J. Complexity*, 8:451–453, (1992).

[113] A. Mekler. Stability of nilpotent groups of class 2 and prime exponent. *J. Symbolic Logic*, 46:781–788, (1981).

[114] G. Melles. An exposition of Shelah's classification theory and classification by set recursive functions (More for Thomas the doubter). Preprint.

[115] M. Messmer. Groups and fields interpretable in separably closed fields. *Trans. Amer. Math. Soc.*, 344:361–377, (1994).

[116] M. Morley. On theories categorical in uncountable powers. *Proc. Nat. Acad. Sc. U.S.A.*, 49:213–216, (1963).

[117] M. Morley. Categoricity in power. *Trans. Amer. Math. Soc.*, 114:514–538, (1965).

[118] M. Morley. *Omitting classes of elements*, pages 265–273. In Addison et al. [1], 1965.

[119] M. Morley. Countable models of $\aleph_1$-categorical theories. *Israel J. Math.*, 5:65–72, (1967).

[120] M. Morley. The number of countable models. *J. Symbolic Logic*, 35:14–18, (1970).

[121] P. Odifreddi. *Classical recursion theory.* North Holland, Amsterdam, 1989.

[122] A. Onshuus. *Thorn-independence in rosy theories.* PhD thesis, Berkeley, 2002.

[123] Y. Peterzil, A. Pillay, and S. Starchenko. Definably simple groups in o-minimal structures. *Trans. Amer. Math. Soc.*, 352:4397–4419, (2000).

[124] Y. Peterzil and S. Starchenko. A Trichotomy Theorem for o-minimal structures. *Proc. London Math. Soc.*, 77:481–253, (1998).

[125] A. Pillay. On groups and fields definable in o-minimal structures. *J. Pure Appl. Algebra*, 53:239–255, (1988).

[126] A. Pillay. *Geometric Stability Theory.* Oxford University Press, Oxford, 1996.

[127] A. Pillay. Model Theory and Diophantine Geometry. *Bull. Amer. Math. Soc.*, 34:405–422, (1997).

[128] A. Pillay. Some model theory of compact complex spaces. *Contemp. Math.*, 270:323–328, (2000).

[129] A. Pillay and C. Steinhorn. Definable sets in ordered structures I. *Trans. Amer. Math. Soc.*, 295:565–592, (1986).

[130] B. Poizat. Une théorie de Galois imaginaire. *J. Symbolic Logic*, 48:1151–1170, (1983).

[131] B. Poizat. *Cours de Théorie des Modèles.* Nur Al - Mantiq Wal - Ma'rifah, Villeurbanne, 1985.

[132] B. Poizat. *Groupes Stables.* Nur Al - Mantiq Wal - Ma'rifah, Villeurbanne, 1987.

[133] B. Poizat. *Les petits cailloux*, volume 3. Nur Al - Mantiq Wal - Ma'rifah, Aléas, Lyon, 1995.

[134] B. Poizat. *Course in Model Theory: An introduction in Contemporary Mathematical Logic.* Springer Verlag, 1999.

[135] B. Poizat. *Stable groups.* American Mathematical Society, 2001.

[136] M. Prest. *Model Theory and Modules.* London Math. Soc. Colloquium Publications. Cambridge University Press, Cambridge, 1988.

[137] M. Prest. Model theory and representation type of algebras. In *Logic Colloquium '86*, pages 219–260, Amsterdam, 1988. North Holland.

[138] M. Rabin. *Decidable Theories*, pages 595–629. In Barwise [10], 1977.

[139] J. P. Ressayre. *Integer parts of real closed exponential fields*, pages 278–288. In Clote and Krajicek [26], 1993.

[140] A. Robinson. *On the metamathematics of Algebra.* North Holland, Amsterdam, 1951.

[141] A. Robinson. *Complete theories.* North Holland, Amsterdam, 1956.

[142] A. Robinson. On the concept of differentially closed field. *Bull. Res. Council Israel*, 8F:113–128, (1959).

[143] A. Robinson. *Introduction to Model Theory and to the Metamathematics of Algebra.* North Holland, Amsterdam, 1963.

[144] H. Rogers. *Theory of recursive functions and effective computability.* MIT Press, Cambridge, 1988.

[145] C. Ryll-Nardzewski. On the categoricity in power $\leq \aleph_0$. *Bull. Acad. Polon. Sciences*, 7:545–548, (1959).

[146] G. E. Sacks. *Saturated model theory.* W. A. Benjamin, Reading, 1972.

[147] T. Scanlon. Diophantine Geometry from Model Theory. *Bull. Symbolic Logic*, 7:37–57, (2001).

[148] S. Shelah. Uniqueness and characterization of prime models over sets for totally transcendental first order theory. *J. Symbolic Logic*, 37:107–113, (1972).

[149] S. Shelah. *Classification theory and the number of non-isomorphic models.* North Holland, Amsterdam, 1978.

[150] S. Shelah. Simple unstable theories. *Ann. Math. Logic*, 19:177–203, (1980).

[151] S. Shelah. *Classification theory.* North Holland, Amsterdam, 1990.

[152] S. Shelah, L. Harrington, and M. Makkai. A proof of Vaught's Conjecture for ω-stable theories. *Israel J. Math.*, 49:259–280, (1984).

[153] J.R. Shoenfield. *Mathematical logic.* Addison-Wesley, Reading, 1967.

[154] T. Skolem. Untersuchungen über die Axiome des Klassen Kalküls and über Produktations-und Summationsprobleme, welche gewisse Klassen von Aussagen betreffen. *Skrifter Vindenskapsakademiet i Kristania*, 3:37–71, (1919).

[155] Z. Sokolovic. *Model theory of differential fields.* PhD thesis, Notre Dame, 1992.

[156] L. Svenonius. $\aleph_0$-categoricity in first order predicate calculus. *Theoria*, 25:82–94, (1959).

[157] A. Tarski. *A decision method for elementary algebra and geometry.* University of California Press, Berkeley and Los Angeles, second edition, 1951.

[158] A. Tarski. Contributions to the Theory of Models I. *Indag. Math.*, 16:572–581, (1954).

[159] A. Tarski. Contributions to the Theory of Models II. *Indag. Math.*, 16:582–588, (1954).

[160] A. Tarski. Foundations of the calculus of systems. In *Logic, semantics, metamathematics. Papers from 1923 to 1938*, pages 342–383. Clarendon Press, Oxford, 1956.

[161] G. Terjanian. Un contre-example a une conjecture d'Artin. *C. R. Acad. Sc. Paris Ser A*, 262:612, (1966).

[162] R. Thom. La stabilité topologique des applications polynomiales. *L'Einseignement Mathématiques*, 8:24–33, (1962).

[163] S. Thomas. On the complexity of the classification problem for torsion-free abelian groups of finite rank. *Bull. Symbolic Logic*, 7:329–344, (2001).

[164] C. Toffalori. Strutture esistenzialmente complete per certe classi di anelli. *Rend. Sem. Mat. Univ. Padova*, 66:51–71, (1981).

[165] C. Toffalori. On pseudo $\aleph_0$-categorical structures. In *Logic Colloquium '84*, pages 303–327, Amsterdam, 1986. North Holland.

[166] L. van den Dries. Remarks on Tarski's problem concerning $(\mathbf{R}, +, \cdot, \mathbf{exp})$. In *Logic Colloquium '82*, pages 97–121. North Holland, 1984.

[167] L. van den Dries. A generalization of the Tarki-Seidenberg Theorem and some nondefinability results. *Bull. Amer. Math. Soc.*, 15:189–193, (1986).

[168] L. van den Dries. Alfred Tarski's elimination theory for real closed fields. *J. Symbolic Logic*, 53:7–19, (1988).

[169] L. van den Dries. *Tame Topology and o-minimal Structures*, volume 248 of *London Math. Soc. Lecture Note Series.* Cambridge University Press, Cambridge, 1998.

[170] L. van den Dries. O-minimal Structures and Real Analytic Geometry. In *Current Developments in Mathematics (B. Mazur ed.)*, pages 105–152. Int. Press, Sommerville MA, 1999.

[171] L. van den Dries and C. Miller. On the real exponential field with restricted analytic functions. *Israel J. Math.*, 85:19–56, (1994).

[172] L. van den Dries and K. Schmidt. Bounds in the polynomial rings over fields. A non standard approach. *Inv. Math.*, 76:77–91, (1984).

[173] B. L. van den Waerden. *Moderne Algebra.* Springer, Berlin, 1930-31.

[174] R. Vaught. Denumerable models of complete theories. In *Infinitistic Methods*, pages 303–321. Pergamon Press, London, 1961.

[175] F. Wagner. *Simple theories.* Kluwer, Dordrecht, 2000.

[176] A. Weil. *Foundations of algebraic geometry*, volume 29 of *Amer. Math. Soc. Colloquium Pubblics.* Amer. Math. Soc., New York, 1946.

[177] A. Wilkie. Model completeness results for expansions of the ordered field of real numbers by restricted Pfaffian functions and the exponential function. *J. Amer. Math. Soc.*, 9:1051–1094, (1996).

[178] A. Wilkie. A general theorem of the complement and some new o-minimal structures. *Selecta Math.(N.S.)*, 5:397–421, (1999).

[179] C. Wood. The model theory of differential fields of characteristic $p \neq 0$. *Proc. Amer. Math. Soc.*, 40:577–584, (1973).

[180] C. Wood. Prime model extensions for differential fields of characteristic $p \neq 0$. *J. Symbolic Logic*, 39:469–477, (1974).

[181] M. Ziegler. Model theory of modules. *Ann. Pure Appl. Logic*, 26:149–213, (1984).

[182] B. Zilber. The structure of models of uncountably categorical theories. In *ICM-Varsavia 1983*, pages 359–368. North Holland, 1984.

[183] B. Zilber. Model Theory and Algebraic Geometry. In *Proceedings 10th Easter Conference*, Berlin, 1993.

Index

TRENDS IN LOGIC

1. G. Schurz: *The Is-Ought Problem.* An Investigation in Philosophical Logic. 1997
ISBN 0-7923-4410-3
2. E. Ejerhed and S. Lindström (eds.): *Logic, Action and Cognition.* Essays in Philosophical Logic. 1997 ISBN 0-7923-4560-6
3. H. Wansing: *Displaying Modal Logic.* 1998 ISBN 0-7923-5205-X
4. P. Hájek: *Metamathematics of Fuzzy Logic.* 1998 ISBN 0-7923-5238-6
5. H.J. Ohlbach and U. Reyle (eds.): *Logic, Language and Reasoning.* Essays in Honour of Dov Gabbay. 1999 ISBN 0-7923-5687-X
6. K. Došen: *Cut Elimination in Categories.* 2000 ISBN 0-7923-5720-5
7. R.L.O. Cignoli, I.M.L. D'Ottaviano and D. Mundici: *Algebraic Foundations of many-valued Reasoning.* 2000 ISBN 0-7923-6009-5
8. E.P. Klement, R. Mesiar and E. Pap: *Triangular Norms.* 2000
ISBN 0-7923-6416-3
9. V.F. Hendricks: *The Convergence of Scientific Knowledge.* A View From the Limit. 2001 ISBN 0-7923-6929-7
10. J. Czelakowski: *Protoalgebraic Logics.* 2001 ISBN 0-7923-6940-8
11. G. Gerla: *Fuzzy Logic.* Mathematical Tools for Approximate Reasoning. 2001
ISBN 0-7923-6941-6
12. M. Fitting: *Types, Tableaus, and Gödel's God.* 2002 ISBN 1-4020-0604-7
13. F. Paoli: *Substructural Logics: A Primer.* 2002 ISBN 1-4020-0605-5
14. S. Ghilardi and M. Zawadowki: *Sheaves, Games, and Model Completions.* A Categorical Approach to Nonclassical Propositional Logics. 2002
ISBN 1-4020-0660-8
15. G. Coletti and R. Scozzafava: *Probabilistic Logic in a Coherent Setting.* 2002
ISBN 1-4020-0917-8; Pb: 1-4020-0970-4
16. P. Kawalec: *Structural Reliabilism.* Inductive Logic as a Theory of Justification. 2002
ISBN 1-4020-1013-3
17. B. Löwe, W. Malzkorn and T. Räsch (eds.): *Foundations of the Formal Sciences II.* Applications of Mathematical Logic in Philosophy and Linguistics, Papers of a conference held in Bonn, November 10-13, 2000. 2003 ISBN 1-4020-1154-7
18. R.J.G.B. de Queiroz (ed.): *Logic for Concurrency and Synchronisation.* 2003
ISBN 1-4020-1270-5
19. A. Marcja and C. Toffalori: *A Guide to Classical and Modern Model Theory.* 2003
ISBN 1-4020-1330-2; Pb 1-4020-1331-0

KLUWER ACADEMIC PUBLISHERS – DORDRECHT / BOSTON / LONDON

Printed in the United States
46878LVS00001BA/133

9 781402 013317